PRENTICE-HALL INTERNATIONAL SERIES IN MANAGEMENT

Athos and Coffey	*Behavior in Organizations: A Multidimensional View*
Baumol	*Economic Theory and Operations Analysis, 2nd ed.*
Boot	*Mathematical Reasoning in Economics and Management Science*
Brown	*Smoothing, Forecasting, and Prediction of Discrete Time Series*
Churchman	*Prediction and Optimal Decision: Philosophical Issues of a Science of Values*
Cohen and Cyert	*Theory of the Firm: Resource Allocation in a Market Economy*
Cyert and March	*A Behavioral Theory of the Firm*
Ewart, Ford, and Lin	*Probability for Statistical Decision Making*
Fabrycky and Torgersen	*Operations Economy: Industrial Applications of Operations Research*
Frank, Massy, and Wind	*Market Segmentation*
Green and Tull	*Research for Marketing Decisions, 2nd ed.*
Hadley and Whitin	*Analysis of Inventory Systems*
Holt, Modigliani, Muth, and Simon	*Planning Production, Inventories, and Work Force*
Ijiri	*The Foundations of Accounting Measurement: A Mathematical, Economic, and Behavioral Inquiry*
Kaufmann	*Methods and Models of Operations Research*
Mantel	*Cases in Managerial Decisions*
Massé	*Optimal Investment Decisions: Rules for Action and Criteria for Choice*
McGuire	*Theories of Business Behavior*
Miller and Starr	*Executive Decisions and Operations Research, 2nd ed.*
Montgomery and Urban	*Management Science in Marketing*
Montgomery and Urban	*Applications of Management Science in Marketing*
Morris	*Management Science: A Bayesian Introduction*
Muth and Thompson	*Industrial Scheduling*
Nelson (ed.)	*Marginal Cost Pricing in Practice*
Nicosia	*Consumer Decision Processes: Marketing and Advertising Decisions*
Peters and Summers	*Statistical Analysis for Business Decisions*
Pfiffner and Sherwood	*Administrative Organization*
Simonnard	*Linear Programming*
Singer	*Antitrust Economics: Selected Legal Cases and Economic Models*
Vernon	*Manager in the International Economy*
Wagner	*Principles of Operations Research with Applications to Managerial Decisions*
Zangwill	*Nonlinear Programming: A Unified Approach*
Zenoff and Zwick	*International Financial Management*
Zionts	*Linear and Integer Programming*

Prentice-Hall, Inc.
Prentice-Hall International, Inc., *United Kingdom and Eire*
Maruzen Company, Ltd., *Far East*

Probability for Statistical Decision Making

PARK J. EWART

JAMES S. FORD

CHI-YUAN LIN

University of Southern California

PRENTICE-HALL, INC., Englewood Cliffs, New Jersey

Library of Congress Cataloging in Publication Data

Ewart, Park J.
Probability for statistical decision making.

(Prentice-Hall international series in management)
1. Probabilities. 2. Statistical decision.
I. Ford, James S., 1926– joint author. II. Lin, Chi-Yuan, 1935– joint author. III. Title.
QA273.E86 519.2 73-21930
ISBN 0-13-711614-4

Printed in the United States of America

10 9 8 7 6 5 4 3 2 1

Prentice-Hall International, Inc., *London*
Prentice-Hall of Australia, Pty. Ltd., *Sydney*
Prentice-Hall of Canada, Ltd., *Toronto*
Prentice-Hall of India Private Limited, *New Delhi*
Prentice-Hall of Japan, Inc., *Tokyo*

To
PAN and VI and MEI-TAI

Contents

Chapter 8

CENTRAL TENDENCY AND VARIABILITY 135

Chapter 9

COMMONLY USED DISCRETE PROBABILITY MODELS 168

Chapter 10

COMMONLY USED CONTINUOUS PROBABILITY MODELS 199

Chapter 11

SOME USEFUL PROBABILITY APPROXIMATIONS 224

Chapter 12

DECISION CONDITIONS AND CRITERIA 248

Chapter 13

THE VALUE OF INFORMATION IN DECISION MAKING 279

Chapter 14

UTILITY IN DECISION MAKING 302

Appendix I

SUMMARIES 319

Appendix II

TABLES 327

ANSWERS TO SELECTED PROBLEMS 369

INDEX 379

Preface

Today's managers, administrators, and policy makers in such fields as business, industry, and government are constantly required to make decisions in the face of uncertainty. As tools for dealing with such decision problems, a body of statistical models and methods has been developed. These statistical tools have been built on the foundation of probability theory. Indeed, a firm grasp of the elements of probability theory is essential to the effective understanding and use of statistical models and methods in making decisions under uncertainty. Hence, the mission of this book is to provide a solid foundation in fundamental probability theory necessary for understanding applications of classical and Bayesian statistics to decision making in business, economics, and public administration.

This volume results from the authors' many years of teaching experience. During these years, students and colleagues have repeatedly expressed a desire for a text that would communicate the essential elements of probability theory more effectively than the generally available texts in probability and statistics. Thus, throughout the book, emphasis has been placed on effective communication of the subject matter in a manner that is conceptually rigorous but easily readable and understandable. Considerable care has been exercised to provide meaningful examples and problems to demonstrate the applicability of the theory and to promote the reader's insight.

As a probability text, this volume is unique in that it develops the fundamentals of probability theory specifically within the context of its application to statistical decision making. In addition to covering the customary topics in probability, the book includes chapters on decision criteria, the value of information, and utility, which provide an interface between probability theory and statistical decision theory. As such, the book is suitable as either a primary text in probability courses or a supplemental text in statistics courses.

In treating certain topics in probability, use of the calculus is necessary. Thus, when unavoidable, the book does employ the calculus, but only the simplest forms of derivatives and integrals are used. Students with an elementary, even though insecure, working knowledge of integral calculus can comfortably handle the entire text. Although calculus is used in treating certain topics, most sections of the book require only high school algebra. For students not having a calculus background, the book may be used without loss of continuity by omitting those parts of Chapters 7 and 8 dealing with continuous functions, and disregarding the occasional appeal to the calculus in Chapters 10 and 12.

During the preparation of this book, we were fortunate to receive valuable assistance from numerous individuals. We are indebted to colleagues and students at the University of Southern California, California State University at Fullerton, California State Polytechnic University at Pomona, and Santa Ana City College for using the preliminary edition of the text. We are particularly grateful to Dr. Wally Blischke, Mr. Joel Greenwald, Mr. Bert Steece, Mr. Tom Wedel, Mr. Bill Heitzman, and Mr. Mike Holland for their critical comments and constructive suggestions.

Special thanks are due to Dr. Robert S. Childress, University of Southern California, Dr. John E. Freund, Arizona State University, and Professor Jay E. Strum, New York University for reviewing the manuscript and providing helpful suggestions. We also wish to extend our gratitude to Dr. George Schick, University of Southern California for his moral support and intellectual stimulation.

Mr. Jeff Bahr and Mr. Bob Hill supplied the computer output for various appendix tables. Miss Chris Kutschinski and Mrs. Toni Graham performed yeoman duty in typing the manuscript. Mr. Andy Lau, Miss Cora Lin, and Miss Mary Liu completed the tedious tasks of proofreading the manuscript and preparing the preliminary art work. To all of these, as well as those who remain unnamed, we express our deep appreciation.

We will warmly welcome any corrections, comments, and suggestions from readers.

Los Angeles, California
March 1974

PARK J. EWART
JAMES S. FORD
CHI-YUAN LIN

Chapter 1

Doubt in Decision Making

From the time men first began to organize into groups to combine their activities and resources in pursuit of common goals, leaders have had to make decisions in order to bring the combined efforts of individuals to successful ends. As the organizational and operational problems of business and government have become increasingly technical, and as the art and science of administration have become more complex and precise, we have witnessed a developing awareness of the practicality of scientifically formal approaches to managerial decision making. One such approach is statistical decision making.

The type of managerial decision problem to which the statistical approach is applicable may be conceptualized as follows:

1. There is a perceived need to attain a certain *goal* or *objective*.
2. There are two or more *alternative courses of action* that might possibly be taken to achieve this goal.
3. Due to *imperfect knowledge* concerning the factors that will determine the *result* or *outcome* of each of the alternative actions, there is *doubt* concerning which course of action will best achieve the goal.
4. The degree of doubt may be reduced by obtaining additional information, usually at some cost.

As an illustration of this type of decision problem, consider the case of George Jones, a successful real estate investor. George holds an option to purchase a large tract of rural land adjacent to a growing town. This land has excellent potential for profitable development *if* the county will build adequate access roads in the near future. Whether or not such roads will be built depends on the outcome of a county-wide bond referendum that will be voted on in four weeks. George's option to buy the land expires in two weeks, and he must therefore make his decision regarding the purchase before the outcome of the election is known. If he buys the land and the roads are built, he anticipates a profit on his investment; but if the roads are not built, he will be forced to dispose of the land at an out-of-pocket loss. If he decides not to buy the land and the roads are built, he will have "missed the boat"; but if the roads are not built, he will be "even" with the world—no better nor worse than at present.

George is a rational man, and his goal is to make the most profitable use of his resources. There are two alternative courses of action available to him as possible means of achieving this goal. We may denote these alternatives as:

a_1: exercise the option
a_2: drop the option

The consequence resulting from either of these courses of action will depend on whether or not the bond issue passes. Such conditions, which determine the specific consequences resulting from particular courses of action, are referred to by modern decision theorists as *states of nature*. Thus, in deciding between the two alternative courses of action, George must consider two possible states of nature:

s_1: bond issue passes
s_2: bond issue fails

Hence, as shown in Table 1.1, the consequence of George's decision will depend not only on which of the two alternative actions he chooses, but also on which of the two states of nature happens to occur. Thus, there are four different consequences, each associated with one of the four possible action-state pairs.

Table 1.1

Possible Consequences for George's Decision Problem

	Alternative Courses of Action	
State of Nature	a_1*: Exercise the Option*	a_2*: Drop the Option*
s_1: Bond Issue Passes	Profit	Miss the Boat
s_2: Bond Issue Fails	Out-of-Pocket Loss	Even

George feels that the "chances" are favorable for the passage of the bond issue, but he is not certain. Since he does not know for sure what the outcome of the voting will be, he is doubtful about which action to take.

To lessen his doubt, George decides to obtain additional information by hiring a local research firm to poll a sample[1] of 100 registered voters in the county. This device of obtaining information from a sample of observations is basic to all statistical decision procedures. If we regard the outcome of George's sample (that is, the result of his poll) as the number of people out of 100 who respond that they favor the bond issue, then there are 101 possible outcomes—the number of favorable responses may be any integral value between 0 and 100, inclusive. Of course, *not all* of these possible outcomes are equally likely. If actually, for instance, a majority of the voting population

[1] The term "sample" will be defined formally in Chapter 4.

favors the bond issue, 60 favorable responses out of a sample of 100 voters would be a more likely outcome than 10 favorable responses. Now, the real question is the following: "If a sample is taken, how can George use his knowledge of the outcome of the sample to lessen his uncertainty about the decision?"

To answer this question, we must resort to the methodology of statistical decision making. This, in turn, requires a basic understanding of the fundamental concepts of probability theory. The present volume is aimed at the development of such an understanding. In other words, the primary purpose of this volume is to present fundamental concepts of probability that are essential to statistical analysis and to illustrate some of their applications to decision making under doubt.

Chapter 2
Meanings of Probability

Although they may never have studied probability, most people have an intuitive feeling for the general concept. This feeling is reflected in such everyday probabilistic statements as "I believe the chances are favorable that my firm will show a tidy profit this year," or "The odds are fair that the Dodgers will win the pennant," or "There is little likelihood that the stock market will display a significant downward turn during the next six months." Implicit in each of these statements is an element of confidence coupled with some uncertainty. Although such indefinite statements of probabilities may be effective in casual communication, the application of probability concepts to decision problems requires that we adopt definitions that will permit us to express the probability of a certain condition or event in specific, quantitative terms.

Even among statisticians, who are greatly concerned with their definitions, there is no single definition of probability that will cover all situations. Before attaching meaning to a probability statement, we must question the *source* of that statement. If a probability is determined by theoretical reasoning, or from actual observation of a sequence of events, we speak of *objective* probability. If an individual states a probability as a measure of the strength of his personal belief in some proposition, we are dealing with *subjective* probability. The distinction between objective and subjective probability involves differences not only in the source of probability statements, but also in the fundamental viewpoint toward the interpretation and application of probability statements.

2.1 OBJECTIVE PROBABILITY

A fundamental notion underlying the objectivistic interpretation is the concept of a *random process* (also called a *stochastic process* or *chance experiment*). A random process is an operation that is (conceptually, at least) *repetitive*. That is, a random process is conceived as a *sequence of trials* conducted in such a manner that the particular outcome of any given trial is the result of unknown or unassignable causes. In other words, the sequence of outcomes produced by a random process is, so far as we know, solely the result of "chance." A simple example of a random process would be the repeated tossing of a coin. Each toss (trial) has two possible outcomes: heads or tails. The particular sequence of heads and tails obtained through repeated tossings is unpredictable and can be explained only in terms of random effects. To the strict objectivist, a probability statement is meaningless unless it can be interpreted with respect to such a series of outcomes produced repeatedly by some operation under essentially the same conditions.

2.1.1 Probabilities Based on Theoretical Reasoning

One source of objective probability statements is the *theoretical analysis* of a random process. As an example, consider the random process of drawing a card from a well-shuffled standard deck of 52 playing cards. We might reason that each card has an *equally likely* opportunity of being drawn, and that the probability of drawing any particular card, say the queen of hearts, is therefore 1/52. Similarly, we know that there are four aces in the deck, and the probability of drawing an ace (without regard to suit) is 4/52. In such cases, wherein each of the possible outcomes is equally likely, we may define the probability of some event E by the ratio:

$$P(E) = \frac{n(E)}{n(T)} \tag{2.1}$$

where $n(T)$ = total number of equally likely outcomes

$n(E)$ = number of those outcomes corresponding to event E.

For example, we might define an event as drawing a spade from a well-shuffled 52-card deck. Since there are 52 possible cards that might be drawn, the total number of equally likely outcomes is 52—that is, $n(T) = 52$. Also, since there are 13 spades in the deck, 13 of these possible outcomes correspond to "drawing a spade." Therefore, $n(E) = 13$, and the probability of drawing a spade is $13/52 = 1/4 = .25$.

2.1.2 Empirical Probabilities

The term *empirical* (from a Latin word meaning "experiment") is frequently applied to probabilities that are derived from actual experimentation or observation of a random process in operation, unlike probabilities derived purely through theoretical analysis. Empirical probabilities result from "experiencing" the outcomes of random processes. On the basis of pure theoretical analysis we might determine that the probability of obtaining heads when tossing a fair coin is 1/2. That is, we might argue that there are two possible outcomes that are equally likely, and that one of these two outcomes corresponds to the event "heads." Empirically, we would determine the probability of obtaining heads with a particular coin by tossing that coin many times and recording the proportion of times it comes up heads.

Assume that a random process is repeated some number of times, n, and that during these n repetitions some event E occurs with a frequency f (that is, the event E occurs f number of times out of n repetitions). Then the empirical probability of the event E may be defined as the ratio

$$P(E) = \frac{f}{n} \tag{2.2}$$

Suppose, for instance, that a distributor of garden seeds wishes to determine the probability that a particular type of seed will germinate when planted under standard conditions. He might estimate this probability empirically by planting a large number of these seeds and then observing the number of seeds that actually germinate. If he planted 1,000 seeds and if 700 of these seeds germinated, then the empirical probability for this type of seed germinating under the standard conditions would be $700/1{,}000 = .70$.

2.1.3 Interpretation of Objective Probability

How do we interpret objective probabilities? What does the objectivist mean, for example, when he says that the probability of drawing a spade at random from a well-shuffled deck is .25, or that the probability for a particular type of seed germinating under standard planting conditions is .70? If we look at a random process as a repeatable operation, the objective probability of an event is the *relative frequency of occurrence* of that event in the long run.

In the example of drawing a spade from a bridge deck, the probability of .25, which was obtained by theoretical analysis, may be interpreted as the relative frequency, in the long run, with which we would expect to draw a spade during an extended series of repeated trials. If all the clubs, diamonds, and hearts were removed from the deck, so that the deck contained only spades, then the probability of drawing a spade would be equal to 1.00; we would be *certain* of drawing a spade every time. If all of the spades were removed from the deck, we would be certain that we could not possibly draw a spade. Under such conditions, the probability of drawing a spade would be equal to 0. Unless the probability of an event is 0 or 1, we are uncertain regarding whether the event will or will not occur in a given instance. If the probability of an event is .999, we may feel fairly "confident" that the event will occur on a specified trial, but we cannot be certain that it will occur, since the probability tells only that we can expect it to occur about 999 times out of 1,000 in the long run.

In contrast to probabilities obtained by theoretical analysis, probabilities obtained by empirical procedures are usually only *estimates.* To obtain the "true" long-run relative frequency of an event would require an impractically large number of observations. In the example of the garden seeds, if the seeds germinated 700 times out of 1,000 attempts, the relative frequency of the event "germination" would be .70. If the seed distributor had planted 10,000 seeds instead of 1,000, he might have found that 7,100 of them germinated, yielding an empirical probability of $7{,}100/10{,}000 = .71$, rather than .70. Although these two figures are very close, neither of them can be considered better than estimates. If he planted 100,000 seeds, he would likely obtain even another value for his empirical probability. To illustrate this point further, suppose that you toss a coin 200 times, and at the end of each 20 tosses you recorded the *cumulative* number of heads obtained *so far.* You could then estimate the probability of obtaining heads by computing the ratio of the total number of heads to the total number of tosses. Your results might look something like those in Table 2.1.

Table 2.1

Results of Coin-Tossing Experiment

Total Number of Tosses	*Total Number of Heads*	*Relative Frequency of Heads*
20	5	.25
40	22	.55
60	27	.45
80	45	.56
100	48	.48
120	60	.50
140	69	.49
160	81	.51
180	88	.49
200	102	.51

Notice in Table 2.1 that the relative frequency of heads fluctuates as the total number of tosses changes. Unless we hold some preconceived conviction that the probability of heads is .50, we have no basis for concluding that any one of these relative frequencies is the "right" figure. Furthermore, even if we argue on logical grounds that the probability of heads is .50, we must realize that we are dealing with one specific coin that is not necessarily "fair," and that the logical probability for coins in general does not necessarily apply to a particular coin that you might have picked at random from your pocket.

A further look at our coin-tossing data reveals that, if the total number of tosses was comparatively small, the relative frequency of heads fluctuated considerably, but as the total number of tosses increased, the relative frequency of heads tended to become more stable. That is, over the second 100 tosses, the relative frequency of heads fluctuated comparatively little around a limiting value of .50, as shown graphically in Figure 2.1.

Figure 2.1

Empirical Estimation of the Probability of Obtaining Heads on the Toss of a Coin

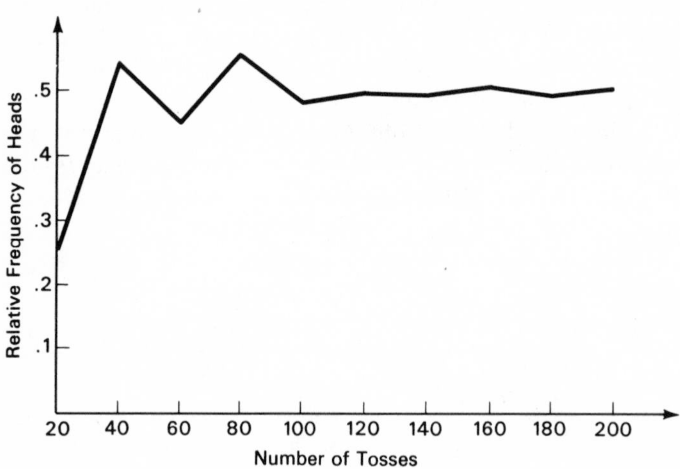

2.2 SUBJECTIVE PROBABILITY

From the subjective viewpoint, a probability is regarded as a measure of an individual's personal confidence in the truth of some proposition, such as the proposition that a business recession will not occur within the next two years. The subjectivist maintains that a "reasonable" person who obtains some practice and experience can learn to use the language of probability to describe his own attitudes about the degree of certainty or uncertainty in a particular situation.

Consider, for example, a pair of subjectivists, Andy Able and Bob Baker, who are partners in a speculative real estate venture. They have the opportunity to acquire several hundred acres of arid land that could have development potential depending on the outcome of a reclamation proposal currently before the state legislature. Andy feels that the probability is about .70 that the reclamation bill will be passed, whereas Bob judges the probability of the bill's passage to be about .40. In other words, both Andy and

Bob feel uncertain about whether the reclamation bill will be passed. Andy has assigned a relative weight of .70 as a measure of his belief that the bill will be passed, and a relative weight of .30 as a measure of his belief that it will not be passed. Similarly, Bob has assigned relative weights of .40 and .60 to indicate the strength of his feelings about these two propositions.

If Andy is a firm subjectivist, he hardly will be distressed by the fact that he and his partner assigned different probabilities to the proposition that the reclamation bill will be passed. The subjectivistic position readily admits—in fact, *assumes*—that two reasonable persons might attach different probabilities to the same event. After all, the beliefs that different people have regarding the likelihood of a particular event are based on their psychological predispositions, which in turn are determined largely by their real-world experiences. If two people with different experiential backgrounds are presented with the same decision problem and are given the same relevant evidence, they may well assign it somewhat different probabilities if their motivational, emotional, and perceptual processes are operating differently at the time.

Because two reasonable individuals may assign different probabilities to the same event, subjective probabilities are frequently described as being *judgmental* or *personalistic*. Nevertheless, a person's subjective probability assessments concerning a particular class of events are usually based on some backlog of experience with similar events, and so usually do not just come from "out of the blue." Indeed, frequently one of the best ways of obtaining information bearing on certain practical decisions is to solicit the personal opinions of individuals who have a great deal of experience with the types of events that concern us. Thus the personalists may cogently argue that the use of subjective probabilities, obtained from experienced persons, have an important function in the process of making practical decisions.

2.3 COMPARISON OF VIEWPOINTS

How does the viewpoint of the subjectivist compare with that of the objectivist? To the objectivist, of course, the personalist's probability statements, based on personal feeling, opinion, or attitude, are unacceptable. The objectivistic approach would assume that two reasonable persons who approach the same problem situation with the same set of assumptions and evidence will arrive at the same probability assessments. The personalist, however, would argue that such an approach is narrow and limited; the objectivist is limited to events that can be interpreted in terms of a random process with resultant relative frequencies, whereas the personalist has no such limitation on his sphere of operation. The personalist is perfectly willing to apply himself to problems involving the probabilities of events produced by random processes, but he is equally willing to work with probabilities of unique events —that is, one-of-a-kind events rather than events that can occur repeatedly. For example, the personalist would feel that the question of the probability that Big Steel will increase its prices during the next six months is a perfectly legitimate problem, but the question would be outside the domain of the objectivist since it concerns a unique event rather than a repeatable process.

The objectivistic and personalistic viewpoints arrive at some degree of rapprochement in their treatments of the role of experience in probability assessment. To the extent that an individual's subjective probability assessment for an event is based on his experience with similar past events, he is more or less empirical. If his experience has been relatively informal, and his probability assessment is affected by memory and judgment, then we would consider his probability statements to be essentially personalistic. If, however, his experience has been gained under relatively standardized, formal conditions, and particularly if he has kept careful records of this experience to form the basis of his probability assessment, then his probability statements may closely approach being empirical in the objectivistic sense. Indeed, in actual business applications of some statistical procedures that are associated primarily with the personalistic approach, the appearance of genuine objectivistic probabilities is not at all unheard of.

PROBLEMS

2.1 Nevada gambling casino operators use probabilities to determine the payoffs on various kinds of bets in such games as craps, blackjack, and roulette. Insurance company managements use probabilities to determine the premiums to be charged for different types of life insurance and annuity policies. Contrast these two industries with respect to the source of probabilities that they use.

2.2 The internationally famous Lloyd's of London will insure almost any risk for a fixed premium for a given period of time. Among the unique risks reported to have been insured by Lloyd's are (a) risk of disabling injury to a race horse, (b) risk of disabling injury to legs of a famous dancer, (c) risk of sufficient rain to cause cancellation of a single scheduled sporting event, such as a specific scheduled track meet, and (d) risk of losing a rare collection of jewelry or paintings. What type of probability is likely to dominate the decision-making process as each underwriter of Lloyd's of London decides the premium at which he will insure a given proportion of the risk?

2.3 What is the probability of each of the following outcomes (events) from a single roll of a single balanced die?

(a) Obtaining a 2?

(b) Obtaining 1, 3, or 5?

(c) Obtaining 2 or 6?

2.4 What is the probability of each of the following outcomes (events) from dealing a single card from a well-shuffled standard deck of 52 cards?
(a) A king?
(b) A face card?
(c) An ace of clubs?
(d) A diamond?
(e) Either a diamond or a heart?
(f) Either a club, spade, diamond, or heart?
(g) A number less than 10 counting an ace as one?
(h) A diamond face card?
(i) A red card?

2.5 The "defective rate" of a production process may be interpreted as the probability that any randomly selected unit produced by the process will be defective. If a sample of 20 bolts is selected at random from the output of a particular production process, and the sample is found to contain one defective, what would be the empirically estimated defective rate of the process?

2.6 Welton Products Company has four separate sales departments. The ages of outstanding invoices in the different departments are summarized below:

	Sales Department			
Age of Invoice	*A*	*B*	*C*	*D*
Under 120 days	78	105	86	64
120–180 days	27	36	28	39
Over 180 days	21	9	6	11

(a) If an outstanding invoice is randomly selected from the pooled central files, what is the probability that it is from Department *B*?
(b) What is the probability that a randomly selected invoice from the pooled files is over 180 days old?
(c) If an outstanding invoice is randomly selected from the pooled files, what is the probability that the invoice is over 180 days old and is from Department *B*?
(d) If an outstanding invoice is selected at random from the files of Department *C*, what is the probability that it is not less than 120 days old?

2.7 The following table was abstracted from the *Statistical Abstract of the United States 1970*.

Number Surviving to Specified Age Per 100,000 Born Live

		White			Negro and Other		
		1949–1951	*1959–1961*	*1967*	*1949–1951*	*1959–1961*	*1967*
Age 20:	Male	95,104	95,908	96,298	91,941	93,108	93,982
	Female	96,454	97,135	97,486	93,544	94,660	95,485
Age 40:	Male	91,173	92,427	92,583	82,832	85,744	85,013
	Female	94,080	95,326	95,662	86,052	89,676	90,529
Age 65:	Male	63,541	65,834	65,990	45,198	51,392	50,180
	Female	76,773	80,739	81,486	52,358	60,825	64,255

(a) Calculate the probability, at birth, based upon 1967 experience, that a white male will live at least to age (i) 20, (ii) 40, (iii) 65.
(b) Calculate the probabilities requested in (a) for Negro and other.
(c) Compare (a) and (b).
(d) Using 1967 experience, compare the probability, at birth, that a white male will live to age 65 with the probability, at birth, that a white female will live to age 65.
(e) Using 1967 experience, compare the probability, at age 40, that a white male will live to age 65 with the probability, at age 40, that a white female will live to age 65.
(f) Are the probabilities derived in this problem empirical, theoretical, or judgmental?

2.8 During the year that the total resident population of the United States was approximately 198 million persons, the U.S. Public Health Service reports that there were 20,120 deaths due to accidental falls and 5,724 deaths due to accidental drowning. Based on these data, what is the empirically estimated probability that a person will die from one or the other of these two causes?

Chapter 3

Counting Possible Outcomes

Many applications of probability theory to decision problems involve random processes that reasonably may be assumed to have a finite number of equally likely outcomes. As an example, consider a manufacturer of electronic equipment who receives transistors from a supplier in lots of sixty. Suppose that the manufacturer wishes to test a sample of ten transistors selected "at random" from each lot. Under these conditions, if a particular lot actually contains two defectives, what is the probability that both defectives will be included among the ten transistors that are selected for testing? To answer this question, we might apply Formula (2.1), which is:

$$P(E) = \frac{n(E)}{n(T)}$$

In this example, E represents the event that both defectives will be in the sample selected for testing, $n(T)$ is the total number of different groups of ten transistors that could possibly be drawn at random from the lot of sixty transistors, and $n(E)$ represents the number of possible groups of ten transistors that contain two defectives.

Applying Formula (2.1) to this problem would be an elementary matter if we could determine the values of $n(E)$ and $n(T)$. One approach might be to identify each transistor by a serial number from 1 to 60. Then we might begin to list all possible groups of ten different numbers from 1 to 60. Some of these would be:

$$\begin{array}{cccccccccc}
1, & 2, & 3, & 4, & 5, & 6, & 7, & 8, & 9, & 10 \\
1, & 2, & 3, & 4, & 5, & 6, & 7, & 8, & 9, & 11 \\
. & . & . & . & . & . & . & . & . & . \\
. & . & . & . & . & . & . & . & . & . \\
41, & 43, & 52, & 53, & 54, & 55, & 56, & 57, & 58, & 59 \\
. & . & . & . & . & . & . & . & . & . \\
. & . & . & . & . & . & . & . & . & . \\
51, & 52, & 53, & 54, & 55, & 56, & 57, & 58, & 59, & 60
\end{array}$$

If a person had unlimited time, patience, pencils, and paper, he eventually could succeed in listing and then counting all the possibilities. This, however, would be an exhausting task, for there are over 75 billion different possible sets of ten items out of sixty (as you will shortly be able to verify for yourself). Obviously, if we are to approach probability problems in a practical manner, we need methods of counting that are more efficient than the tedious process of listing and enumerating. We would do well to defer answering our transistor problem until the end of this chapter, after we have considered some of these counting methods.

3.1 FACTORIAL NOTATION

In our discussion of "short-cut" counting procedures, it will be helpful to employ *factorial* notation. The expression $n!$ is read "n factorial" and is used to denote the product of all whole numbers from 1 to n. That is,

$$n! = n(n-1)(n-2)\ldots(3)(2)(1)$$

For example:

$$7! = 7 \times 6 \times 5 \times 4 \times 3 \times 2 \times 1 = 5{,}040$$

As a special case, we must define $0! = 1$. In working with factorials, it is also helpful to note that for $n \geq 1$,

$$n! = n(n-1)!$$

As examples:

$$6! = 6 \times 5!$$

$$50! = 50 \times 49!$$

$$100! = 100 \times 99 \times 98!$$

3.2 THE FUNDAMENTAL PRINCIPLE OF COUNTING

Most counting procedures are based on the following fundamental principle:

> If a first task can be conducted in n_1 different ways and, after it is done in any one of those ways, a second task can be conducted in n_2 different ways and, after the first two tasks, a third can be conducted in n_3 different ways, and so on for r tasks, then the number of different ways in which the r tasks can be accomplished in the given order is
>
> $$n_1 \times n_2 \times n_3 \times \ldots \times n_r \tag{3.1}$$

Thus, if a first task can be conducted in five different ways, after any one of which a second task can be conducted in six different ways, after any one of which a third task can be conducted in four different ways, then together the three tasks can be accomplished in the stated order in $5 \times 6 \times 4 = 120$ different ways.

Example 1:

A real estate developer is planning a new housing tract. He has four different floor plans, three different exterior styles (modern, colonial, and rustic), and two different types of garages (attached and detached). How many different house designs may he offer to the public?

For each of the four floor plans, three different exterior styles are possible, making $4 \times 3 = 12$ different combinations of floor plans and exterior styles. For each of these twelve combinations there are two possible types of garages, making a total of $12 \times 2 = 24$ different designs. By direct application of Formula (3.1), $4 \times 3 \times 2 = 24$. This result may also be obtained by means of the tree diagram *shown in Figure 3.1.*

Figure 3.1

Tree Diagram Showing Number of Different House Designs

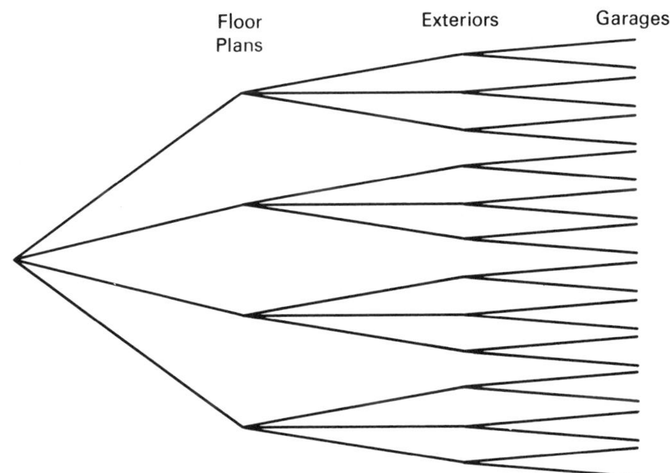

Example 2:

An office manager has four girls—Alice, Betty, Cora, and Dorothy—to whom she must make work assignments. She needs one girl to operate the telephone switchboard, one to take dictation, one to type, and one to file. Only Alice and Betty can operate the switchboard, and only Cora and

Dorothy can take dictation. All four girls can type and file. In how many different ways can the tasks be assigned?

There are two ways to fill the switchboard position (either Alice or Betty), and two ways to fill the dictation position (either Cora or Dorothy). After these two positions have been filled, there are two possible ways to fill the typing position; for instance, if Alice is assigned to the switchboard and Cora takes dictation, then either Betty or Dorothy may do the typing. Once the switchboard, dictation, and typing assignments have been made, only one girl remains to do the filing. Thus there are $2 \times 2 \times 2 \times 1 = 8$ *different possible ways of making the work assignments. The tree diagram in Figure 3.2 illustrates this result.*

Figure 3.2

Tree Diagram Illustrating Assignment of Office Tasks

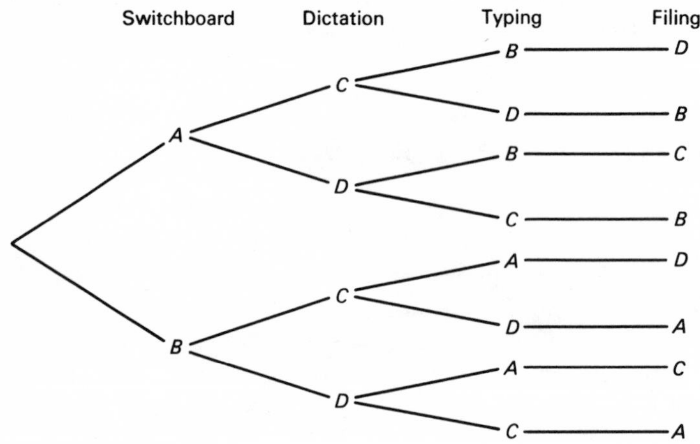

Example 3:

At no extra cost, Joe's Hamburger Haven will, at the customer's request, add any or all of the following ingredients to a plain hamburger: lettuce, tomato, onion, catsup, pickle, and mustard. In how many different ways may a hamburger be concocted?

In this example, there are six optional ingredients. To apply the fundamental counting principle, the choice of including or omitting each of these ingredients can be considered as a separate task. Thus, there are six successive tasks to be performed in preparing a hamburger. Since each of these tasks can be performed in two different ways, the total number fo ways in which a hamburger may be prepared is $2 \times 2 \times 2 \times 2 \times 2 \times 2 = 2^6 = 64$. *Notice that in this example,* $n_1 = n_2 = \ldots = n_6 = 2$.

3.3 PERMUTATIONS

When we talk about the number of ways in which a set of objects may be *arranged in order*, we are speaking of the number of possible *permutations* of the set. Depending on conditions, there are various permutation formulas, all of which may be derived from the fundamental principle of counting.

3.3.1 Permuting *n* Distinguishable Objects

Suppose that you are placing three books on a shelf. In how many orders, from left to right, could you arrange them on the shelf? Designating the books as A, B, and C, we can list the orders:

$$\begin{matrix} ABC & BAC & CAB \\ ACB & BCA & CBA \end{matrix}$$

These are the six possible permutations of the set of three books. We could obtain this same result, without actually listing, by directly applying the fundamental counting principle, reasoning as follows: Any one of the three books may be placed in the first position; that is, there are three ways in which the first choice may be made. Assuming that the first choice has been made, either of the two remaining books may be placed in the second position, so there are two ways in which the second choice may be made. Whichever of the two books is placed in the second position, the one remaining book will go in the third position. Thus the total number of permutations may be obtained from the calculation $3 \times 2 \times 1 = 6 = 3!$ To generalize, if we have a set consisting of n different objects, the *total number of ways in which these n objects, taken all together, may be arranged in order*, is given by the expression

$$_nP_n = n! \tag{3.2}$$

Example:

A door-to-door salesman has five different kitchen gadgets that he demonstrates to housewives. He would like to determine an order of presentation of the five gadgets that will create an optimal "impact" on the housewife. How many different orders of presentation are there?

From Formula (3.2), the total number of permutations of five objects, taken all together, is:

$$_5P_5 = 5! = 120$$

3.3.2 Permutations Involving Indistinguishable Objects

So far, in discussing the permutations of a set of n objects, we have assumed that all n objects in the set were distinguishably different. Suppose, however, that some of the objects in the set are indistinguishable from others. For instance, we might have a twelve-item set composed of three x's, four 0's, and five +'s. If all twelve symbols were distinguishably different, we would have $12! = 479{,}001{,}600$ possible permutations of the elements of the set. However, since some of the elements are identical to others, not all of these 12! permutations are distinguishably different. For any given permutation of the twelve items, there are 3! ways in which the x's may be permuted among themselves, 4! ways in which the 0's may be permuted among themselves, and 5! ways that the +'s may be permuted among themselves. Thus, for every distinguishably different permutation of the twelve items, there are 3!4!5! identical permutations that may be obtained by permuting identical items among themselves. Hence, to obtain the number of distinguishably different permutations of the twelve items, we would perform the following division:

$$\frac{12!}{3!\,4!\,5!} = 27{,}720$$

To generalize this procedure: If we are given a set of n objects, of which n_1 objects of one kind are identical, n_2 objects of a second kind are identical, n_3 objects of a third kind are identical, and so on for k kinds of objects, then *the number of possible distinguishably different permutations of the n objects, taken all together*, is:

$${}_nP_n(n_1,n_2,\ldots,n_k) = \frac{n!}{n_1!n_2!\ldots n_k!} \tag{3.3}$$

where $\sum_{i=1}^{k} n_i = n$

Example:

A costume jeweler is designing a new necklace. The necklace will consist of a string of twenty-one beads arranged in a pattern. Of the twenty-one beads, there are three identical red beads, four identical blue beads, six identical green beads, and eight identical yellow beads. How many distinguishably different patterns are possible?

From Formula (3.3), the total number of distinguishably different permutations is:

$${}_{21}P_{21}(3,4,6,8) = \frac{21!}{3!\;4!\;6!\;8!} = 12{,}221{,}609{,}400$$

A special, but important, case of Formula (3.3) is the situation in which there are only two kinds of objects. Suppose that we have a set of just two kinds of objects. If there are r objects of one kind, then there must be $(n - r)$ objects of the other kind. Then Formula (3.3) reduces to

$$ {}_nP_n(r,n-r) = \frac{n!}{r!\,(n-r)!} \tag{3.4} $$

Example:

During the month of June, a space exploration facility conducts seven successive attempts to launch satellites. Of these attempts, four were successes and three were failures. In how many different orders might these four successes and three failures have occurred?

From Formula (3.4)

$$ {}_7P_7(4,3) = \frac{7!}{4!\,3!} = 35 $$

3.3.3 Selecting and Permuting *r* Objects from *n* Objects

Formula (3.2) is concerned with the permutations of n objects *taken all together*. Suppose, however, that we have a group of n objects from which we are to select a sub-group of r objects that are to be placed in order. In how many different ways may r objects be selected from n objects and then arranged in order? For example, we might have a group of five books from which we are to select three books and place them on a shelf in some order. How many different possible arrangements may result? Designating the five books as A, B, C, D, and E, we might list the permutations:

ABC	*BAC*	*CAB*	*DAB*	*EAB*
ABD	*BAD*	*CAD*	*DAC*	*EAC*
ABE	*BAE*	*CAE*	*DAE*	*EAD*
ACB	*BCA*	*CBA*	*DBA*	*EBA*
ACD	*BCD*	*CBD*	*DBC*	*EBC*
ACE	*BCE*	*CBE*	*DBE*	*EBD*
ADB	*BDA*	*CDA*	*DCA*	*ECA*
ADC	*BDC*	*CDB*	*DCB*	*ECB*
ADE	*BDE*	*CDE*	*DCE*	*ECD*
AEB	*BEA*	*CEA*	*DEA*	*EDA*
AEC	*BEC*	*CEB*	*DEB*	*EDB*
AED	*BED*	*CED*	*DEC*	*EDC*

From our list we can count a total of sixty permutations. With much less labor we could have obtained the same result by applying the fundamental counting principle: The first position may be filled by any one of five books, after which the second position may be filled by any one of four remaining books, after which the third position may be filled by any one of three remaining books. Thus the total number of permutations is $5 \times 4 \times 3 = 60$. Another way of expressing this computation would be

$$\frac{5 \times 4 \times 3 \times 2 \times 1}{2 \times 1} = \frac{5!}{(5-3)!}$$

Notice that the numerator of this ratio is the factorial of the total number of books from which the selection of three books was made, and the denominator is the factorial of the number of books remaining after the selection. Generalizing this result, we can say that the *total number of possible permutations resulting from selecting and arranging r objects from a group of n objects* is given by the expression

$$_nP_r = \frac{n!}{(n-r)!} \tag{3.5}$$

Example:

A personnel manager has received requisitions for one typist each from the Production Department, Marketing Department, and Research Department. There are seven applicants available from which these three positions may be filled. In how many ways may three typists be selected from the seven applicants and assigned to the three different openings?

Applying the fundamental counting principle, there are seven ways to fill the first position, after which there are six ways to fill the second position, after which there are five ways to fill the third position. This gives $7 \times 6 \times 5 = 210$ permutations. From Formula (3.5) we obtain the same result.

$$_7P_3 = \frac{7!}{(7-3)!} = \frac{7 \times 6 \times 5 \times 4!}{4!} = 7 \times 6 \times 5 = 210$$

3.3.4 Permutations Allowing Repetitions

Formula (3.5) assumes that all n elements are distinguishably different and that any particular element may appear only once in any given permutation of r elements. Suppose, however, that any element may appear more than once in an arrangement—i.e., elements may be repeated. For instance, in the preceding discussion of the number of possible book arrangements, we did not allow for sequences containing repetitions—such as AAA, ABA, and BCC. If such repetitions are allowed, any one of the n elements may appear at any point in the sequence regardless of whether it has previously occurred. Consequently, at any point in the sequence there are n possible ways of selecting an element. Thus, *if repetitions are allowed, the total number of possible ways of forming a sequence of r elements from a set of n elements* is $n \times n \times \ldots \times n$, which is n multiplied by itself r times. This can be considered as a special case of the fundamental counting principle in the sense that $n_1 = n_2 = \ldots = n_r = n$. Thus, in terms of a formula,

$$_nP^r = n^r \tag{3.6}$$

Observe that in this formula the notation $_nP^r$ is used to denote permutations in which repetitions are allowed, whereas in Formula (3.5) the notation $_nP_r$ denotes permutations in which repetitions are not allowed.

Example:

*As a secret agent, you are preparing to embark on a mission into enemy territory and wish to design a baffling code to smuggle messages through the underground. You decide to use an "alphabet" consisting of seven symbols: ¢, #, @, *, &, %, $. Each message will consist of an arrangement of three symbols, and repetitions will be allowed. Thus, some of the possible messages would be as follows:*

¢@%   @¢%   #&#   **$ ***

For each of the possible three-symbol arrangements, your code book would indicate an associated meaning. For instance, the message "%%@" might mean "The enemy is planning to cross the frontier at midnight." Allowing repetitions of symbols, how many different three-character messages may be constructed from the seven characters available? Applying Formula (3.6), we obtain the answer:

$$_7P^3 = 7^3 = 343$$

3.4 COMBINATIONS

In practical business situations wherein mathematical models involving the counting of possible outcomes are appropriate, the order or arrangement of objects is often inconsequential. If, on a given day, a floor salesman sells five TV sets out of twenty models in stock, the particular order in which he made the sales is not likely to be of great importance, although knowing the brands of the five models he sold can be of some value for such purposes as inventory control. If a quality control inspector randomly selects a sample of ten items from the output of an automatic production process and finds that three of the items are defective, the particular positions in which the three defectives occurred in the sequence of the ten items is hardly of any consequence, although the fact that there were three defectives out of ten may have an important bearing on a production control decision. In such cases we are interested only in the objects selected and not their order or position. Such a collection of objects, considered *without regard to order or arrangement,* is called a *combination.*

If we consider a set of n objects from which we select a group of r objects, then the combination of r objects may be regarded as a subset of the set of n objects. Suppose again that we have a set of five books that we designate as A, B, C, D, and E. Let us select a subset of three books and place them on a shelf. If we are concerned only with the identities of the three books selected, without regard to the order in which they are selected or placed on the shelf, how many different three-book groupings or collections are possible? Listing the combinations, we have

$$\begin{array}{ccccc} ABC & ABD & ABE & ACD & ACE \\ ADE & BCD & BCE & DBE & CDE \end{array}$$

From our list we can see that there are only ten combinations of three books selected from a set of five, as compared to sixty permutations. Recalling Formula (3.2), we are reminded that for a group of three objects taken all together, there are $3! = 6$ permutations. Thus we have a total of $10 \times 6 = 60$ possible permutations of three books selected from a set of five books. Conversely, if we divide the total number of permutations by the total number of permutations per combination, we obtain the total number of combinations. That is,

$$\frac{60 \text{ permutations}}{6 \text{ permutations/combination}} = 10 \text{ combinations.}$$

More generally, if there are $r!$ permutations of a subset of r objects that have been selected from a set of n objects, then the total number of possible combinations of r objects selected from n objects may be obtained by dividing Formula (3.5) by $r!$. Thus we may obtain *the number of possible combinations of r objects selected from a set of n objects* from

$${}_nC_r = \binom{n}{r} = \frac{n!}{(n-r)!\,r!} \tag{3.7}$$

Notice that the notation $\binom{n}{r}$ is used to denote the combinations of r objects selected from a set of n objects. This notation is more convenient than ${}_nC_r$ in the more complex formulas to be presented in later chapters.

Example:

A sales manager has ten field representatives working under him. A local consulting firm, at a fee of $150 per man, is conducting a three-day sales clinic to which the sales manager would like to send all ten of his field representatives. However, his budget will allow him to send only three men. How many different ways are there for him to compose this group of three men?

From Formula (3.7) the number of possible combinations of three men selected from a set of ten men is

$$_{10}C_3 = \binom{10}{3} = \frac{10!}{7!\,3!} = 120$$

A comparison of Formula (3.7) with Formula (3.4) reveals an interesting and important relationship. Specifically,

$$_nP_n(r, n - r) = \frac{n!}{r!\,(n - r)!} = \binom{n}{r} = {_nC_r}$$

Expressed in words, the number of distinguishable permutations in the special case handled by Formula (3.4) is equal to the number of combinations given by Formula (3.7). We shall make use of this relationship during our discussion of the binomial distribution in Chapter 9.

3.5 DEBRIEFING

We are now ready to return to the transistor testing problem posed at the outset of this chapter. To rephrase the problem: If a lot of sixty transistors contains two defectives and a sample of ten transistors is selected at random from the lot, what is the probability that both defectives will be included in the sample? From Formula (3.7), the total number of different groups of ten transistors that can possibly be drawn from the lot of sixty transistors is $\binom{60}{10}$. If the sample of ten contains two defectives, then it must contain eight good transistors. The total number of ways that both defectives can be selected from the two available is $\binom{2}{2}$. The total number of ways that eight good transistors can be selected from the fifty-eight good ones in the lot is $\binom{58}{8}$. Applying the fundamental counting principle, the total number of possible groups of ten transistors that contain two defectives becomes $\binom{2}{2}\binom{58}{8}$. Therefore, from Formula (2.1) we obtain the ratio

$$\frac{\binom{2}{2}\binom{58}{8}}{\binom{60}{10}}$$

designating the probability that the sample will contain both defectives. Evaluating this ratio yields a probability of approximately .025. The objective interpretation of this probability would be that in taking samples of ten transistors from sixty-transistor lots that contain two defectives, we would expect to obtain both defectives in the sample 2.5 per cent of the time in the long run. You will later recognize the solution to this problem as an example of the application of the hypergeometric probability function, which is considered in detail in Chapter 9.

PROBLEMS

3.1 An automobile purchaser has decided to buy a Gazell XC-2000. He has the following options:

Body color: Red, blue, citron, sand, white, green
Interior color: Red, lime, mauve
Air conditioning: Yes or no
Power steering: Yes or no

In how many ways may he specify his options?

3.2 A production supervisor has 4 men—Arthur, Bill, Charlie, and Don—to whom he must make work assignments. There are 4 tasks to be performed, but not all men can perform all tasks. Specifically:

Arthur can perform tasks 1, 3
Bill can perform tasks 1, 2, 4
Charlie can perform tasks 1, 2, 3, 4
Don can perform tasks 2, 4

In how many ways can the tasks be assigned if each task is given to a different man?

3.3 An investment club has 4 candidates for president, 2 candidates for vice-president, and 3 candidates for secretary-treasurer. In how many different ways may a slate of officers be elected?

3.4 In a marketing study, a subject is asked to taste 4 different brands of ginger ale and place them in order of preference. For a given subject, how many different outcomes are possible?

3.5 Seven job applicants in a personnel office are waiting to be interviewed by the personnel manager. If the personnel manager decides to interview them in random order, in how many different orders may the 7 interviews be conducted?

3.6 In 1952, the Graduate School of Business of Siwash University conferred MBA degrees on 12 graduates. At the twentieth reunion of this group, all 12 graduates met to recall old times and to brag about their economic accomplishments since graduation. Assuming that the dollar value of each graduate's net accomplishment can be determined and that none of them is lying, in how many different ways may they be placed in rank order with respect to net accomplishment?

3.7 In a job shop, 6 different operations must be performed on a particular part. These operations may be performed in any sequence. In how many different orders may these operations be performed?

3.8 At a sales clinic, individual presentations are to be given by 3 men (Art, Bob, and Chuck) and 3 women (Dorothy, Eva, and Fanny). In how many different orders might these 6 presentations be arranged if:
(a) The order of speakers is completely random?
(b) The first speaker must be a man?
(c) The first and last speakers must be women?
(d) The first speaker must be a man, and the successive presentations must alternate between men and women?
(e) The first speaker may be either a man or a woman, but the successive presentations must alternate between men and women?

3.9 Twelve books are scattered on a professor's desk. Among these, there are 2 identical copies of *Wealth of Nations*, 3 identical copies of *Das Kapital*, and 4 identical copies of *The Mentality of Apes*. The remainder of the books are all different. In how many ways may these 12 books be arranged on a shelf in distinguishably different orders?

3.10 A Scrabble player has 8 tiles with the following letters: A, A, B, E, H, S, S, S. In how many distinguishably different ways may these tiles be arranged in a row?

3.11 From a set of nine 1's and five 0's, how many different 14-digit binary numbers can be formed (assuming that the first digit must be 1)?

3.12 A wide-mouthed jar contains 10 blue and 15 green discs of equal size and weight. If all 25 of the discs are drawn randomly in sequence, how many distinguishably different sequences of blue and green discs may be drawn?

3.13 A standard deck of cards contains 26 red and 26 black cards. If all 52 cards are dealt in sequence from a well-shuffled deck, how many distinguishably different arrangements of red and black cards may be dealt?

3.14 For a class of 12 students, a professor has a list of 15 topics for class reports. For next week, he plans to assign a different one of these topics to each of the students. The assignment will be made at random, with 3 remaining topics left unassigned. In how many different ways may the assignments be made?

3.15 At a regional track meet, 18 contestants are entered in the mile run. In how many ways may 5 different prizes (first, second, third, fourth, and fifth places) be awarded?

3.16 A television variety show is to consist of 4 different commercial segments and 3 different entertainment segments. The show must begin with a commercial, and each entertainment segment must be followed by a commercial.
(a) If the sponsor has already selected the 4 commercials and 3 entertainment segments, in how many different ways may the 7 segments of the show be arranged?
(b) Suppose that the producer of the show has 7 different commercials and 5 different entertainment segments from which to choose. In how many different ways may the show's 7 segments be selected and arranged?

3.17 In a particular state, auto license plates currently consist of a sequence of three digits followed by a sequence of two letters. It is predicted that, in a few years, there will be more vehicles than the number of distinguishably different license plates which are possible under this system. If the state adopts a new system of using three digits followed by three letters, what will be the increase in the number of distinguishably different license plates which are possible?

3.18 A real estate developer has just won a contract to build an apartment complex. In order to complete this prime contract, the developer plans to place 6 (distinguishable) subcontracts with one or more of 3 different subcontractors (Allan, Bob, and Charles). The contracts are labeled 1 through 6 and will be placed with subcontractors in the order 1 through 6.
(a) In how many different ways can the developer place the subcontracts?
(b) In how many distinguishable ways can it place 3 with contractor Allan, 1 with contractor Bob, and 2 with contractor Charles?

3.19 Three managers of different departments of a large U.S. company are on European vacations. They have made a date to meet at the Tivoli Hotel in Copenhagen, quite unaware that there are 5 hotels in the city with that name. Suppose that each manager independently chooses one of these 5 hotels at random.
(a) What is the probability that all 3 managers choose the same hotel?
(b) What is the probability that all 3 managers choose different hotels?

3.20 On a Caribbean trip of the cruise ship *Princess Gertrude*, the Activities Director has established tournaments in the following activities: badminton, shuffleboard, chess, bridge, and volleyball.
(a) Assuming that a passenger may sign up for any, none, or all of these activities, in how many ways can he establish the list of tournaments in which he wishes to participate?
(b) After interviewing the passengers, who are largely of retirement age, the Activities Director decides to eliminate badminton and volleyball, but to add backgammon to the list of tournament games. However, due to scheduling problems, a passenger cannot sign up for both backgammon and bridge. However, if a person signs up for either backgammon or bridge, he can still sign up (if he wishes) for shuffleboard and chess. Under these new conditions, in how many ways can a passenger establish the list of tournaments in which he wishes to participate?
(c) Under the conditions of (b) above, suppose that a passenger definitely wishes to participate in the bridge tournament. Under these conditions, in how many ways can he establish his list of tournaments?
(d) Under the conditions of (b) above, suppose that a passenger definitely wishes to participate in the shuffleboard tournament. Under these conditions, in how many ways can he establish his list of tournaments?

3.21 A retailing chain has 20 stores. Management decides to try out a new product in a sample of 5 of those stores. How many different possible 5-store samples are there?

3.22 Twelve members of the board of directors of the Homestake Savings and Loan Association attend a board meeting. If each board member shakes hands with each other board member once and only once, how many handshakes will each board member make? What will be the total number of handshakes exchanged?

3.23 A group of 25 nurses are to be assigned to work shifts so that there are 12 on the first shift, 9 on the second shift, and 4 on the third shift. In how many different possible ways might the shift assignments be made?

3.24 A retailer receives a shipment of 20 stereo sets, of which 3 are defective. The retailer plans to select 4 sets from the shipment to place on the showroom floor. In how many ways may this random selection of 4 sets be made such that:
(a) None of the 4 sets is defective;
(b) exactly 1 of the 4 sets is defective;
(c) exactly 2 of the 4 sets are defective;
(d) exactly 3 of the 4 sets are defective;
(e) at least 2 of the 4 sets are defective;
(f) no more than 2 of the 4 sets are defective?

3.25 A bridge hand consists of 13 cards dealt from a deck of 52 cards. How many different possible bridge hands are there?

3.26 Jay Inc. has entered into a contract with television station KTN to purchase 30 seconds of commercial time on the Noontime News five days a week for fifteen weeks. Each 30-second spot will consist of a cartoon.
(a) Assume that no cartoon may be used more than once during any given week, although any cartoon shown in one week may be shown again in any other week. However, the group of 5 cartoons selected for any week must be different from that for any other week during the 15-week period. What is the minimum number of cartoons required?
(b) Assume that no cartoon can be repeated during the entire 15-week period. What is the minimum number of cartoons needed?

3.27 The Welfax Manufacturing Company is making preparations to establish a 14-member company grievance committee. Five seats on the committee must be held by supervisory personnel, 5 other seats must be held by nonsupervisory personnel, and each of the remaining 4 seats may be held by either supervisory or nonsupervisory personnel. There are 10 supervisors and 12 nonsupervisors who are candidates for selection to the committee. In how many different ways may the committee be formed from these 22 candidates?

3.28 At the horse races, a quinella bet involves betting on 2 horses, which are selected to finish first and second regardless of order of finish. For example, if a bettor buys a quinella ticket on horse No. 3 and horse No. 6, he wins if either (i) horse No. 3 wins and horse No. 6 comes in second; or (ii) horse No. 6 wins and horse No. 3 comes in second. An exacta bet, on the other hand, involves betting on 2 horses that are selected to finish first and second in an indicated order. For example, if a bettor buys an exacta ticket on horse No. 3 to win and horse No. 6 to come in second, he will win only if the two horses finish in the designated order.
(a) If 10 horses are entered in a race, how many different quinella bets are possible? Is this question concerned with combinations or permutations? Explain.
(b) If 10 horses are entered in a race, how many different exacta bets are possible?
(c) Suppose that a racetrack bettor makes 2 different bets on the same race: (i) a \$5.00 quinella bet on horses No. 3 and No. 6, and (ii) a \$5.00 exacta bet on horse No. 3 to win and horse No. 6 to finish second. Suppose also that he is so lucky ("scientific"?) that horse No. 3 wins and horse No. 6 comes in second. On which of these 2 bets would you think he would make the bigger profit? Explain.

3.29 CTO, Inc., is a prime contractor for a complicated electronic assembly. It plans to place 7 different subcontracts for subassemblies with 4 different subcontractors —Adams, Blue, Carter, and Dogwood. The contracts are numbered 1 through 7. All contracts are to be placed at the same time.
(a) If the company places 2 contracts with Adams, 3 with Carter, and 1 with

each of the other 2, in how many different ways can these contracts be assigned?

(b) If the company has decided to place all 7 contracts with a single contractor among the 4, in how many distinguishable ways can these contracts be assigned?

3.30 A governmental research unit consists of a chief scientist, 50 clerks, and 200 research scientists. The unit is divided into 3 groups. The chief scientist is employed at the GS-13 level. Among the research scientists at the GS-10 level, there are 20 in Group I, 70 in Group II, and 30 in Group III; at the GS-11 level, there are 15 in Group I, 20 in Group II, and 25 in Group III; at the GS-12 level, there are 5 in Group I, 10 in Group II, and 5 in Group III. Among the clerks, there are 10 in Group I, 25 in Group II, and 15 in Group III.

(a) Answer each of the following questions using *combinatorial* notation:

(i) The chief scientist wishes to appoint a 7-member scientific advisory board consisting entirely of GS-12 employees. Of these, he wishes 2 from Group I, 3 from Group II, and 2 from Group III. In how many different ways might this board be constituted?

(ii) The chief scientist wishes to appoint a 10-member proposal evaluation board consisting of 2 research scientists from Group I, 5 research scientists from Group II, and 3 research scientists from Group III. In how many different ways might this board be constituted?

(iii) The chief scientist wishes to appoint a 12-member professional review board consisting of 6 GS-10's, 4 GS-11's, and 2 GS-12's. He has already decided that Dr. Graymatter, a GS-11 scientist, will be chairman of this board. In how many different ways might this board be constituted?

(iv) A 9-member grievance committee is to be selected by lot from the total group of 200 research scientists and 50 clerks. What is the probability that this committee will be composed of 5 research scientists and 4 clerks?

(b) Answer the following questions using *factorial* notation:

(i) An 8-member recreation committee is to be chosen by lot from all employees (except the chief scientist). What is the probability that this committee will consist of 3 clerks and 5 research scientists?

(ii) An 11-member welfare committee is to be chosen by lot from all employees (except the chief scientist). What is the probability that this committee will consist of 2 employees from Group I, 6 employees from Group II, and 3 employees from Group III?

(iii) Dr. Honcho, the chairman of Group II, wishes to have a 10-member advisory council selected by lot from all of his subordinates. What is the probability that this council will contain at least 3 clerks?

3.31 In playing poker, the set of all possible hands is customarily partitioned into the following subsets:

Royal flush
Straight flush (excluding royal flush)
Four of a kind
Full house (3 of a kind plus 1 pair)
Flush (excluding straight and royal flushes)
Straight (excluding straight and royal flushes)
Three of a kind (excluding full houses)
Two pairs (excluding 4 of a kind)
One pair (excluding full houses and 2 pairs)
All other possible hands.

How many possible hands are contained in the following subsets selected from the list above:

(a) One pair (excluding full houses and 2 pairs)
(b) Four of a kind
(c) Full house
(d) Three of a kind (excluding full houses)

Chapter 4
Sets and Sample Spaces

Probability theory had its beginnings in the analysis of gambling problems, from which seventeenth century philosophers and mathematicians derived their basic insights concerning chance mechanisms. Today, however, our approach to these problems is more sophisticated than the methods employed in the mid 1600s. Specifically, our approach will be based on the theory of sets, which was not formally introduced into mathematics until late in the nineteenth century.

4.1 BASIC SET CONCEPTS

The modern approach to the study of probability, as well as to other branches of "pure" and "applied" mathematics, is rooted in the theory of sets. Although the formal development of set theory is both elegant and elaborate, the fundamental idea of a set is simple and intuitive. The term *set* refers in a very general way to any *well-defined collection* of "objects." Each of the following is an example of a set:

1. The cans of stewed tomatoes on the shelves of a particular supermarket.
2. The men who served as secretary of state under the presidential administration of James Madison.
3. The common stock issues traded on the New York Stock Exchange on August 25, 1972.
4. The symphonies composed by Wolfgang Amadeus Mozart.
5. The manned trips to the moon carried out by the United States during the calendar years 1970 and 1971.

4.1.1 Ways of Defining Sets

In set theory, an important requirement is that a set be well-defined. Clear definition is necessary in order to declare whether or not any given object belongs, or does not belong, to a particular set. Thé objects belonging to a given set are called the *elements* or *members* of that set. For example, under the administration of James Madison there were two, and only two, secretaries of state—Robert Smith and James Monroe. Thus, if a set is defined as the men who served as secretary of state under Madison, Robert Smith and James Monroe clearly are members of that set and Thomas Jefferson clearly is not a member of that set.

The symbol $\in$ is used to denote that an object "is an element of," "is a member of," or "belongs to" a particular set. The expression "James Monroe $\in$ James Madison's secretaries of state" denotes that James Monroe belongs to the set of Madison's secretaries of state. Similarly, the symbol $\notin$ denotes that an object "is not an element of," "is not a member of," or "does not belong to" a particular set. Hence the expression "Thomas Jefferson $\notin$ James Madison's secretaries of state" indicates that Thomas Jefferson does not belong to the set of Madison's secretaries of state.

There are two common ways of defining sets—*listing* and *describing*. In either case, the set is defined within braces: $\{\ldots\}$. By the *listing* method, a set is specified by listing all of its elements. For example, using this method to define the set of Madison's secretaries of state, we would have

$$S = \{\text{Robert Smith, James Monroe}\}$$

This expression may be read "S is a set consisting of the elements Robert Smith and James Monroe." It then follows that Robert Smith $\in S$, James Monroe $\in S$, whereas Florence Nightingale $\notin S$.

By the *describing* method, a set is specified by describing the elements of the set in terms of some common property. By this method, the set of Madison's secretaries of state would be denoted as:

$$S = \{x \mid x \text{ is one of James Madison's secretaries of state}\}$$

This expression is read "S is the set of elements x such that x is one of James Madison's secretaries of state." Notice that the symbol $\mid$ stands for the phrase "such that."

In the set consisting of Madison's secretaries of state, it makes no substantive difference whether the set is defined by the listing method or the describing method; either method provides a clear, unambiguous definition of the set. Why, then, should there be two methods? The answer is that, with many important kinds of sets, either one method or the other is not feasible or is impossible. For instance, if a set should contain a very large number of elements, such as the set of all voters who cast their ballots for Richard Nixon in the 1972 presidential election, it would not be practical to list every member of that set. Thus, the only practical way of defining such a set would be the describing method. In contrast, consider the set consisting of the symbols %, #, k, and 7. For this set, there is no common property that could be used to describe membership in the set unambiguously, and the listing method then would be necessary.

In conceptualizing sets, it is important to distinguish between "finite" and "infinite" sets. A *finite* set is one that consists of a fixed, definite number of elements. If a set is finite, it should be possible to count all of its elements. In an *infinite* set, however, the number of elements is limitless or unspecifiable, and therefore cannot be counted. Examples of finite sets would be all the legal residents of the state of California, or all the industrial firms incorporated within the state of New Jersey, or all the transactions posted in a particular ledger, or all the articles of merchandise stored in a particular warehouse. Examples of infinite sets would include all the possible tourists who might visit Yellowstone National Park, all the possible flips of a coin, or all the transistors that could be produced by a particular manufacturer. Many sets of practical concern actually are finite but are so large, such as the population of the United States, that they may be considered as infinite for purposes of analysis. We might regard such very large finite sets as being *effectively infinite.*

4.1.2 Subsets

Consider the following two sets:

$$X = \{r,s,t,u\} \qquad Y = \{r,s,t\}$$

We readily see that every element contained in Y is also contained in X. In this situation, we may say that Y is a "subset" of X. More formally, any set A is a *subset* of set B if each element of A is also an element contained in B. Symbolically, this situation is denoted as $A \subseteq B$, which is read "A is a subset of B." Thus, for the two sets listed above, we may write $Y \subseteq X$.

In set theory, any set is commonly regarded as a subset of itself. Thus, in the above example, X may be considered as a subset of itself, which is denoted as $X \subseteq X$. Similarly, we may write $Y \subseteq Y$.

A set that contains no elements is called the *empty* or *null* set. The empty set is denoted as $\emptyset$. It is customary to regard the empty set as a subset of any set. Therefore, in the above example we have $\emptyset \subseteq X$ and $\emptyset \subseteq Y$.

If A is a subset of B such that A contains at least one element, but not all the elements, contained in B, then A is said to be a *proper subset* of B, which is denoted as $A \subset B$. Thus, although the empty set and the entire set are regarded as subsets of a given set, they are not proper subsets of that set. In our example, expressions such as $Y \subseteq X$ and $Y \subset X$ are both correct. However, $X \subset X$ and $\emptyset \subset X$ are invalid expressions, since the entire set and the empty set are not proper subsets (although they are subsets).

For any set A, the expression $n(A)$ denotes the number of elements contained in the set. In our example, we have $n(X) = 4$ and $n(Y) = 3$. It is important to note that $n(\emptyset) = 0$.

In applying set theory to probability problems, it often is useful to obtain a count of the total number of different subsets that may be formed from a particular set. In our example, we have seen that Y, $\emptyset$, and X are all subsets of X. However, these are not the only subsets of X, as we can see from Table 4.1, which lists all the possible subsets of X. Notice that the numbers of subsets of different sizes, shown in the far-right column of Table 4.1, are obtained by applying Formula (3.7) for determining the number of possible combinations of r objects selected from a set of n objects. Thus, the number of ways of forming the empty set from a set of four elements is $\binom{4}{0}$. Similarly, the number of ways of forming a subset of one element from a set of four elements is $\binom{4}{1}$. Continuing this procedure for subsets of two elements, three elements, and four elements, we obtain:

$$\binom{4}{0} + \binom{4}{1} + \binom{4}{2} + \binom{4}{3} + \binom{4}{4} = 16.$$

Sixteen is the total number of subsets that can be formed from a set containing four elements.

Table 4.1

All Possible Subsets of $X = \{r,s,t,u\}$

Number of Elements in the Subset	*Subsets of X*	*Number of Subsets*
0	$\emptyset$	$\binom{4}{0} = 1$
1	$\{r\}, \{s\}, \{t\}, \{u\}$	$\binom{4}{1} = 4$
2	$\{r,s\}, \{r,t\}, \{r,u\}, \{s,t\}, \{s,u\}, \{t,u\}$	$\binom{4}{2} = 6$
3	$\{r,s,t\}, \{r,s,u\}, \{s,t,u\}, \{r,t,u\}$	$\binom{4}{3} = 4$
4	$\{r,s,t,u\}$	$\binom{4}{4} = 1$
		16

In general, if a set consists of n elements, the total number of subsets of that set is

$$\sum_{r=0}^{n} \binom{n}{r}$$

It may be demonstrated mathematically that the above expression is equal to 2^n. Therefore, the total number of subsets of a set containing n elements is 2^n. Notice in Table 4.1, for example, that the total number of subsets is 16, which is 2^4.

4.1.3 Complementary and Universal Sets

In considering the elements of a set A, we frequently may wish to consider the collection of objects that is not contained in A. This set of objects that is not in A is called the *complement* of A, denoted as A'. For example, suppose we define the set $V = \{x \mid x$ is a utilities issue currently listed on the New York Stock Exchange$\}$. In this case, how do we define the complementary set V'? Our first inclination might be to say that V' is the set of all non-utilities issues currently listed on the New York Stock Exchange. But this would not necessarily be correct, because V' could just as well be the set of all utilities stocks that are not listed on the New York Exchange, the set of all non-utilities stocks that are not listed on the New York Exchange, and so on. It becomes obvious that in order for the complement of a set to be well-defined, it is necessary to specify some reference set that limits or bounds the entire collection of elements under consideration in a given situation. This delimiting set is called the *universal* set, denoted as $\mathcal{U}$. Thus, if we define $\mathcal{U}$ as the set of all stock issues currently listed on the New York Stock Exchange, then V' is the set of all non-utilities issues currently listed on the New York Stock Exchange. But if we should define $\mathcal{U}$ as the set of all utilities stocks in the U.S., then V' would be the set of all utilities stocks in the U.S. that are not currently listed on the New York Exchange.

In dealing with decision problems, statisticians frequently refer to the *universe* or *population* under consideration. In common usage, the term "population" generally is employed to refer to the total number of people in some geographical area or political subdivision, such as the population of the United States or the population of Cook County, Illinois. To the statistician, however, "population" is a much broader concept. The term "population" refers to any *entire* aggregate of elements under consideration in a given situation. Thus a population may be a group of human beings, as in the popular sense of the term; but it may just as well refer to any other universal set of objects or phenomena, such as articles produced by a particular manufacturing process, dwelling units in a particular city, or industrial firms belonging to a particular industry.

As we pointed out in Chapter One, sampling from a population is an important statistical device for obtaining information to reduce uncertainty in decision making. We should recognize that a population is simply a universal set. In this context, a *sample* may be defined as a subset of the universe.

4.2 DEFINING EVENTS ON SAMPLE SPACES

Let us consider the random process of drawing a card from a well-shuffled bridge deck. In this case, we might regard the deck of cards as a universe, and the card that is drawn from the deck is a sample taken from that universe. Since the deck contains fifty-two cards, each distinguishably unique, we might say that the process has fifty-two possible *elementary outcomes*. The set of all possible elementary outcomes of such a random process is called the *sample space*. Each elementary outcome is referred to as a *sample point* or *element* of the sample space. Thus each element of the sample space represents an elementary outcome of the random process, and each trial of the process must result in an outcome corresponding to exactly one element of the sample space. Diagrammatically, we might represent the sample space of our card-drawing "experiment" as in Figure 4.1.

Figure 4.1

Sample Space for Drawing a Single Card from a 52-Card Bridge Deck

	Spades	Hearts	Diamonds	Clubs
Ace	♠ A •	♥ A •	♦ A •	♣ A •
King	♠ K •	♥ K •	♦ K •	♣ K •
Queen	♠ Q •	♥ Q •	♦ Q •	♣ Q •
Jack	♠ J •	♥ J •	♦ J •	♣ J •
10	♠ 10 •	♥ 10 •	♦ 10 •	♣ 10 •
9	♠ 9 •	♥ 9 •	♦ 9 •	♣ 9 •
8	♠ 8 •	♥ 8 •	♦ 8 •	♣ 8 •
7	♠ 7 •	♥ 7 •	♦ 7 •	♣ 7 •
6	♠ 6 •	♥ 6 •	♦ 6 •	♣ 6 •
5	♠ 5 •	♥ 5 •	♦ 5 •	♣ 5 •
4	♠ 4 •	♥ 4 •	♦ 4 •	♣ 4 •
3	♠ 3 •	♥ 3 •	♦ 3 •	♣ 3 •
2	♠ 2 •	♥ 2 •	♦ 2 •	♣ 2 •

Until now we have used the term "event" intuitively. We may now define an *event* as any subset of a sample space. We use the term *simple event* to designate an event that corresponds to a single sample point. Thus the event "drawing the queen of spades" is a simple event. A *compound event* is an event that is defined by more than a single sample point. That is, a compound event is composed of two or more simple events. The occurrence of any one of the simple events that comprise a compound event permits us to say that the compound event has occurred. For example, the event "drawing an ace" is a compound event composed of four simple events, as illustrated by the shaded area in Figure 4.2. If a single draw of a card turns up either the ace of hearts, or ace of diamonds, or ace of clubs, or ace of spades, we may say that the event "drawing an ace" has occurred.

Figure 4.2

Subset of the Sample Space Defining the Event "Drawing an Ace"

	Spades	Hearts	Diamonds	Clubs
Ace	♠ A •	♥ A •	♦ A •	♣ A •
King	♠ K •	♥ K •	♦ K •	♣ K •
Queen	♠ Q •	♥ Q •	♦ Q •	♣ Q •
Jack	♠ J •	♥ J •	♦ J •	♣ J •
10	♠ 10 •	♥ 10 •	♦ 10 •	♣ 10 •
9	♠ 9 •	♥ 9 •	♦ 9 •	♣ 9 •
8	♠ 8 •	♥ 8 •	♦ 8 •	♣ 8 •
7	♠ 7 •	♥ 7 •	♦ 7 •	♣ 7 •
6	♠ 6 •	♥ 6 •	♦ 6 •	♣ 6 •
5	♠ 5 •	♥ 5 •	♦ 5 •	♣ 5 •
4	♠ 4 •	♥ 4 •	♦ 4 •	♣ 4 •
3	♠ 3 •	♥ 3 •	♦ 3 •	♣ 3 •
2	♠ 2 •	♥ 2 •	♦ 2 •	♣ 2 •

It is important to realize that the definition of particular events is an arbitrary matter, depending on the particular characteristic of the outcomes that may be of interest. Suppose, for example, that you propose the following game to a friend: You will shuffle a standard deck, and he will draw a single card from the shuffled deck. If he draws a spade, you will pay him $1.00; but he will pay you 50¢ if the card is a heart, 30¢ if it is a diamond, and 20¢ if it is a club. In this case, you are interested in the suit of the card drawn, and so you are concerned with four events—drawing a spade, heart, club, or diamond. Each of these four events is a compound event that is made up of thirteen simple events, since each suit is defined by thirteen sample points representing the thirteen denominations of that suit. Similarly, in a different game, you might be interested in the color of the card drawn, so that you would be concerned with two events ("red" and "black"), each defined by twenty-six sample points. Other events of common concern to card players might be drawing an ace (four sample points), or drawing a face card (twelve sample points).

When you think about it, the number of different events that can be defined on a fifty-two-point sample space is staggering. The total number of events (subsets) is $2^n = 2^{52} = 4{,}503{,}599{,}627{,}370{,}496$. Of these, the fifty-two simple events (sets composed of a single sample point) and the null event (empty set) account for fifty-three of the total, leaving a remainder of 4,503,599,627,370,443 compound events. Of course, most of these combinations of sample points define only events that are logically possible but would not be of the slightest interest for any practical purpose.

In dealing with probability problems, the critical considerations in defining an event are (1) to define the event such that the definition includes all those sample points, and only those sample points, that represent the elementary outcomes of concern, and (2) to define the event in an unambiguous manner such that each and every point in the sample space may be identified as being included or excluded in the set defining the event. As we shall see in later chapters, these considerations are crucial, since whether or not a particular sample point is included in the definition of an event can materially affect the probability assigned to the occurrence of that event.

4.3 INTERSECTIONS OF EVENTS

For variety's sake, let us consider another random process—the ancient pastime of throwing a pair of dice. In fact, it was a dice problem, presented by the Chevalier de Meré to Blaise Pascal in 1654, which led to the formal origin of mathematical probability theory. Assuming that each die is "fair," each of the six sides, marked by spots representing the numbers 1 through 6, is equally likely to come up when the die is cast. To keep the two dice distinct in our thinking, let us identify one as a red die and the other as a green die. For each of the six faces that might come up on the red die, any of the six faces may come up on the green die when the two dice are thrown. Thus we might regard the random process of rolling a pair of dice as having $6 \times 6 = 36$ possible elementary outcomes. This thirty-six-element sample space is represented diagrammatically in Figure 4.3. In this sample space, each sample point represents an elementary outcome that is defined by assigning a value to each of two distinguishable dice in a specified order. That is, in sequence, the value of the red die is specified first, and the value of the green die is specified second. Thus, each sample point is defined as an *ordered pair* of values. Hence, if we let r = the number of dots on the upper face of the red die when it is cast and g = the number of dots on the face of the green die, we may then denote an elementary outcome by the ordered pair (r,g). For example, (5,3) indicates the outcome that the red die comes up 5 and the green die comes up 3, whereas (3,5) indicates that the red die comes up 3 and the green die comes up 5.

Figure 4.3

Sample Space of Outcomes for Tossing a Pair of Dice

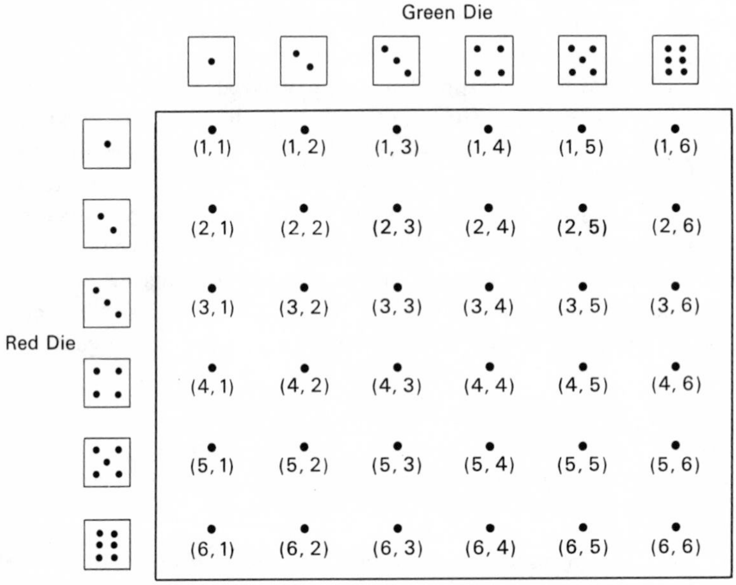

Red Die \ Green Die	1	2	3	4	5	6
1	(1, 1)	(1, 2)	(1, 3)	(1, 4)	(1, 5)	(1, 6)
2	(2, 1)	(2, 2)	(2, 3)	(2, 4)	(2, 5)	(2, 6)
3	(3, 1)	(3, 2)	(3, 3)	(3, 4)	(3, 5)	(3, 6)
4	(4, 1)	(4, 2)	(4, 3)	(4, 4)	(4, 5)	(4, 6)
5	(5, 1)	(5, 2)	(5, 3)	(5, 4)	(5, 5)	(5, 6)
6	(6, 1)	(6, 2)	(6, 3)	(6, 4)	(6, 5)	(6, 6)

In the popular dice game of craps, interest is focused on the sum of the two values when a shooter rolls the dice. An event that might be of concern to a player is the event that the sum equals 10; i.e., $r + g = 10$. We can define this event (call it E_1) on the sample space by listing the sample points for which $r + g = 10$. Thus,

$$E_1 = \{(4,6), (5,5), (6,4)\}$$

Another event that might be of concern to a player would be shooting a 10 "the hard way" (double 5). This is a simple event that corresponds to the single sample point (5,5). If we denote this event by E_2, we have

$$E_2 = \{(5,5)\}$$

Notice that the sample point (5,5) is common to both events E_1 and E_2. If a player does shoot a double 5, we may say that both E_1 and E_2 have occurred. The occurrence of one of these two events, however, does not necessarily imply that the other event has occurred. If a player shoots either (6,4) or (4,6), we can say that E_1 has occurred, but not E_2. The set of those elements that are common to two events is referred to as the *intersection* of the two events. In our example, the intersection of E_1 and E_2 is the common element (5,5). In set theory, the symbol $\cap$ is used to denote an intersection. Thus we may write $E_1 \cap E_2 = \{(5,5)\}$. The expression $E_1 \cap E_2$ is read "E_1 cap E_2" or "E_1 intersect E_2."

Suppose we define another event, E_3, as "rolling a sum greater than 9"; i.e., $r + g > 9$. Symbolically,

$$E_3 = \{(4,6), (5,5), (5,6), (6,4), (6,5), (6,6)\}$$

As they have been defined, the three events E_1, E_2, and E_3 all intersect with one another. Specifically,

$$\begin{aligned} E_1 \cap E_2 &= \{(5,5)\} \\ E_1 \cap E_3 &= \{(4,6)(5,5)(6,4)\} \\ E_2 \cap E_3 &= \{(5,5)\} \\ E_1 \cap E_2 \cap E_3 &= \{(5,5)\} \end{aligned}$$

If a player rolls a (6,4) or a (4,6), we may say that both events E_1 and E_3 have occurred. If he rolls a double 5, we may say that all three events have occurred.

4.4 UNIONS OF EVENTS

We have seen that the intersection of two events A and B, denoted by $A \cap B$, refers to the set of all sample points which are contained in *both* A and B. A related concept is the *union* of two events, which is the set of all sample points contained in *either* A *or* B. That is, the union of A and B consists of all the sample points that are contained either in A but not B, or in B but not A, or in both A and B. The union of A and B is denoted by $A \cup B$, which is read "A cup B" or "A union B." Thus the set $A \cup B$ consists of the three sets $A \cap B'$, $A' \cap B$, and $A \cap B$. Generally we shall refer to the union of A and B as "A or B," bearing in mind that the word "or" is used inclusively to mean "A and/or B."

As an example, consider the following two sets:

$$A = \{a,e,i,o,u\}$$
$$B = \{a,b,c,d,e\}$$

The elements contained in A but not in B form the set

$$A \cap B' = \{i,o,u\}$$

Similarly, the elements contained in B but not in A form the set

$$A' \cap B = \{b,c,d\}$$

Also, the elements contained in both A and B form the set

$$A \cap B = \{a,e\}$$

By combining the elements contained in these three intersections, we obtain the union

$$A \cup B = \{a,e,i,o,u,b,c,d\}$$

As this example illustrates, if an element is contained in either A or B, it is included in the union of A and B. This is in contrast to the intersection of A and B, which consists of only those elements which are common to both A and B.

For another illustration, let us return to our dice example. We have already seen that

$$E_1 \cap E_3 = \{(4,6), (5,5), (6,4)\}$$

We may further observe that

$$E_1' \cap E_3 = \{(5,6), (6,5), (6,6)\}$$

and

$$E_1 \cap E_3' = \emptyset$$

Thus the union of E_1 and E_3 is

$$E_1 \cup E_3 = \{(4,6), (5,5), (6,4), (5,6), (6,5), (6,6)\}$$

Notice in this example that $E_1 \cup E_3$ is identical to E_3. This is because E_1 is a subset of E_3, so that all of the elements contained in E_1 are also contained in E_3.

4.5 MUTUALLY EXCLUSIVE EVENTS

Let us define one more event, E_4 as "rolling an odd-numbered sum." Inspection of the sample space in Figure 4.3 indicates that there are eighteen sample points defining this event. Specifically,

$$E_4 = \{(1,2), (1,4), (1,6), (2,1), (2,3), (2,5), (3,2), (3,4), (3,6), (4,1), (4,3), (4,5), (5,2), (5,4), (5,6), (6,1), (6,3), (6,5)\}$$

We may note that $E_3 \cap E_4 = \{(5,6), (6,5)\}$. Thus if a player should roll either (5,6) or (6,5), both events E_3 and E_4 will have occurred. Notice, however, that E_4 has no sample points in common with either E_1 or E_2. In order for either E_1 or E_2 to occur, the player must roll a 10, which is an even number rather than an odd number. In such cases, wherein two events have no common sample points, the occurrence of one event prohibits the simultaneous occurrence of the other event. Such events, which by nature of their definitions occupy no common part of the sample space and therefore cannot occur simultaneously, are said to be *mutually exclusive*. Because two mutually exclusive events have no sample points in common, their intersection is an empty set. Thus we may write:

$$E_1 \cap E_4 = \emptyset$$

and

$$E_2 \cap E_4 = \emptyset$$

Consequently, if we wish to know whether two events are mutually exclusive, we need simply to determine whether their intersection is empty. If two events are mutually exclusive, they cannot occur simultaneously.

4.6 PARTITIONING SAMPLE SPACES

We have defined the event E_1 as rolling a sum equal to 10. The complement of E_1 is the event defined by all sample points that are not included in the definition of E_1. Thus the complement of E_1 denoted by E_1' represents the event of rolling any sum other than 10. Similarly, since E_4 represents the event of rolling an odd-numbered sum, its complement E_4' represents an event of rolling an even-numbered sum. Obviously, since an event and its complement have no sample points in common, the intersection of an event and its complement is empty. For example, $E_4 \cap E_4' = \emptyset$; it is impossible to roll an odd number and an even number simultaneously.

From the various events that we have defined on the sample space of Figure 4.3, let us now consider E_1 and E_4 and their complements E_1' and E_4'. We could then define the following events:

$E_5 = E_4 \cap E_1$ Throwing an odd-numbered sum equal to 10

$E_6 = E_4' \cap E_1$ Throwing an even-numbered sum equal to 10

$E_7 = E_4 \cap E_1'$ Throwing an odd-numbered sum not equal to 10

$E_8 = E_4' \cap E_1'$ Throwing an even-numbered sum not equal to 10

Obviously, $E_5 = \emptyset$, since it is impossible to throw an odd-numbered sum that is equal to 10. Each of the other three events—E_6, E_7, and E_8—is a possible event defined by particular sample points, as shown in Table 4.2. Notice that all of these four events are mutually exclusive—none of the four events is defined by any sample points in common with any other of the four events. Furthermore, these four events are *collectively exhaustive* in the sense that every element in the sample space is represented among the four events. In other words, each and every element of the sample space is included in the definition of one and only one of the four events: E_5, E_6, E_7, and E_8. Such a division of a sample space into a group of events that are mutually exclusive and collectively exhaustive is called a *partition* of the sample space. In approaching many of the problems that are presented in the following pages, you will find that the technique of partitioning a sample space may lead to a much faster and more meaningful solution than the application of standard probability formulas.

Table 4.2

Partition of Sample Space for Tossing of Two Dice

	E_1: Sum = 10	E_1': Sum ≠ 10
E_4: Odd Sum	$E_5 = \emptyset$	$E_7 =$ {(1,2), (1,4), (1,6), (2,1), (2,3), (2,5), (3,2), (3,4), (3,6), (4,1), (4,3), (4,5), (5,2), (5,4), (5,6), (6,1), (6,3), (6,5)}
E_4': Even Sum	$E_6 =$ {(4,6), (5,5), (6,4)}	$E_8 =$ {(1,1), (1,3), (1,5), (2,2), (2,4) (2,6), (3,1), (3,3), (3,5), (4,2), (4,4), (5,1), (5,3), (6,2), (6,6)}

Example 1:

An electronic assembly requires one each of Type A and Type B tubes. At a particular point in the assembly process, an assembler reaches with one hand into a bin containing Type A tubes, and with the other hand reaches *into a bin containing Type B tubes. He randomly selects a tube from each bin and inserts them into the assembly. Suppose, at a particular moment when the assembler is going through this operation, there are only four tubes remaining in the Type A bin (call them A_1, A_2, A_3, and A_4) and five tubes remaining in the Type B bin (call them B_1, B_2, B_3, B_4, and B_5). If we view this operation of selecting two tubes as a random process, with four ways of selecting a tube from the Type A bin and five ways of selecting a tube from the Type B bin, there are $4 \times 5 = 20$ elementary outcomes, as illustrated in the tree diagram of Figure 4.4. That is, we have a twenty-element sample space, as shown in Figure 4.5.*

Figure 4.4

Tree Diagram Illustrating Elementary Outcomes

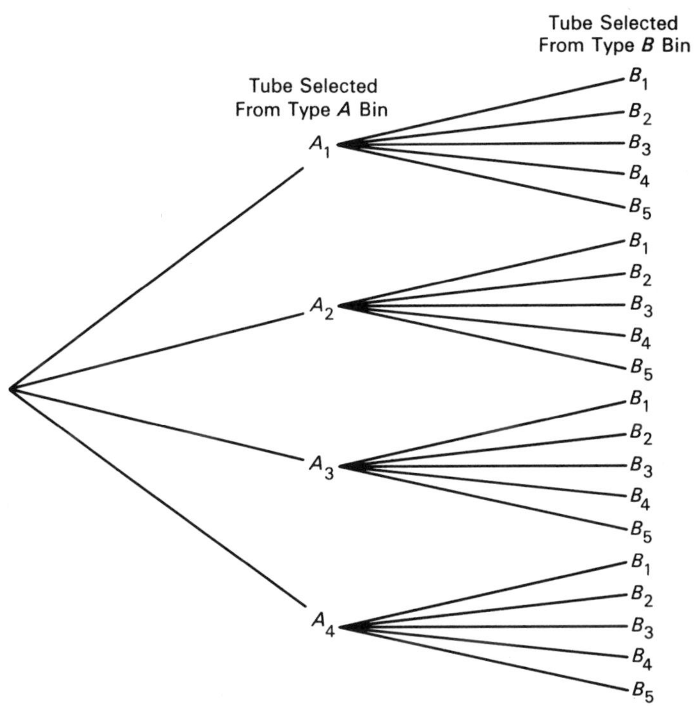

Assume that, of the total of nine tubes (four of Type A and five of Type B), A_4, B_2, and B_5 are defective, whereas the remainder are good. We might then define some events in terms of the total number of defective tubes that the assembler draws from the two bins. Let E_0 be the event that neither of the two tubes selected is defective, E_1 be the event that just one of the two

tubes is defective, and E_2 be the event that both tubes are defective. In terms of the sample points in Figure 4.5:

$$E_0 = \{(A_1B_1), (A_1B_3), (A_1B_4), (A_2B_1), (A_2B_3), (A_2B_4), (A_3B_1), (A_3B_3), (A_3B_4)\}$$

$$E_1 = \{(A_1B_2), (A_1B_5), (A_2B_2), (A_2B_5), (A_3B_2), (A_3B_5), (A_4B_1), (A_4B_3), (A_4B_4)\}$$

$$E_2 = \{(A_4B_2), (A_4B_5)\}$$

Notice that these three events are mutually exclusive and exhaust the sample space. In other words, the three events—E_0, E_1, and E_2—constitute a partition of the sample space.

Figure 4.5

Sample Space for Selecting Two Tubes

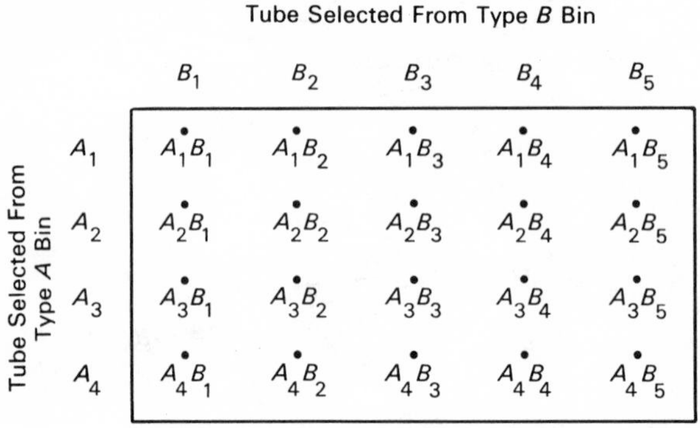

Tube Selected From Type *B* Bin

Tube Selected From Type *A* Bin	B_1	B_2	B_3	B_4	B_5
A_1	• A_1B_1	• A_1B_2	• A_1B_3	• A_1B_4	• A_1B_5
A_2	• A_2B_1	• A_2B_2	• A_2B_3	• A_2B_4	• A_2B_5
A_3	• A_3B_1	• A_3B_2	• A_3B_3	• A_3B_4	• A_3B_5
A_4	• A_4B_1	• A_4B_2	• A_4B_3	• A_4B_4	• A_4B_5

Example 2:

A manufacturer produces steel shafts as components for a precision machine. Each shaft is carefully inspected to determine whether it is within tolerance limits with respect to the dimensions of its diameter and length. If a shaft is within tolerance limits with respect to these two characteristics, it is accepted. If it is outside the tolerance limits with respect to either or both characteristics, it is scrapped. Focusing our attention on the diameters of the shafts, we might consider two events, D (in tolerance with respect to diameter) and D' (out of tolerance with respect to diameter). These two events—D and D'—form a partition of the sample space, since they are mutually exclusive and collectively exhaustive; each shaft must be either in or out of tolerance with respect to diameter. We might also consider two other events, L (in tolerance with respect to length) and L' (out of tolerance with respect to length). Just as D and D' form a partition, so do the two events

L and L' form a partition (but a different partition) of the sample space. Considering both diameter and length together, we might define certain other events in terms of intersections:

$W = D \cap L$	*In tolerance with respect to both dimensions*
$X = D \cap L'$	*Diameter in tolerance but length out of tolerance*
$Y = D' \cap L$	*Diameter out of tolerance but length in tolerance*
$Z = D' \cap L'$	*Both diameter and length out of tolerance*

These four events—W, X, Y, and Z—form another partition of the sample space, since they meet the requirements that they are mutually exclusive and collectively exhaustive. This partitioning is illustrated in Table 4.3.

Table 4.3

Partition of Sample Space for Inspection of Shafts

		Length	
		In Tolerance L	*Out of Tolerance* L'
Diameter	*In Tolerance* D	$W = D \cap L$	$X = D \cap L'$
	Out of Tolerance D'	$Y = D' \cap L$	$Z = D' \cap L'$

4.7 VENN-EULER DIAGRAMS

We have seen how events may be defined on a sample space in terms of the language and symbolism of set theory. The set operations and relations expressed in this language and symbolism may also be portrayed pictorially by *Venn-Euler diagrams.* Developed by the English logician John Venn (1834–1923) and the Swiss geometrician Leonhard Euler (1707–1783), Venn-Euler diagrams are also simply referred to as Venn diagrams or Euler diagrams.

In a Venn-Euler diagram, the universal set is represented by the set of points contained in a rectangle. For example, consider a universal set consisting of the five vowels and the first five consonants of the English alphabet. That is,

$$\mathcal{U} = \{a,b,c,d,e,f,g,i,o,u\}$$

This universal set is represented pictorially by the Venn-Euler diagram in Figure 4.6.

Figure 4.6

Universal Set

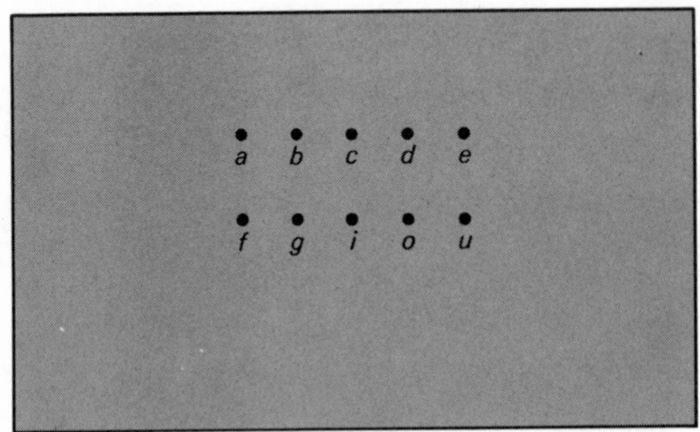

A set is represented in a Venn-Euler diagram by some geometric shape, such as a circle, which encompasses all of the elements belonging to the set. For example, let us define a set A as the set of the five English vowels. Symbolically,

$$A = \{a,e,i,o,u\}$$

This set is represented in a Venn-Euler diagram by the shaded area in Figure 4.7. Recalling that the complement of a given set consists of all elements in the universe which are not contained in that set, the complement of A is

$$A' = \{b,c,d,f,g\}$$

which is represented by the shaded area of the Venn-Euler diagram in Figure 4.8.

Figure 4.7

A Set Within the Universal Set

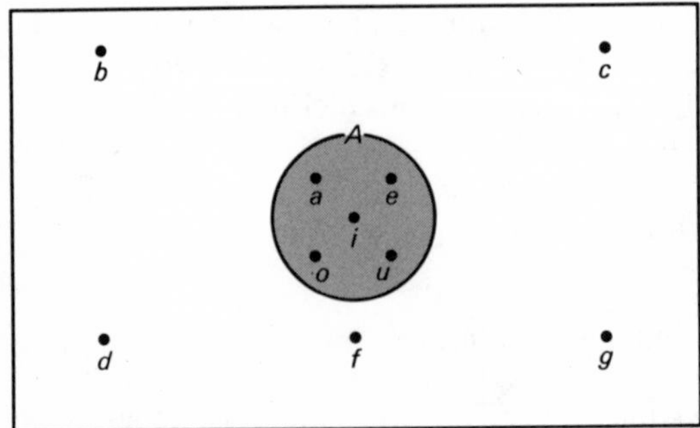

Figure 4.8

The Complement of a Set

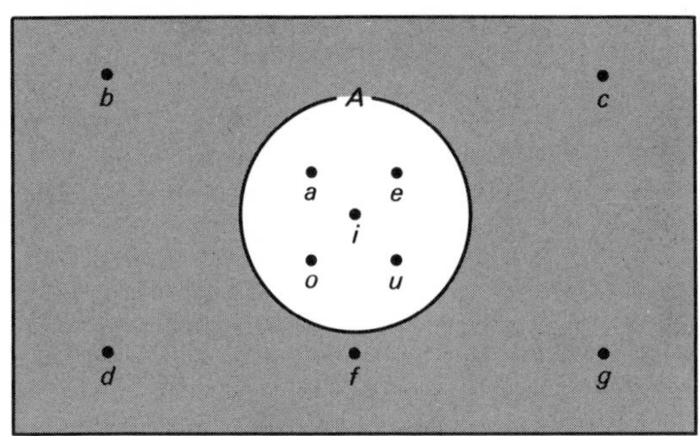

The intersection of two sets in a Venn-Euler diagram is represented by two geometric shapes which overlap in such a way that the elements contained in the overlapping area are those elements, and only those elements, which are common to the two sets. To illustrate, let us define another set, B, consisting of the first five letters of the alphabet. That is,

$$B = \{a,b,c,d,e\}$$

By inspection, the two elements a and e, and only those two elements, are common to sets A and B. Thus, $A \cap B = \{a,e\}$. This intersection is shown by the shaded area in Figure 4.9. Similarly, there are three elements— i, o, u— which are contained in A but not in B. Therefore, $A \cap B' = \{i,o,u\}$, as shown by the shaded area in Figure 4.10. Likewise, the three elements b, c, and d are contained in B but not in A, so that $A' \cap B = \{b,c,d\}$, which is represented by the shaded area in Figure 4.11.

Figure 4.9

The Intersection of A and B

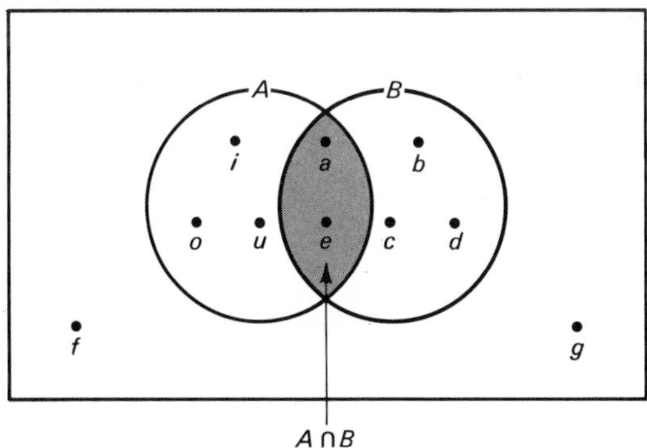

Figure 4.10

The Intersection of A and B′

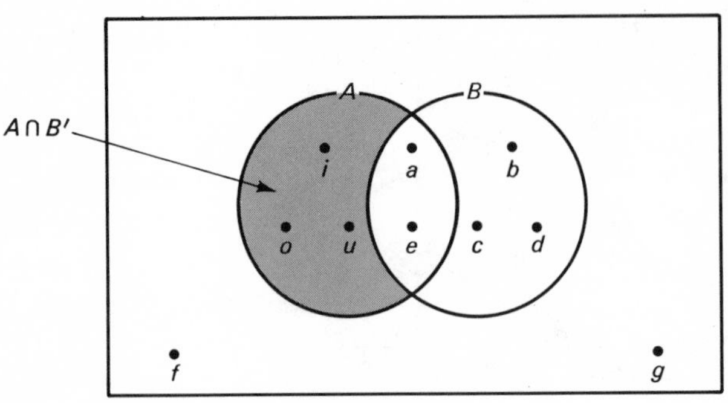

Figure 4.11

The Intersection of A′ and B

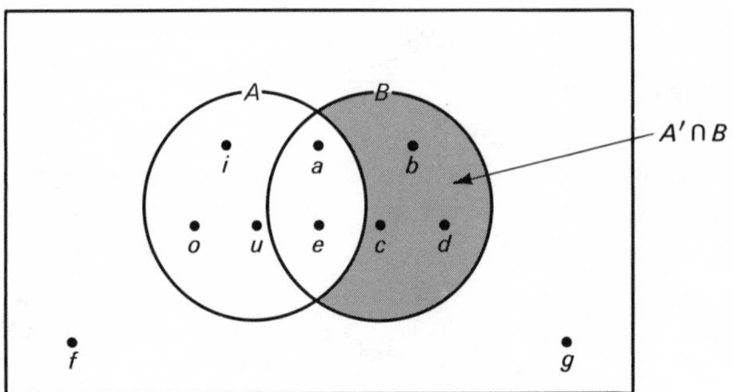

As we noted in Section 4.4, the union of two events A and B consists of all of the elements contained in the three intersections $A \cap B'$, $A' \cap B$, and $A \cap B$. Thus, by combining the shaded areas in Figures 4.9, 4.10, and 4.11, we obtain the shaded area of Figure 4.12, which represents the union of A and B. From Figure 4.12 we see that

$$A \cup B = \{a,b,c,d,e,i,o,u\}$$

In Figure 4.9, events A and B are represented by two circles which overlap in such a way that the elements which are common to both events are contained in the region of overlap. This overlapping reflects the fact that the intersection of A and B is non-empty. In contrast, if two events are mutually exclusive (i.e., if their intersection is the empty set), they are represented in a Venn-Euler diagram by two separate shapes that do not overlap. To illustrate this principle, let us define a new event

$$C = \{i,o,u\}$$

Note that event C has no elements in common with event B. Thus, B and C are mutually exclusive. This is demonstrated by the non-overlapping circles in Figure 4.13. The absence of overlap indicates that $B \cap C = \emptyset$, $B \cap C' = B$, and $B' \cap C = C$. Thus, the union of B and C simply consists of all of the elements contained in B and C. This is illustrated by the shaded regions of Figure 4.13.

Figure 4.12

The Union of A and B

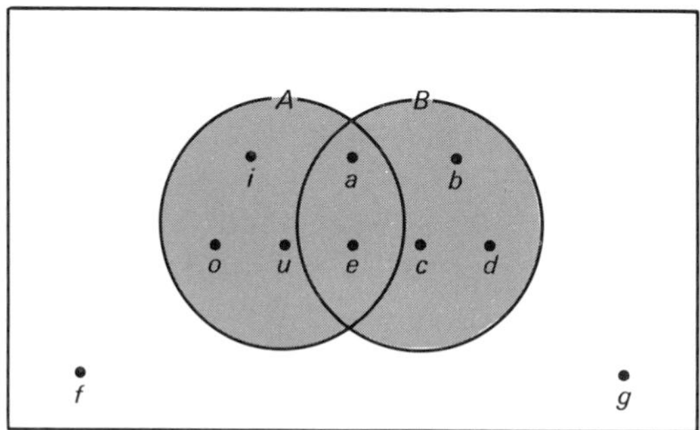

Figure 4.13

Union of Two Mutually Exclusive Events

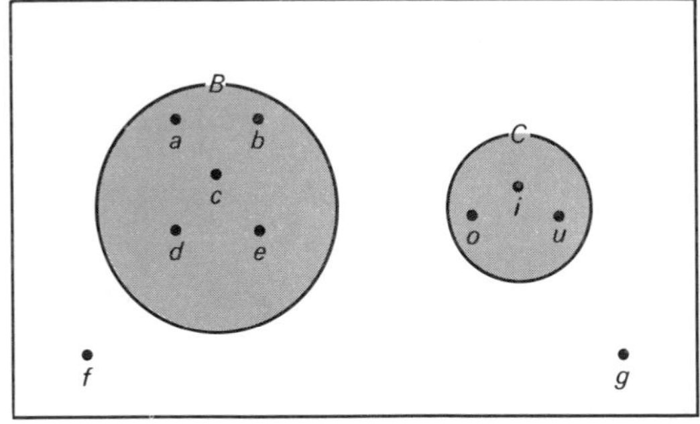

While event C has no elements in common with event B, we may note that all of the elements in C are included among the elements of the event A. That is, C is a subset of A. This is illustrated in Figure 4.14 by two circles drawn in such a way that the circle representing C is entirely contained inside the circle representing A. Since C is completely contained in A, the union of A and C consists of all of the elements in A. That is, since $C \subseteq A$, then $A \cup C = A$. This is depicted by the shaded area of Figure 4.15.

Figure 4.14

C is a Subset of A

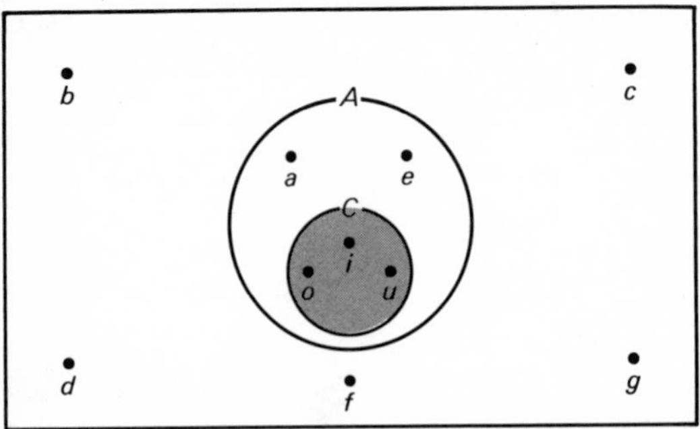

Figure 4.15

The Union of A and C

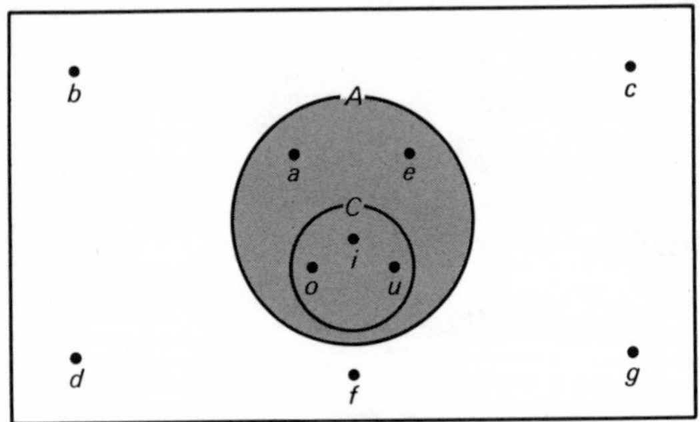

Up to this point in our discussion of Venn-Euler diagrams, we have not considered the depiction of more than two events. However, the principles employed in depicting two events may be extended to three or more events. To illustrate this process, consider three new events

$$D = \{a,b,c\}$$
$$E = \{o,u\}$$
$$F = \{d,e,f\}$$

None of these three events contains any elements in common with either of the other events. That is, events D, E, and F are mutually exclusive. Recalling that the universe consists of the five vowels and the first five consonants of the English alphabet, we also can see that D, E, and F do not include all of the elements of the universe. In other words, the events D, E, and F are not collectively exhaustive. These three events are depicted in Figure 4.16 by three non-overlapping circles which do not exhaust the total area of the rectangle.

Figure 4.16

Three Mutually Exclusive Events

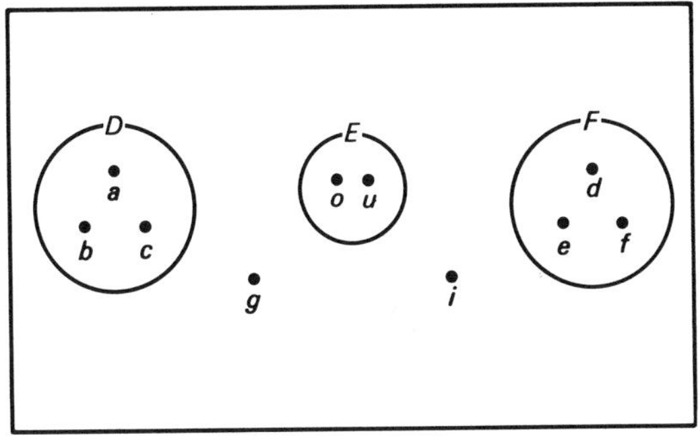

Let us now consider the three events D, E, and G, where

$$G = \{d,e,f,g,i\}$$

As the reader may verify, the events D, E, and G are not only mutually exclusive but also collectively exhaustive. In other words, these three events form a partition of the universe. This partition is depicted by the Venn-Euler diagram in Figure 4.17, in which three non-overlapping regions exhaust the total area of the rectangle.

Figure 4.17

A Partition of the Universe

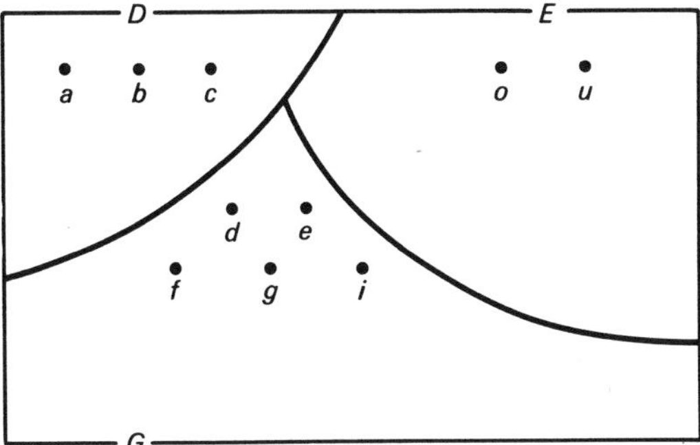

As a final example of Venn-Euler diagrams, consider the three events A, B, and G. From the definitions of these three events, we may observe that each one has a non-empty intersection with each of the others. That is, none of these three events is mutually exclusive of either of the other events. However, the three events do exhaust all of the elements of the universe. The Venn-Euler diagram for these three events is shown in Figure 4.18 by three overlapping shapes which exhaust the total area of the rectangle.

Figure 4.18

Three Collectively Exhaustive Events

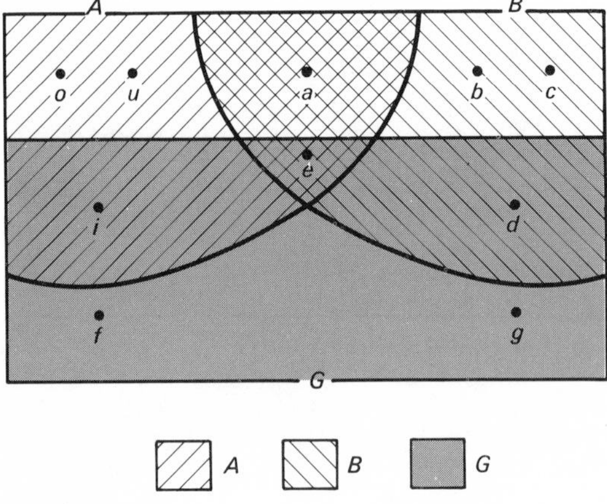

A B G

PROBLEMS

4.1 Below are four sets that are defined by the describing method

$S_1 = \{x|x \text{ is one of the first four months of the year}\}$
$S_2 = \{x|x \text{ is a prince consort of England since 1800}\}$
$S_3 = \{x|x \text{ is one of the United States' three most populous cities}\}$
$S_4 = \{x|x \text{ is a non-negative integer}\}$

Define each of the above sets by the listing method.

4.2 Below are three sets that are defined by the listing method.

$S_1 = \{\text{October, November, December}\}$
$S_2 = \{a,e,i,o,u\}$
$S_3 = \{2,4,6, \ldots\}$

Define each of the above sets by the describing method.

4.3 Consider the following two sets:

$$A = \{2,3,4\}$$
$$B = \{b|2 \leq b \leq 4\}$$

(a) Explain how these two sets differ.
(b) Define Set A by the describing method.
(c) Explain why it is impossible to define Set B by the listing method.

4.4 A particular sample space is composed of 10 sample points.
(a) How many different events, each encompassing 5 sample points, may be defined on this sample space?
(b) Remembering that the empty set and the universal set are both regarded as events, how many different events of all possible sizes may be defined on this sample space?

4.5 A box contains 7 circular tags individually numbered from 1 through 7. A tag is drawn randomly from the box, its number is recorded, and the tag is laid aside. Then a second tag is drawn from the 6 tags remaining in the box, and its number is recorded.
(a) Assume that an elementary outcome of this experiment consists of an ordered pair of values; e.g., the ordered pair (3,5) indicates the outcome that tag #3 is drawn first and tag #5 is drawn second. Diagram the sample space of the experiment.
(b) From analysis of the sample space, and assuming that all elementary outcomes are equally likely, determine the probability that:
(i) The first tag is odd-numbered;
(ii) The second tag is even-numbered;
(iii) Both tags are even-numbered;
(iv) Both tags are odd-numbered;
(v) The first tag is odd and the second tag is even;
(vi) One tag is odd and the other tag is even;
(vii) The value on the first tag is greater than the value on the second tag;
(viii) The value on the first tag is equal to the value on the second tag;
(ix) The value on the first tag is equal to or greater than the value on the second tag;
(x) The sum of the values on the 2 tags is equal to 10;
(xi) The sum of the values on the 2 tags is greater than 10;
(xii) The sum of the values on the 2 tags is equal to or greater than 10.

4.6 Repeat Exercise 4.5 with the following change: After the number of the first tag is recorded, the tag is returned to the box and the contents of the box are stirred. Thus, all 7 tags are equally available on both the first and second draws.

4.7 Consider the following sets:

$$S_1 = \{c,r,e,a,m\}$$
$$S_2 = \{c,h,o,r,e\}$$
$$S_3 = \{d,a,m,p\}$$

Using the listing method, define each of the following sets:

(a) $S_1 \cap S_2$
(b) $S_1 \cap S_3$
(c) $S_2 \cap S_3$
(d) $S_1 \cup S_2$
(e) $S_1 \cup S_3$
(f) $S_2 \cap S_3$

4.8 Consider the following sets:

$$A = \{3,4,5,6,7\}$$
$$B = \{5,6,7,8,9\}$$
$$C = \{c|3 \leq c \leq 7\}$$
$$D = \{d|3 < d < 7\}$$
$$E = \{e|7 \leq e \leq 8\}$$
$$F = \{f|7 < f < 8\}$$
$$G = \{g|4 \leq g \leq 9\}$$

Using either the listing or describing method, whichever seems more appropriate in each case, define each of the following sets so that its elements are clearly specified.

$H = A \cap B$	$N = C \cap E$
$I = A \cap C$	$O = C \cap F$
$J = A \cap D$	$P = F \cap G$
$K = C \cap D$	$Q = A \cup B$
$L = D \cap E$	$R = D \cup E$
$M = E \cap F$	$S = C \cup G$

4.9 As the president of the XYZ Company, Mr. Anderson must decide whether or not to establish a branch office in a foreign country. This decision depends primarily on the market potential of the company's products in that country. He advised the vice-president, Mr. Baker, to perform an analysis and to submit his recommendation. In addition, he has hired two consultants, Mr. Clark and Mr. Decker, to study the problem separately and make their respective recommendations. Unfortunately, none of the three reports will be perfectly reliable (each has a .10 chance of being wrong). Thus, the president has decided to take the action recommended by a majority of these reports, rather than any single report. He further realizes that such a decision rule does not guarantee that he will make a correct decision.

We may recognize that, under various situations, Mr. Anderson will make a wrong decision. To specify these situations precisely, we let

$$B = \{\text{Baker's report is right}\}$$
$$C = \{\text{Clark's report is right}\}$$
$$D = \{\text{Decker's report is right}\}$$

Use B, C, D, and their complements to express each of the possible events that will lead the president to make a wrong decision.

4.10 An investment club has 100 members, consisting of 50 men between the ages of 25 and 60, inclusive, and 50 women between the ages of 30 and 55, inclusive. At a meeting of the club, the names of the 100 members have been written on separate slips of paper and placed in an urn. These slips have been thoroughly mixed, and one slip is to be drawn from the urn. The person whose name is drawn will win one share of Great Caspian Tea Company stock. Consider the following possible events:

E_1: {The winner is a man}
E_2: {The winner is a woman}
E_3: {The winner is a person less than 57 years old}

E_4: {The winner is a person more than 50 years old}
E_5: {The winner is a woman less than 40 years old}
E_6: {The winner is a woman at least 40 years old}
E_7: {The winner is a person no more than 55 years old}
E_8: {The winner is a man over 55 years old}
E_9: {The winner is a woman who is 40 years old}
E_{10}: {The winner is a man who is 55 years old}
E_{11}: {The winner is a man no more than 55 years old}

Which of the following sets of events constitute partitions of the sample space for the drawing?

(a) E_1 and E_2
(b) E_3 and E_4
(c) E_5 and E_6
(d) E_1, E_5, and E_6
(e) E_1, E_5, E_6, and E_9
(f) E_7 and E_8
(g) E_7, E_8, and E_{10}
(h) E_2, E_8, and E_{11}
(i) E_2, E_8, E_{10}, and E_{11}

4.11 Let a universal set be defined as $\mathcal{U} = \{x|1 \leq x \leq 12\}$. Consider the following subsets of $\mathcal{U}$:

$S_1 = \{x|1 \leq x \leq 3\}$
$S_2 = \{x|4 \leq x \leq 6\}$
$S_3 = \{x|x = 7, 8, 9\}$
$S_4 = \{x|x = 10, 11, 12\}$
$S_5 = \{x|7 \leq x \leq 9\}$
$S_6 = \{x|10 \leq x \leq 12\}$
$S_7 = \{x|1 \leq x \leq 6\}$
$S_8 = \{x|7 \leq x \leq 12\}$
$S_9 = \{x|1 \leq x \leq 2\}$
$S_{10} = \{x|3 \leq x \leq 4\}$
$S_{11} = \{x|5 \leq x \leq 6\}$
$S_{12} = \{x|3 \leq x \leq 6\}$

Which of the following groups of subsets constitute partitions of the universal set?

(a) S_1, S_2, S_3, and S_4
(b) S_1, S_2, S_4, and S_5
(c) S_{10}, S_{11}, and S_8
(d) S_7 and S_8
(e) S_8, S_{11}, and S_{12}
(f) S_8, S_{10}, and S_{11}
(g) S_8, S_9, S_{10}, and S_{11}
(h) S_1, S_2, S_5 and S_6
(i) S_5, S_6, S_9, S_{10}, and S_{11}

4.12 Suppose the universal set $\mathcal{U}$ is the set of positive integers: {1,2,3,4,5,6,7,8,9}. Consider the following subsets:

$$X = \{x|x \text{ is a positive integer less than } 4\}$$
$$Y = \{y|y \text{ is an even integer in } \mathcal{U}\}$$
$$Z = \{z|z \text{ is an integer in } \mathcal{U} \text{ not evenly divisible by } 4\}$$

Using the listing method, define each of the following subsets:

(a) $X \cap Y$
(b) $X' \cap Y'$
(c) $X' \cap Z'$
(d) $X \cap Y \cap Z'$
(e) $(X \cap Y')'$
(f) $(Y \cap Z)'$
(g) $(X \cap Y \cap Z)'$
(h) $(X' \cap Y' \cap Z)'$

4.13 Referring to problem 4.12 above:

(a) For each of the events X, Y, and Z, draw a Venn-Euler diagram showing the specified event as a shaded area.
(b) Draw a Venn-Euler diagram for each of the expressions in (a) through (h).

4.14 Mr. Howard Sellmore, proprietor of the Bon Chance Boutique, has decided to conduct a sales contest among his eight employees. He will award a bonus to the clerk who makes the largest total gross sales during the month of January. The eight clerks employed by Mr. Sellmore are described in the table below:

Clerk	Sex	Hair Color	Eye Color	Married?
Art	Male	Blond	Brown	Yes
Bob	Male	Black	Brown	Yes
Chuck	Male	Blond	Blue	No
Doug	Male	Brown	Brown	No
Ella	Female	Red	Green	No
Fay	Female	Black	Blue	Yes
Gwen	Female	Brown	Brown	No
Helen	Female	Blond	Blue	Yes

Consider the following possible events:

J: The winner is a male.
K: The winner has red hair.
L: The winner has blond hair.
M: The winner has green eyes.
N: The winner has blue eyes.
O: The winner is married.

(a) Denoting the eight clerks by the letters a, b, c, d, e, f, g, h, use the listing method to define the events J, K, L, M, N, and O.
(b) Express each of the following events in terms of J, K, L, M, N, O and their complements:

P: The winner is a blue-eyed blond.
Q: The winner is a married male.
R: The winner is an unmarried female.
S: The winner has green eyes or blond hair.
T: The winner is a female or has brown eyes.

(c) List the elements (clerks) contained in each of the events defined in (b) above.
(d) For each of the following events, draw a Venn-Euler diagram showing the specified event as a shaded area:
  (i) $J \cap L$
  (ii) $J \cap N$
  (iii) $J \cup O$
  (iv) $L \cup M$
  (v) $L \cap N \cap O$
(e) Describe in words each of the events in (d) above.
(f) List the elements contained in each of the events in (d) above.

4.15 The Venn-Euler diagram in Figure 4.18 shows

$$\mathcal{U} = \{a,b,c,d,e,f,g,i,o,u\}$$
$$A = \{a,e,i,o,u\}$$
$$B = \{a,b,c,d,e\}$$
$$G = \{d,e,f,g,i\}.$$

From this diagram, list the elements contained in each of the following expressions:

(a) $A \cap B \cap G$
(b) $A \cap B \cap G'$
(c) $A \cap B' \cap G$
(d) $A' \cap B \cap G$
(e) $A' \cap B' \cap G'$
(f) $A \cup B \cup G$
(g) $A \cup B \cup G'$
(h) $(A \cap B \cap G)'$
(i) $A' \cup B' \cup G'$
(j) $(A \cup B \cup G)'$

Chapter 5

Probabilities of Events

In Chapter Four we were concerned with defining events on sample spaces. In this chapter we will begin to consider how probabilities of such events are determined.

5.1 ASSIGNING PROBABILITIES TO EVENTS

In assigning probabilities to a set of events, we must satisfy three requirements:

1. The probability of an event must be equal to or greater than zero. In other words, an event cannot have a negative probability.
2. The probability of the occurrence of one *or* the other of any two mutually exclusive events must equal the sum of the separate probabilities of these two events.
3. The sum of the probabilities of *all* the separate events comprising a partition must equal 1.00.

Example:

A speculator purchases 1,000 shares of a common stock, which he plans to sell at the end of ten trading days, hopefully at a profit. He considers three possible events—that the stock will go up in price, that it will decline, and that it will remain unchanged. Viewing the short-term behavior of stock prices as a random process, these three events form a partition of the sample space; they are mutually exclusive and collectively exhaustive. Indicating the three events by the letters U (price goes up), D (price goes down), and N (price remains unchanged), he assigns the following subjective probabilities: $P(U) = .70$, $P(D) = .10$, $P(N) = .20$. Notice, first, that none of these probabilities is negative. Second, the probability that the speculator will fail to sell at a profit is given by $P(U') = P(D) + P(N) = .10 + .20 = .30$; that is, the probability that the stock will not go up (i.e., either go down or remain unchanged) is the sum of the probabilities of two separate, mutually exclusive events. Finally, we see that $P(U) + P(D) + P(N) = .70 + .10 + .20 = 1.00$, indicating it is certain that one of these three events will occur.

Let us return to the random process of rolling a pair of dice, illustrated by the sample space of Figure 4.3. Let us consider a first event, O, that the sum of the two uppermost faces is an odd number, and a second event, S, that the sum is greater than 7. We might then partition the sample space in terms of four events:

$O \cap S$	an odd number greater than 7
$O \cap S'$	an odd number equal to or less than 7
$O' \cap S$	an even number greater than 7
$O' \cap S'$	an even number equal to or less than 7

If we assume all simple events to be equally likely, Figure 4.3 indicates that we may make probability assignments as shown in Table 5.1. Thus we can see from Table 5.1 that if we are interested, for instance, in the probability of throwing an odd number greater than 7, then $P(O \cap S) = 6/36$. Similarly, $P(O \cap S') = 12/36$, $P(O' \cap S) = 9/36$, and $P(O' \cap S') = 9/36$. Notice that each of these four probabilities is the probability of an intersection of two events. Since the probability of an intersection of two events is the probability that the two events will occur jointly, such a probability is called a *joint probability.*

Table 5.1

Probability Table for Throwing Dice

	S	S'	Total
O	6/36	12/36	18/36
O'	9/36	9/36	18/36
Total	15/36	21/36	1

Notice from Table 5.1 that the event S (throwing a sum greater than 7) is composed of the two mutually exclusive events $(O \cap S)$ and $(O' \cap S)$. Thus we may obtain the probability of rolling a sum greater than 7 (regardless of whether it is odd or even) by the addition $P(O \cap S) + P(O' \cap S) = 6/36 + 9/36 = 15/36$. This result is shown in the "Total" cell for Column S in Table 5.1. Similarly, $P(S') = P(O \cap S') + P(O' \cap S') = 12/36 + 9/36 = 21/36$. Notice that since S and S' are mutually exclusive and collectively exhaustive, $P(S) + P(S') = 15/36 + 21/36 = 1.00$. Thus, an alternative way of calculating the probability of throwing a sum equal to 7 or less would be $P(S') = 1 - P(S) = 1 - 15/36 = 21/36$. Probabilities such as $P(S)$ and P(S'), which appear in the margins of a probability table like Table 5.1, are sometimes referred to as *marginal* probabilities. The marginal probability of an event is the probability that the event will occur regardless of any other conditions that might be imposed. For instance, the probability of rolling a sum greater than 7 is 15/36 if we disregard whether the sum is odd or even. Thus, marginal probabilities also are called *unconditional* probabilities.

Example:

The president of a retailing chain is trying to decide whether to present a new luxury item to the public at the present time. His decision depends partly on the likelihood of the onset of a business recession during the next twelve months. He estimates a .20 probability of a recession during the coming year. He also estimates that, regardless of whether a recession occurs, there is a .70 probability that a marketing campaign would be successful if it were launched at the present time. In addition, he feels that there is a .10 probability of the joint occurrence of a recession and a successful marketing campaign.

If we let R represent the occurrence of a recession during the next twelve months, and S represent the successful outcome of a marketing campaign launched at the present time, we can express the probability estimates given above in the following form:

$$P(R) = .20$$
$$P(S) = .70$$
$$P(S \cap R) = .10$$

Since the occurrence and non-occurrence of a recession are mutually exclusive and collectively exhaustive events (a recession will either occur or not occur), we can compute the probability that a recession will not occur as $P(R') = 1 - P(R) = 1 - .20 = .80$. *Similarly, if we disregard whether a recession will or will not occur, the probability of the failure of a marketing campaign launched at the present time is* $P(S') = 1 - P(S) = 1 - .70 = .30$. *Recognizing the probability of a successful campaign (regardless of the*

Table 5.2

Probability Table for Recession Example

		Recession		
		Will Occur R	*Won't Occur* R'	*Total*
Outcome of Campaign	*Successful* S	.10	.60	.70
	Unsuccessful S'	.10	.20	.30
	Total	.20	.80	1

occurrence of a recession) as a marginal probability, we obtain the expression $P(S) = P(S \cap R) + P(S \cap R')$. *Thus, as the probability of the joint occurrence of a successful campaign and no recession, we may compute* $P(S \cap R') = P(S) - P(S \cap R) = .70 - .10 = .60$. *Likewise,* $P(S' \cap R) = P(R) - P(S \cap R) = .20 - .10 = .10$; *and* $P(S' \cap R') = P(R') - P(S \cap R') = .80 - .60 = .20$. *These calculations are summarized in Table 5.2.*

5.2 CONDITIONAL PROBABILITY

Returning to Table 5.1, we see that the marginal probability of obtaining a sum greater than 7, when two fair dice are rolled, is $P(S) = 15/36$. That is, if your back is turned when a player rolls the dice, you may say that the chances are 15 out of 36 that he threw a sum greater than 7. Suppose, however, that you are given the information that the sum was an odd number. From Figure 4.3 and Table 5.1, you can ascertain that, among those sample points representing odd-numbered sums, there are six elements corresponding to sums greater than seven and twelve sample points corresponding to sums equal to or less than 7. Thus, in light of the information that an odd-numbered sum was thrown, you might want to revise your initial statement that the probability is 15/36 that the sum was greater than 7. Indeed, you might now say that the probability that the sum was greater than 7 is 1/3. Let us investigate how you might arrive at this conclusion.

The assessment of 15/36 as the probability of rolling a sum greater than 7 was based on our definition of the event S on the sample space of Figure 4.3. However, on being informed that the sum was an odd number, we can immediately eliminate from consideration eighteen of those thirty-six sample points—that is, those eighteen elements corresponding to even-numbered sums. Thus, in determining our revised probability assessment, we are working with a *reduced sample space* consisting of only those eighteen elements corresponding to odd-numbered sums. That is, with reference to Table 5.1, we may limit our attention to those events represented in row O of the table. Whereas originally $P(O)$ was equal to 18/36, $P(O)$ must now equal 1, since we are certain (having faith in the truth of the information) that an odd-numbered sum was thrown. Since $P(O) = P(O \cap S) + P(O \cap S')$, we must revise the values of $P(O \cap S)$ and $P(O \cap S')$ so that they will add to 1. That is, we want to change the values of these two probabilities so that the proportion between them remains the same but so that their sum will equal 1. This is easily achieved by dividing each value by their sum. If $P(O) = P(O \cap S) + P(O \cap S')$, then we can divide both sides of the equation to obtain:

$$\frac{P(O)}{P(O)} = \frac{P(O \cap S)}{P(O)} + \frac{P(O \cap S')}{P(O)}$$

In terms of the values in Table 5.1, we have:

$$\frac{18/36}{18/36} = \frac{6/36}{18/36} + \frac{12/36}{18/36}$$

$$1 = 1/3 + 2/3$$

Thus, *given* that an odd-numbered sum was thrown, we obtain 1/3 as the probability that the sum was greater than 7, and 2/3 as the probability that the sum was equal to or less than 7. These probabilities, obtained by re-evaluating the original probabilities after reducing the sample space in accordance with the given information, may be written:

$$P(S|O) = 1/3$$

$$P(S'|O) = 2/3$$

where the symbol | stands for "given."[1] Such probabilities are called *conditional* probabilities. The conditional probability of an event is the probability of that event's occurrence, *given* some specified *condition*. In our example, $P(S|O)$ represents the probability of the occurrence of the event S, given the condition that the event O has occurred. In applying the concept of conditional probability to statistical decision making, the "conditions" either represent information that is obtained from empirical observation, or indicate certain assumptions or judgments that arise during the decision-making process.

Example:

The credit manager of a particular department store in a major city defines a "good credit risk" as a person who has a history of customarily paying his bill within thirty days of billing. One of the items that appears on the information form that he uses to screen credit applicants is whether or not the applicant has established credit at any other department store in the city. From his analysis of several hundred credit records, the credit manager has discovered that there is a .40 probability that an applicant does have established credit at another department store. He also has determined that, considering all applicants, there is a .65 probability that the applicant will be a good credit risk. His records further show that there is a joint probability of .30 that an applicant will have established credit at another store and will be a good credit risk.

From the information given above, we may construct the probability table shown in Table 5.3.

[1] Notice that the symbol | is read "such that" when used to define a set, but "given" when used to express a conditional probability.

Table 5.3

Probability Table for Credit Evaluation

		Established Credit?		
		Yes	*No*	*Total*
Risk	*Good*	*.30*	*.35*	*.65*
	Poor	*.10*	*.25*	*.35*
	Total	*.40*	*.60*	*1.00*

From the credit manager's position, a significant question is, Does the fact that an applicant has established credit at another department store provide useful information to help predict whether or not he will be a good credit risk? If we reduce the sample space to only those applicants who indicate that they do have established credit, the conditional probability that an applicant will be a good credit risk is .30/.40 = .75. This may be compared with the marginal probability of .65 that any applicant will be a good risk, regardless of whether or not he has established credit elsewhere. In other words, if we know that a person does have other established credit, we may make an upward revision from .65 to .75 in the probability that he will be a good risk. We may interpret this finding as suggesting that a person's response to the item on the application form does provide predictive information.

Given any two events, A and B, which are defined on a sample space in such a way that $P(B)$ is not equal to zero, the conditional probability of A given B is mathematically defined by the formula:

$$P(A|B) = \frac{P(A \cap B)}{P(B)} \tag{5.1}$$

Notice that $P(A|B)$ is defined only if $P(B)$ is greater than zero. Obviously, if $P(B) = 0$, the statement "given that B occurs" would be meaningless. Multiplying both sides of this equation by $P(B)$, we obtain the general formula for the joint probability of two events:

$$P(A \cap B) = P(A|B)P(B) \tag{5.2}$$

Similarly,

$$P(B|A) = \frac{P(B \cap A)}{P(A)} \tag{5.3}$$

Multiplying both sides of this equation by $P(A)$ yields:

$$P(B \cap A) = \mathrm{P}(B|A)P(A) \tag{5.4}$$

Since $P(B \cap A) = P(A \cap B)$, we see that:

$$P(B|A)P(A) = P(A|B)P(B) \tag{5.5}$$

Example:

From a computer tally based on employee records, the personnel manager of a large manufacturing firm finds that 15 per cent of the firm's employees are supervisors and 25 per cent of the firm's employees are college graduates. He also discovers that 5 per cent of the firm's employees are both supervisors and college graduates. Suppose that an employee is selected at random from the firm's personnel roster. Let S denote the event that the person selected is a supervisor and C denote the event that the person selected is a college graduate. We then could state the following probabilities: $P(S) = .15$; $P(C) = .25$; $P(C \cap S) = .05$. From these facts we can construct Table 5.4.

Table 5.4

Probability Table from Employee Records

	C	*C′*	*Total*
S	*.05*	*.10*	*.15*
S′	*.20*	*.65*	*.85*
Total	*.25*	*.75*	*1.00*

From Table 5.4 we may arrive at the following probability statements:

$P(S) = .15$	*Probability of selecting a person who is a supervisor*
$P(S') = .85$	*Probability of selecting a person who is not a supervisor*
$P(C) = .25$	*Probability of selecting a person who is a college graduate*
$P(C') = .75$	*Probability of selecting a person who is not a college graduate*
$P(S \cap C) = .05$	*Probability of selecting a person who is both a college graduate and a supervisor*

$P(S \cap C') = .10$	*Probability of selecting a person who is a supervisor but not a college graduate*
$P(S' \cap C) = .20$	*Probability of selecting a person who is not a supervisor but is a college graduate*
$P(S' \cap C') = .65$	*Probability of selecting a person who is neither a supervisor nor a college graduate*
$P(S\|C) = \frac{.05}{.25} = .2000$	*Probability of selecting a supervisor if the selection is confined to college graduates*
$P(S'\|C) = \frac{.20}{.25} = .8000$	*Probability of selecting a non-supervisor if the selection is confined to college graduates*
$P(S\|C') = \frac{.10}{.75} = .1333$	*Probability of selecting a supervisor if the selection is confined to non-college graduates*
$P(S'\|C') = \frac{.65}{.75} = .8667$	*Probability of selecting a non-supervisor if the selection is confined to non-college graduates*
$P(C\|S) = \frac{.05}{.15} = .3333$	*Probability of selecting a college graduate if the selection is confined to supervisors*
$P(C'\|S) = \frac{.10}{.15} = .6667$	*Probability of selecting a non-college graduate if the selection is confined to supervisors*
$P(C\|S') = \frac{.20}{.85} = .2353$	*Probability of selecting a college graduate if the selection is confined to non-supervisors*
$P(C'\|S') = \frac{.65}{.85} = .7647$	*Probability of selecting a non-college graduate if the selection is confined to non-supervisors*

The various joint and conditional probabilities given above may be confirmed by applying formulas (5.1) through (5.5). For example:

$$\begin{aligned} P(S' \cap C) &= P(S'|C)P(C) = (.8000)(.25) \\ &= P(C|S')P(S') = (.2353)(.85) \\ &= .20 \end{aligned}$$

For his own satisfaction, the reader is left to make formula confirmations of the other probabilities on his own.

5.3 INDEPENDENCE

Many statistical decision problems are concerned with establishing the *relevance* of data to a particular decision. Far too often in everyday life—whether the matter be business or personal—we base our decisions on facts that are irrelevant to our problems but that we mistakenly assume to be relevant, in which case we might as well make our decisions by drawing numbers out of a hat.

A striking example of the failure to recognize the necessity of establishing the relevance of data that are used in making business decisions may be drawn from the history of industrial employment testing. The American public's first widespread exposure to mental tests occurred during World War I, when recruits were administered the Army Alpha, which was a paper-and-pencil test of "general intelligence." Impressed by the results obtained by the Army in using this test as a personnel classification instrument, men returned to their businesses after the war and began using similar tests more or less indiscriminantly as personnel hiring devices. More often than not, these testing programs were utter failures, and the practice of personnel testing fell into disrepute in most companies by the end of the 1920s. What was not understood then, as psychologists have since established, is that specific jobs require specific mental abilities, and that scores on a general intelligence test may be entirely unrelated to the level of employee performance or output on a particular job. In order to justify the use of a particular personnel test as a hiring device for a particular job, scores on the test must be relevant to the hiring decision with respect to that job. This process of establishing the relevance of test scores with respect to a particular job is called *test validation*, which amounts essentially to determining that the probability of a person's successful performance on the job is conditional on his test score. For example, it has been demonstrated, for many managerial jobs, that persons with high scores on certain vocabulary tests have a greater probability of job success than persons with low scores.

When knowledge of the occurrence or non-occurrence of one event has no influence on the probability assessment of the occurrence of another event, the two events are said to be *independent*. That is, an event A is independent of another event B if the conditional probability of A, given B, is equal to the marginal (unconditional) probability of A. In terms of a formula, the event A is independent of the event B if:

$$P(A|B) = P(A) \tag{5.6}$$

Thus, *in the special case in which A and B are independent*, we may revise Formula (5.2) to read:

$$P(A \cap B) = P(A)P(B) \tag{5.7}$$

That is, the probability of the joint occurrence of two independent events is equal to the product of the marginal probabilities of those two events. Therefore, to determine whether or not two events A and B are independent, we simply check to see if either Formula (5.6) or Formula (5.7) holds. If the two events are independent, then the knowledge that one event has or has not occurred will not influence the probability that the other event will occur. Thus, when two events are independent, knowledge of the occurrence of one event is irrelevant to a decision that depends on the probability of occurrence of the other event.

Example:

For the past several months, a would-be investor has been keeping careful records of the stock-market behavior of a particular issue. His records indicate that, for a randomly selected day, there is a .60 probability that the stock will go up in price. Furthermore, there is a .50 probability that the volume of shares traded on this stock will be high (200,000 shares or more). He also notices that, on those days when the trading volume of the stock is high, the price of the stock goes up 60 per cent of the time.

Let U indicate the event that the price of the stock goes up, and H indicate the event that the trading volume on the stock is high. Then, from the above information:

$$P(U) = .60$$

$$P(H) = .50$$

$$P(U|H) = .60$$

Notice that $P(U|H) = P(U)$. That is, the probability that the stock will go up in price on a randomly selected trading day is independent of the day's trading volume on that issue. From Formula (5.2) we see that $P(H \cap U) = P(U|H)P(H) = (.60)(.50) = .30$. This, by virtue of Formula (5.7), confirms that U and H are independent. From the data given, we can construct the following probability table:

Table 5.5

Probability Table for Stock Behavior

	U	U'	*Total*
H	.30	.20	.50
H'	.30	.20	.50
Total	.60	.40	1

Notice that the probability shown in each of the four cells in the body of the table is equal to the product of the marginal probabilities for the row and column in which the cell resides. Notice, also, the constant proportionalities across rows and down columns. That is, reading across:

$$\frac{.30}{.20} = \frac{.30}{.20} = \frac{.60}{.40}$$

Similarly, reading down:

$$\frac{.30}{.30} = \frac{.20}{.20} = \frac{.50}{.50}$$

To gain a clear understanding of the concept of independence, the reader must be very careful to avoid confusing independent events with mutually exclusive events. These two concepts are *not* synonymous. Indeed, they are incompatible in the sense that two possible events cannot be both mutually exclusive and independent. Consider, for example, tossing a coin, with the two possible outcomes—heads and tails. We may view these two outcomes as mutually exclusive events. If the coin is fair, $P(H) = .50$ and $P(T) = .50$. If someone tosses the coin, and we are told that it came up heads, then the probability that it came up tails is reduced to zero; that is, $P(T|H) = 0$. Since the occurrence of one event precludes the occurrence of the other, $P(H \cap T) = 0$, so that $P(H \cap T) \neq P(H)P(T)$, which fails to meet the definition of independence given by Formula (5.7).

For any two events, A and B, the condition that $P(A \cap B) > 0$ *is necessary but not sufficient* for independence. That is, in order to be independent, A and B must be defined in such a way that they overlap in some common portion of the sample space. Such an overlap (i.e., non-empty intersection) does not, however, necessarily guarantee that the events are independent. More specifically:

1. If $P(A \cap B) = 0$, then A and B must be dependent. Two mutually exclusive events cannot be independent.
2. If $P(A \cap B) > 0$, then A and B may or may not be independent, depending on whether the requirement of Formula (5.7) is satisfied. Of course, if Formula (5.7) is satisfied, Formula (5.6) also is necessarily satisfied.
3. If A and B are not independent, $P(A \cap B)$ may or may not be greater than zero.

Examples:

	B	B′	
A	0	.30	.30
A′	.40	.30	.70
	.40	.60	1

A and B are mutually exclusive and therefore dependent.

$P(A \cap B) = 0$; *hence*, $P(A \cap B) \neq P(A)P(B)$

	B	B′	
A	.20	.10	.30
A′	.20	.50	.70
	.40	.60	1

A and B are neither mutually exclusive nor independent.

$P(A \cap B) > 0$, *and* $P(A \cap B) \neq P(A)P(B)$.

	B	B′	
A	.12	.18	.30
A′	.28	.42	.70
	.40	.60	1

A and B are not mutually exclusive but they are independent.

$P(A \cap B) > 0$, *but* $P(A \cap B) = P(A)P(B)$.

5.4 PROBABILITIES OF UNIONS OF EVENTS

In Chapter 4 we observed that the union of two events A and B, denoted by $A \cup B$, is the set of all sample points contained in A or in B or in both A and B. Thus, by the notation $P(A \cup B)$ we mean the probability that either A occurs or B occurs or both A and B occur. The calculation of $P(A \cup B)$ depends on whether or not A and B are mutually exclusive.

5.4.1 Unions of Mutually Exclusive Events

We have mentioned in our discussion of partitioning that if two events defined on the same sample space are mutually exclusive, then the probability of the occurrence of one or the other must equal the sum of the probabilities of the separate events. That is, *if A and B are mutually exclusive events*, then:

$$P(A \cup B) = P(A) + P(B) \tag{5.8}$$

Of course, in this special case in which A and B are mutually exclusive, their joint occurrence is impossible.

Example:

A particular mail-order house usually receives between 2,000 and 4,000 orders per day. Consider the following events:

A: less than 2,000 orders received
B: at least 2,000 but less than 3,000 orders received
C: at least 3,000 but less than 4,000 orders received
D: 4,000 or more orders received

Obviously, these are mutually exclusive events. From records maintained over a period of years, the following probabilities have been determined for order volume on a randomly selected day:

$$P(A) = .10 \qquad P(B) = .40 \qquad P(C) = .30 \qquad P(D) = .20$$

From these data, we may draw the following conclusions:

Probability of receiving less than 3,000 orders	$P(A \cup B)$ $= P(A) + P(B)$ $= .10 + .40 = .50$
Probability of receiving less than 4,000 orders	$P(A \cup B \cup C)$ $= P(A) + P(B) + P(C)$ $= .10 + .40 + .30 = .80$
Probability of receiving at least 3,000 orders	$P(C \cup D)$ $= P(C) + P(D)$ $= .30 + .20 = .50$
Probability of receiving at least 2,000 orders	$P(B \cup C \cup D)$ $= P(B) + P(C) + P(D)$ $= .40 + .30 + .20 = .90$
Probability of receiving less than 2,000 orders or 4,000 or more orders	$P(A \cup D)$ $= P(A) + P(D)$ $= .10 + .20 = .30$
Probability of receiving at least 2,000 but less than 4,000 orders	$P(B \cup C)$ $= P(B) + P(C)$ $= .40 + .30 = .70$

5.4.2 Unions of Non-Mutually Exclusive Events

So far we have restricted our discussion of unions to mutually exclusive events. Consider, however, the case where events are not mutually exclusive. Suppose, for example, that several hundred small-business executives are gathered at a trade show. A drawing is to be held, and a door prize awarded. Of the total number of executives present, 40 per cent are under forty and 80 per cent are college graduates. Of the men under forty, 75 per cent are college graduates. What is the probability that the door prize will be won by either a college graduate and/or a man under forty? Let A represent the event that the winner is under forty and B represent the event that he is a college graduate. In this case, A and B are not mutually exclusive; in fact, from the above data, $P(B|A) = .75$. If we were (wrongly) to apply Formula (5.8) to this problem, we would obtain $P(A \cup B) = P(A) + P(B) = .40 + .80 =$

1.20, which is not only incorrect but ridiculous. The key to the error, of course, lies in the fact that A and B are not mutually exclusive; rather than being equal to zero, the joint probability of A and B is $P(A \cap B) = P(B|A)P(A) = (.75)(.40) = .30$. To diagnose the error, we might prepare the probability table shown in Table 5.6:

Table 5.6

Probability Table for Small-Business Executives

	B	B'	
A	.30	.10	.40
A'	.50	.10	.60
	.80	.20	1

From the table it is clear that $P(A \cup B) = P(A \cap B) + P(A \cap B') + P(A' \cap B) = .30 + .10 + .50 = .90$. Since $P(A \cap B)$ is included in both $P(A)$ and $P(B)$, the error was one of adding $P(A \cap B)$ to the sum twice instead of only once. To obtain the correct answer, we must amend Formula (5.8) to read:

$$P(A \cup B) = P(A) + P(B) - P(A \cap B) \tag{5.9}$$

This is the general formula for the probability of the union of two events. If the two events are mutually exclusive, then $P(A \cap B) = 0$, and Formula (5.9) reduces to Formula (5.8).

Example 1:

A brewery maintains a large panel of beer tasters. The panel is equally divided among men and women. The panel has just been asked to compare three new brews (call them A, B, C), and of the total panel 50 per cent preferred Brew A, 30 per cent preferred Brew B, and 20 per cent preferred Brew C. Assuming that taste preference is independent of sex, what is the probability that a randomly selected panel member will be either a man and/or a person who prefers Brew B?

Since brew preference is assumed to be independent of sex, we may prepare Table 5.7, in which the probability shown in each cell is the product of the marginal probabilities for that particular row and column. From the table, the desired probability may be obtained by adding .25 + .15 + .10 + .15 = .65. The same result may be obtained from Formula (5.9): .50 + .30 − .15 = .65.

Table 5.7

Probability Table for Brewery Panel

	Preference			
	A	*B*	*C*	*Total*
Men	.25	.15	.10	.50
Women	.25	.15	.10	.50
Total	.50	.30	.20	1

Example 2:

The reliability of a system or of a component of a system might be regarded as the probability that the system or component will function correctly for a specified period of time. Assume that a particular satellite system is composed of two components, A and B, as shown in Figure 5.1. Each component is composed of two identical parts connected in parallel, and the two components are connected in series. Component A will function properly as long as A_1 and/or A_2 functions properly, and Component B will function properly as long as B_1 and/or B_2 functions properly. Parts A_1 and A_2 each has a reliability of .70; parts B_1 and B_2 each has a reliability of .80. Assuming that whether a particular part of the system functions is independent of whether any other part functions, what is the reliability of the entire system?

Figure 5.1

Schematic of a Hypothetical Satellite System

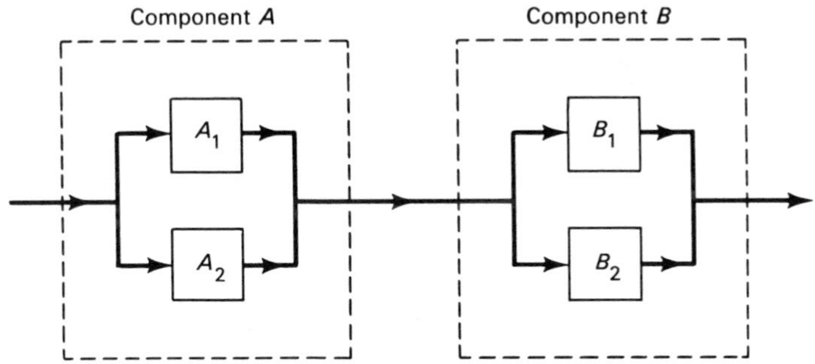

From Formula (5.7), the probability that both A_1 and A_2 will function properly for the specified length of time is $(.70)(.70) = .49$. From Formula (5.9), the probability that A_1 and/or A_2 will function is $.70 + .70 - .49 = .91$. Thus the reliability of Component A is .91. Similarly, the probability

that both B_1 and B_2 will function is $(.80)(.80) = .64$, *and the reliability of component B is* $.80 + .80 - .64 = .96$. *Then, from Formula (5.7), the probability that both components A and B will function properly for the specified time is* $(.91)(.96) = .8736$, *which is the reliability of the entire system.*

5.5 BAYES' THEOREM

In his *Essay Towards Solving a Problem in the Doctrine of Chance,* published in 1763, the Reverend Thomas Bayes presented a now well-known formula for computing conditional probabilities that has come to be known as Bayes' Theorem. Today this formula occupies such a prominent position in modern statistical decision theory that it demands our particular attention.

5.5.1 Intuitive Demonstration

As an informal introduction to our discussion of Bayes' Theorem, let us consider the following random process: There are three urns that contain mixtures of red and green balls as follows:

Urn	*Number of Red Balls*	*Number of Green Balls*	*Total*
A	3	7	10
B	5	15	20
C	6	6	12

Two fair coins are tossed. If they both come up heads (HH), a ball is selected randomly from Urn A; if one comes up heads and the other comes up tails (either HT or TH), a ball is selected randomly from Urn B; if they both come up tails (TT), a ball is selected randomly from Urn C. Thus there is a .25 probability that the selection will be made from Urn A, a .50 probability that the selection will be made from Urn B, and a .25 probability that the selection will be made from Urn C. If a ball is selected according to this process, and we are told that the ball is red, what is the probability that it was selected from Urn A or B or C? Likewise, if the ball so selected is green, what is the probability that it came from each urn?

On the basis of the given information, we may state three marginal probabilities: $P(A) = .25$, $P(B) = .50$, $P(C) = .25$. Also, knowing the particular mixture in each urn, we may state conditional probabilities for drawing a ball of a given color if we know which urn the ball is selected from. Thus, if R

denotes the event of drawing a red ball and G denotes the event of drawing a green ball, we have:

$$P(R|A) = 3/10 = .30 \qquad P(G|A) = 7/10 = .70$$
$$P(R|B) = 5/20 = .25 \qquad P(G|B) = 15/20 = .75$$
$$P(R|C) = 6/12 = .50 \qquad P(G|C) = 6/12 = .50$$

Applying Formula (5.2) we may compute certain joint probabilities:

$$P(A \cap R) = P(R|A)P(A) = (.30)(.25) = .075$$
$$P(B \cap R) = P(R|B)P(B) = (.25)(.50) = .125$$
$$P(C \cap R) = P(R|C)P(C) = (.50)(.25) = .125$$
$$P(A \cap G) = P(G|A)P(A) = (.70)(.25) = .175$$
$$P(B \cap G) = P(G|B)P(B) = (.75)(.50) = .375$$
$$P(C \cap G) = P(G|C)P(C) = (.50)(.25) = .125$$

From these calculations, we may construct the probability table shown in Table 5.8.

Table 5.8

Probability Table for Urn Experiment

	R	G	Total
A	.075	.175	.25
B	.125	.375	.50
C	.125	.125	.25
Total	.325	.675	1

From Table 5.8, we obtain two more marginal probabilities:

$$P(R) = P(A \cap R) + P(B \cap R) + P(C \cap R) = .075 + .125 + .125 = .325$$
$$P(G) = P(A \cap G) + P(B \cap G) + P(C \cap G) = .175 + .375 + .125 = .675$$

Now, by applying Formula (5.1), we may answer the original problem. If we know that a *red* ball was drawn, the probabilities that it was selected from each of the urns are given by the following calculations:

$$P(A|R) = \frac{P(A \cap R)}{P(R)} = \frac{.075}{.325} = .230$$

$$P(B|R) = \frac{P(B \cap R)}{P(R)} = \frac{.125}{.325} = .385$$

$$P(C|R) = \frac{P(C \cap R)}{P(R)} = \frac{.125}{.325} = .385$$

If we know that a *green* ball was drawn, the probabilities that it was selected from each of the urns are calculated as follows:

$$P(A|G) = \frac{P(A \cap G)}{P(G)} = \frac{.175}{.675} = .259$$

$$P(B|G) = \frac{P(B \cap G)}{P(G)} = \frac{.375}{.675} = .556$$

$$P(C|G) = \frac{P(C \cap G)}{P(G)} = \frac{.125}{.675} = .185$$

Thus, if we know that a *red* ball was drawn, there is a .230 probability that it was drawn from Urn A, a .385 probability that it was drawn from Urn B, and a .385 probability that it was drawn from Urn C. Notice that these three probabilities add up to 1, since the ball must have been drawn from one of the three urns. Similarly, if we know that a *green* ball was drawn, the probabilities are .259 that it came from Urn A, .556 that it was drawn from Urn B, and .185 that it came from Urn C.

Notice in this problem that one set of marginal probabilities—$P(A)$, $P(B)$, and $P(C)$—was given at the outset. Such probabilities are sometimes called *prior probabilities,* since they are the probabilities that exist prior to obtaining any information about the outcome of the process. The conditional probabilities that constitute the solution to the problem—$P(A|R)$, $P(B|R)$ . . . —are called *posterior probabilities,* since they represent probability assessments that have been revised on the basis of later information.

Another method of arriving at the same solution to the problem—a method that is particularly convenient in solving many applied decision problems—is to use a computational table. Such a table is illustrated in Table 5.9 for the condition that a red ball was drawn.

Table 5.9

Computation of Probabilities Posterior to Observing a Red Ball

Urn	*(1)* *Prior Probability*	*(2)* *Conditional Probability*	*(3)* *Joint Probability*	*(4)* *Posterior Probability*
A	$P(A) = .25$	$P(R\|A) = .30$	$P(A \cap R) = .075$	$P(A\|R) = .230$
B	$P(B) = .50$	$P(R\|B) = .25$	$P(B \cap R) = .125$	$P(B\|R) = .385$
C	$P(C) = .25$	$P(R\|C) = .50$	$P(C \cap R) = .125$	$P(C\|R) = .385$
	1.00		$P(R) = .325$	1.000

In Table 5.9, figures in Columns (1) and (2) are given in the statement of the problem. Column (1) contains the prior (marginal) probabilities that each of the three urns will be selected. Column (2) gives the conditional probabilities that a red ball will be drawn if the indicated urn is selected. Column (3) is obtained by multiplying the corresponding values from columns (1) and (2). The resulting figures in Column (3) are the joint probabilities that the indicated urn will be selected and a red ball drawn from it. The sum of the values in Column (3) is the marginal probability that a red ball will be drawn, regardless of which urn is selected. The figures in Column (4) are obtained by dividing the figures in Column (3) by their sum. Careful study of this computational table will reveal that the calculations are exactly the same as those that were made in solving the problem by formulas.

The solution to the problem under the condition of drawing a green ball is shown in Table 5.10.

Table 5.10

Computation of Probabilities Posterior to Observing a Green Ball

Urn	*Prior Probability*	*Conditional Probability*	*Joint Probability*	*Posterior Probability*
A	$P(A) = .25$	$P(G\|A) = .70$	$P(A \cap G) = .175$	$P(A\|G) = .259$
B	$P(B) = .50$	$P(G\|B) = .75$	$P(B \cap G) = .375$	$P(B\|G) = .556$
C	$P(C) = .25$	$P(G\|C) = .50$	$P(C \cap G) = .125$	$P(C\|G) = .185$
	1.00		$P(G) = .675$	1.000

5.5.2 Formal Statement

We may now state Bayes' Theorem more formally. Let $E_1, E_2, \ldots, E_i, \ldots, E_n$ be n mutually exclusive and collectively exhaustive events with known probabilities $P(E_1), P(E_2), \ldots, P(E_i), \ldots, P(E_n)$. Let C be an event for which the conditional probabilities $P(C|E_i), i = 1, \ldots, n$ are known. Then for any specific event E_j out of the set of n events, the conditional probability $P(E_j|C)$ may be computed from the formula:

$$P(E_j|C) = \frac{P(C \cap E_j)}{P(C)} = \frac{P(C|E_j)P(E_j)}{\sum_i [P(C|E_i)P(E_i)]} \qquad (5.10)$$

Formula (5.10) is the celebrated Bayes' Theorem. Careful examination of the formula will reveal that it specifies exactly the calculations that we have been making in solving the problem of the urns.

Example 1:

A popular example of the application of Bayes' Theorem is provided by the mathematical brain-teaser known as the "diagnosis problem." Suppose that a new diagnostic test has been developed for the detection of psittacosis in humans. Of the people with the disease, the technique correctly gives a positive diagnosis 95 per cent of the time. However, among people who do not have the disease, the test incorrectly gives a positive diagnosis 10 per cent of the time. From a large population in which two-tenths of 1 per cent of the people actually have the disease, a randomly selected person is given the test, and a positive diagnosis results. What is the probability that the person actually has the disease?

Let E_1 represent the event that the person has the disease and E_2 represent the event that the person is free of the disease. These two events are mutually exclusive and collectively exhaustive. If we let C indicate the event of a positive diagnosis, we may write the following probabilities as given in the statement of the problem:

$$P(E_1) = .002 \qquad P(C|E_1) = .95$$
$$P(E_2) = .998 \qquad P(C|E_2) = .10$$

From direct application of Formula (5.10), we obtain:

$$\begin{aligned} P(E_1|C) &= \frac{P(C|E_1)P(E_1)}{P(C|E_1)P(E_1) + P(C|E_2)P(E_2)} \\ &= \frac{(.95)(.002)}{(.95)(.002) + (.10)(.998)} = \frac{.0019}{.1017} \\ &= .0187 \end{aligned}$$

This same result also may be obtained by using Table 5.11.

Table 5.11

Probability Table for Diagnosis Problem

	C (*positive diagnosis*)	*C′* (*negative diagnosis*)	*Total*
E_1 (*Has disease*)	(.95)(.002) = .0019	(.05)(.002) = .0001	.002
E_2 (*Free of disease*)	(.10)(.998) = .0998	(.90)(.998) = .8982	.998
Total	.1017	.8983	1.000

From Table 5.11, using Formula (5.1), we may compute:

$$P(E_1|C) = \frac{P(C \cap E_1)}{P(C)} = \frac{.0019}{.1017} = .0187$$

This result may surprise the reader. Although there is a .95 probability that the test will give a correct diagnosis if a person has the disease, there is only a .0187 probability that a person has the disease if he receives a positive diagnosis. However, we should point out that without any diagnostic information, there is only a .002 probability that a person has the disease. But, given the information that a person has received a positive diagnosis, the probability of his having the disease is .0187. The conditional *(posterior) probability of .0187 is more than nine times as great as the* marginal *(prior) probability of .002.*

Example 2:

A manufacturer produces a product that involves an intricate assembly of a large number of components with different reliabilities. The life of a unit depends partly on the soundness of its components and partly on the integrity of the assembly. The product is marketed in three grades: Grade A, with a one-year warranty; Grade B, with a six-month warranty; and Grade C, with no warranty. As each unit comes off the assembly line, it is processed by an inspector who, according to his judgment, classifies it as Grade A, B, or C. The production manager estimates that, in the long run, 20 per cent of the units are actually of Grade A quality, 50 per cent are Grade B, and 30 per cent are Grade C. His estimates of the probabilities that a unit of specified actual quality will be classified by the inspector as Grade A, B, or C are shown in Table 5.12. If the inspector classifies a unit as Grade A, what is the probability that the unit actually is of Grade A quality? Grade B? Grade C?

Table 5.12

Conditional Probabilities for Product Quality Classification

If Actual	*Probability that it will be classified:*		
Quality is:	*Grade A*	*Grade B*	*Grade C*
Grade A	*.50*	*.30*	*.20*
Grade B	*.20*	*.50*	*.30*
Grade C	*.10*	*.30*	*.60*

Let A, B, and C, represent the events that actual quality of a unit is Grade A, Grade B, or Grade C, respectively. Furthermore, let a, b, and c

represent the events that the inspector classifies a unit as Grade A, Grade B, or Grade C, respectively. From the statement of the problem, we have the following probabilities:

$$\begin{array}{llll} P(A) = .20 & P(a|A) = .50 & P(b|A) = .30 & P(c|A) = .20 \\ P(B) = .50 & P(a|B) = .20 & P(b|B) = .50 & P(c|B) = .30 \\ P(C) = .30 & P(a|C) = .10 & P(b|C) = .30 & P(c|C) = .60 \end{array}$$

Applying Bayes' Formula:

$$P(A|a) = \frac{P(a|A)P(A)}{P(a|A)P(A) + P(a|B)P(B) + p(a|C)P(C)}$$

$$= \frac{(.50)(.20)}{(.50)(.20) + (.20)(.50) + (.10)(.30)} = \frac{.10}{.23} = .435$$

$$P(B|a) = \frac{P(a|B)P(B)}{P(a|A)P(A) + P(a|B)P(B) + P(a|C)P(C)}$$

$$= \frac{(.20)(.50)}{.23} = \frac{.10}{.23} = .435$$

$$P(C|a) = \frac{P(a|C)P(C)}{P(a|A)P(A) + P(a|B)P(B) + P(a|C)P(C)}$$

$$= \frac{(.10)(.30)}{.23} = \frac{.03}{.23} = .130$$

Thus, if the inspector classifies a unit as Grade A, there is a .435 probability that the unit is actually Grade A quality, and an equal probability of .435 that it is actually Grade B quality. There is also a .130 probability that a unit that is classified as Grade A is actually Grade C quality. The reader is urged to use the figures in Table 5.13 to verify the above results.

Table 5.13

Joint Probability Table for Product Classification Problem

	a	*b*	*c*	*Total*
A	.10	.06	.04	.20
B	.10	.25	.15	.50
C	.03	.09	.18	.30
Total	.23	.40	.37	1.00

PROBLEMS

5.1 For two events X and Y it is known that $P(X) = .80$ and $P(X \cap Y) = .60$. Determine $P(Y|X)$.

5.2 There is a .60 probability that a randomly identified housewife in a supermarket will buy bread. There is a .15 probability that she will buy both bread and cheese. If a housewife buys bread, what is the probability that she will also buy cheese?

5.3 A job applicant has just taken a vocabulary test and an arithmetic test. He estimates that there is a .60 probability that he obtained a passing score on the vocabulary test, and a .30 probability that he obtained a passing score on both tests. If he learns that he passed the vocabulary test, what is the probability that he passed the arithmetic test?

5.4 For two events G and H, it is known that $P(H) = .50$ and $P(G \cap H) = .20$. Determine $P(G'|H)$.

5.5 For two events S and T it is known that $P(S) = .40$ and $P(T|S) = .25$. Determine $P(S \cap T)$.

5.6 A salesman is demonstrating a vacuum cleaner to Mrs. Smith and Mrs. Jones. The two ladies have withdrawn to the next room to discuss whether each of them should purchase one of these appliances. The salesman estimates a .60 probability that Mrs. Jones will buy one of his cleaners. He also estimates that if Mrs. Jones buys a vacuum, there is a .40 probability that Mrs. Smith will buy one. What is the probability that both women will buy vacuum cleaners?

5.7 In a large company, 60% of the employees are males and 25% of the employees are management trainees. Furthermore, 80% of the management trainees are males. Suppose an employee is to be selected at random from the company's personnel roster. Let M denote the event that the person selected is a male, and T denote the event that the person selected is a management trainee.

(a) Express each of the following probability statements in words:
- (i) $P(T')$
- (ii) $P(M \cap T)$
- (iii) $P(M \cap T')$
- (iv) $P(T|M)$
- (v) $P(M'|T)$

(b) Compute each of the probabilities in (a).

(c) Express each of the following probability statements in symbols:
- (i) Probability of selecting a female who is a management trainee.
- (ii) Probability of selecting a female who is not a management trainee.
- (iii) Probability of selecting a male if the person selected is not a management trainee.
- (iv) Probability of selecting a person who is not a management trainee if the person selected is a male.
- (v) Probability of selecting a female if the person selected is not a management trainee.

(d) Compute each of the probabilities in (c).

5.8 For two events K and L it is known that $P(K) = .70$ and $P(L) = .40$. Furthermore, it is known that K and L are independent. Determine:

(a) $P(K|L)$

(b) $P(L|K)$

(c) $P(K \cap L)$

5.9 Consider two *independent* events, G and H, for which $P(G) = .80$ and $P(H) = .40$. Determine each of the following probabilities:

(a) $P(G \cap H)$
(b) $P(G \cap H')$
(c) $P(G' \cap H)$
(d) $P(G' \cap H')$
(e) $P(G|H)$
(f) $P(G|H')$
(g) $P(G'|H)$
(h) $P(G'|H')$

5.10 The probability that a customer entering a supermarket buys eggs is .10. If the customer buys eggs, the probability that he will also buy bacon is .40. Whether or not the customer buys eggs, the probability is .50 that he will buy milk.

(a) What is the probability that a customer will buy both eggs and bacon?
(b) What is the probability that the customer will buy both eggs and milk?

5.11 For a particular population of consumers, 60 per cent of the total are males and 30 per cent of the total prefer Brand A toothpaste. If toothpaste preference is independent of sex, and a random sample of 100 consumers is selected, how many consumers in the sample would you expect to be males who prefer Brand A toothpaste?

5.12 The probability that a USC student was born in California is .50. The probability that a USC student owns a car is .40. The probability that a USC student both owns a car and also was born in California is .25. Within the USC student population, are these two events—owning a car and being born in California—independent events? Explain.

5.13 Of the executives at a convention, 60 per cent attended luncheon and 70 per cent attended dinner. Of the total, 40 per cent attended both luncheon and dinner.

(a) If it is known that an executive attended luncheon, what is the probability that he also attended dinner?
(b) Were these two acts—attending luncheon and attending dinner—independent events? Explain.

5.14 The probability that Stock A will rise in price on any given trading day is .40. The probability that Stock B will rise in price on any trading day is .20. If these two stocks behave independently, what is the probability that, on a particular trading day:

(a) Both stocks will rise in price?
(b) Neither stock will rise in price?
(c) Stock A will rise in price and Stock B will not rise?
(d) Stock B will rise in price and Stock A will not rise?
(e) One of the stocks will rise in price and the other will not rise?

5.15 The Guessright Employment Agency administers a Verbal Comprehension test and a Verbal Reasoning test to each of its applicants. On the Verbal Comprehension test, a score above 14 is considered passing, and on the Verbal Reasoning test, a score of 19 is considered passing. From the agency's records, it has been determined that 10 per cent of the applicants fail the Verbal Comprehension test, 12 per cent fail the Verbal Reasoning test, and 2 per cent fail both.

(a) What is the probability that a randomly selected applicant passed both tests?
(b) Is the event "pass Verbal Comprehension test" independent of the event "pass Verbal Reasoning test"? Explain.

5.16 Marlo Associates, a marketing research firm, has been retained by the Lipgram Tea Co. to conduct a marketing research study of tea-buying behavior in Greentree County. A randomly selected sample of 200 housewives in Cedar City, the county seat, were asked the question, "The last time you purchased tea, what brand did you buy?" In a discreet manner, each respondent was also queried concerning her age. The number of respondents in each age preference category are shown below:

	Last Brand Purchased		
Age	*Lipgram*	*Other*	*Total*
Under 30	10	50	60
30 and over	90	50	140
Total	100	100	200

(a) If a respondent is selected at random from this sample, what is the probability that the last brand of tea she purchased was Lipgram?
(b) If a respondent under 30 is selected at random from this sample, what is the probability that the last brand of tea she purchased was Lipgram?
(c) If a respondent is selected at random from those whose last tea purchase was Lipgram, what is the probability that she is at least 30 years old?
(c) Within this sample, is "Last Brand Purchased" independent of the purchaser's age? Explain.
(e) What reservations might you have about extrapolating these Cedar City results to the greater population of Greentree County? Explain.

5.17 For two events C and D it is known that $P(C) = .25$ and $P(D) = .35$. It is also known that C and D are mutually exclusive. Determine $P(C \cup D)$.

5.18 Consider two *mutually exclusive* events F and G. It is known that $P(F) = .40$ and $P(G) = .50$.
(a) Compute the following probabilities:
(i) $P(F \cup G)$
(ii) $P(F \cup G')$
(iii) $P(F|G')$
(iv) $P(G|F)$
(v) $P(G'|F)$
(b) Are F and G independent events? Explain.

5.19 Events X, Y, and Z are mutually exclusive and collectively exhaustive. If $P(X') = .60$ and $P(Y) = 2P(Z)$, what is $P(X)$? $P(Y)$? $P(Z)$?

5.20 A child's piggy bank contains a large number of coins, each of which is either a 1968 penny, a 1969 penny, or a 1971 penny. We know that 20 per cent of the pennies are dated 1968, and 25 per cent of the pennies are dated either 1968 or 1969. If the child shakes the bank until a penny falls out, what is the probability that it is dated:
(a) 1969?
(b) 1971?

5.21 Of the entrants in a regional contest sponsored by a soap company, 80 per cent were women and 20 per cent were men. Of the total number of entrants, 40 per cent were Californians. Of the Californians, 75 per cent were women.
(a) What percentage of the entrants were either Californians and/or women?
(b) If an entrant was a man, what is the conditional probability that he was *not* a Californian?

5.22 On two successive days, a job-seeker is interviewed by Alkol Industries and by Benson Associates. In both cases he is told, "Don't call us, we'll call you." He estimates that the probability of receiving a favorable response from Alkol is .70, that the probability he will receive an unfavorable response from Benson is .50, and that the probability of at least one of the firms' replying unfavorably. What is the probability that he will receive a favorable response from at least 1 of the firms?

5.23 A social-climbing young executive has applied for membership to two exclusive clubs—Les Chevaliers de Vin Aigre and the Sons of the Establishment. He estimates that the probability of being accepted by Les Chevaliers is .40, that he will be accepted by at least 1 of the 2 clubs is .60, but that he has only a .10 probability of being accepted by both clubs. What is the probability that he will be accepted by the Sons of the Establishment?

5.24 In a certain business school, 10 per cent of the senior class were elected to Beta Gamma Sigma, 40 per cent of the class held part-time jobs, and 20 per cent of those holding part-time jobs were elected to Beta Gamma Sigma. What is the probability that a senior chosen at random either held a part-time job and/or was elected to Beta Gamma Sigma?

5.25 Of the 160 supervisors in a company, a total of 100 have college degrees and a total of 55 are veterans. Of the total, 35 of the supervisors are veterans who have college degrees.

(a) If a supervisor is selected at random, what is the probability that he is either a veteran and/or has a college degree?

(b) Suppose the supervisors who are veterans are separated from the non-veterans and a man is randomly selected from the veteran supervisors. What is the probability that the man has a college degree?

5.26 A political analyst estimates that, in an upcoming statewide election, the probability that Goodheart will be elected governor is .70, and the probability that Morris will be elected Treasurer is .60. He also estimates a .56 probability that both men will be elected. Symbolically, these three probabilities may be written:

$$P(G) = .70$$
$$P(M) = .60$$
$$P(G \cap M) = .56$$

(a) Express each of the following probability statements in words:
(i) $P(G \cup M)$; (ii) $P(M|G')$; (iii) $P(G|M)$; (iv) $P(G \cap M')$; (v) $P(G \cup M')$; (vi) $P(G'|M)$.

(b) Express each of the following probability statements symbolically:
(i) Probability that Morris will be elected if Goodheart is elected.
(ii) Probability that Goodheart will not be elected but Morris will be elected.
(iii) Probability that either Goodheart and/or Morris will be defeated.
(iv) Probability that both Goodheart and Morris will be defeated.
(v) Probability that Morris will be defeated if Goodheart is defeated.
(vi) Probability that Goodheart will be defeated if Morris is defeated.

(c) Compute each of the probabilities in (a) and (b) above.

5.27 Two *independent* events, A and B, are defined on the same sample space such that the marginal probability of A is .70 and the marginal probability of B is .80. Determine each of the following probabilities:

(a) $P(A \cap B)$; (b) $P(A' \cap B')$; (c) $P(A \cup B)$;
(d) $P(A' \cup B')$; (e) $P(A|B')$; (f) $P(B'|A)$;
(g) $P(A \cup B')'$.

5.28 Two *independent* events, A and B, have the marginal probabilities $P(A) = .30$ and $P(B) = .60$. Determine the following probabilities:
(a) $P(A \cap B)$; (b) $P(A|B)$; (c) $P(A \cap B')$; (d) $P(A|B')$;
(e) $P(A \cup B)$; (f) $P(A' \cup B')$; (g) $P(A' \cup B)$; (h) $P(A \cup B')$.

5.29 Two events are defined on the same sample space. Let us designate these events A and B. Suppose that $P(B) = .60$, $P(A' \cap B) = .10$, and $P(A \cup B) = .70$. Determine the following:
(a) $P(A \cap B')$; (b) $P(A \cup B')$; (c) $P(A' \cap B')$; (d) $P(A \cap B)$;
(e) $P(A|B)$; (f) $P(B|A)$; (g) $P(B'|A')$.

5.30 Consider the two different events, A and B, both defined on the same sample space. If $P(A \cap B) = .40$ and $P(A' \cap B') = .10$, determine the following:
(a) $P(A' \cup B')$
(b) $P[(A \cap B') \cup (A' \cap B)]$

5.31 The Cryolex Manufacturing Company is composed of 3 divisions, as indicated below:

Division	*Number of Employees*
Production	700
Accounting	100
Sales	200

An opinion survey is to be conducted with a randomly selected sample. The number of employees from each division to be included in the sample is as follows:

Division	*Number in Sample*
Production	160
Accounting	10
Sales	30

(a) If it is known only that a person is a Cryolex employee, what is the probability that he will be included in the survey sample?
(b) If an employee is a member of the production division, what is the probability that he will be included in the sample?
(c) If an employee is not in the accounting division, what is the probability that he will not be included in the sample?
(d) If an employee is included in the sample, what is the probability that he is in the production division?
(e) If an employee is not in the sample, what is the probability that he is not in the sales division?
(f) If an employee is selected at random, what is the probability that he is either a member of the accounting division and/or will be included in the sample?
(g) If an employee is selected from the combined membership of the accounting and sales divisions, what is the probability that he is either a member of the sales division and/or will not be in the sample?

5.32 A group of investors has hatched a plan to organize 3 different firms, A, B, and C. Before rushing headlong into the project, they pause to consider the circumstances that might confront them during the first year of operation. Let A represent the event that Firm A survives its first year (so A' is the event that it fails), B be the event that Firm B survives the first year, and C be the event that Firm C survives the first year. Upon contemplation, the investors estimate the following probabilities:

$$P(A) = .75 \quad P(B) = .70 \quad P(C) = .60 \quad P(C|B) = .80 \quad P(A|B) = .75$$

(a) Express in words the event represented by the following expression:

$$(A \cap B \cap C)'$$

(b) Justify the following expression by stating how the 7 terms on the right are related to the meaning of the term on the left:

$$(A \cap B \cap C)' = (A' \cap B \cap C) \cup (A \cap B' \cap C) \cup (A \cap B \cap C') \cup (A' \cap B' \cap C) \cup (A' \cap B \cap C') \cup (A \cap B' \cap C') \cup (A' \cap B' \cap C')$$

(c) Which of the following is a correct statement? Events B and C are (i) independent and mutually exclusive; (ii) independent but not mutually exclusive; (iii) mutually exclusive but not independent; (iv) neither mutually exclusive nor independent.

(d) Which of the following is a correct statement? Events A and B are (i) mutually exclusive but not independent; (ii) independent but not mutually exclusive; (iii) neither mutually exclusive nor independent; (iv) both mutually exclusive and independent.

(e) Which of the following is a correct statement? (i) A is a subevent of B; (ii) B is a subevent of A; (iii) A is not a subevent of B.

(f) If Firm A should fail during the first year, what is the probability that Firm B will fail during the first year?

(g) What is the probability that both Firm A and Firm B will fail during the first year?

(h) What is the probability that either Firm A and/or Firm B will survive the first year?

5.33 As manager of the international division of a firm, you must decide whether or not to market your product in a certain foreign country. Your three subordinates —David, Edward, and Fred—have analyzed this matter, and each of them has prepared a report for you. Experience has shown that each of them is capable of making a faulty analysis and thus might recommend a wrong decision. Furthermore, because of similar backgrounds and experience, the probabilities of their making right recommendations are not necessarily independent.

Denote D: {David's report is right}
E: {Edward's report is right}
F: {Fred's report is right}

The following probabilities have been estimated on the basis of the past performances of these men in similar situations:

$$P(D) = .70 \qquad P(E) = .80 \qquad P(F) = .90$$
$$P(D|E) = .75 \qquad P(F|E) = .90 \qquad P(D \cup F) = .95$$

(a) What is the probability that both Edward's and Fred's reports are wrong?

(b) What is the probability that either David's and/or Fred's reports are wrong?

(c) If we have complete confidence that Edward's report is right, what is the probability that David's report is right?

(d) If David submits a wrong report, what is the probability that Fred will submit a correct report?

(e) What is the probability that both David and Edward will submit correct reports?
(f) What is the probability that either David and/or Edward will submit a correct report?
(g) What is the probability that either Edward or Fred (but not both) will submit a wrong report?

5.34 In the two-person Antediluvian game of Glak, players take turns tossing 10 colored stones onto a marked playing surface. Depending on the way the stones fall in the various areas of the surface, a play is scored as an Ank (4 points), a Blit (3 points), a Chaw (2 points), or a Dud (1 point). The probability of making each type of play is:

Play	*Probability*
Ank	.10
Blit	.20
Chaw	.30
Dud	.40

If each player makes 1 toss independently, what is the probability that:
(a) Both players toss Blits?
(b) Neither player tosses an Ank?
(c) Player I tosses a Blit and Player II tosses a Chaw?
(d) One player tosses a Blit and the other player tosses a Chaw?
(e) At least 1 of the players tosses a Dud?
(f) Each of the 2 players makes at least 3 points?
(g) The sum of the points scored by the 2 players is 6?
(h) The sum of the points scored by the 2 players is greater than 6?
(i) The sum of the points scored by the 2 players is at least 6?
(j) The sum of the points scored by the 2 players is less than 6?
(k) The sum of the points scored by the 2 players is no greater than 6?

5.35 An electronic subsystem is composed of 3 components—A, B, and C—as shown in the diagram below. Component A will function only if both parts A_1 and A_2 function. Component B will function as long as either B_1 and/or B_2 functions. Component C will function as long as at least 1 of its 3 parts—C_1, C_2, and C_3—

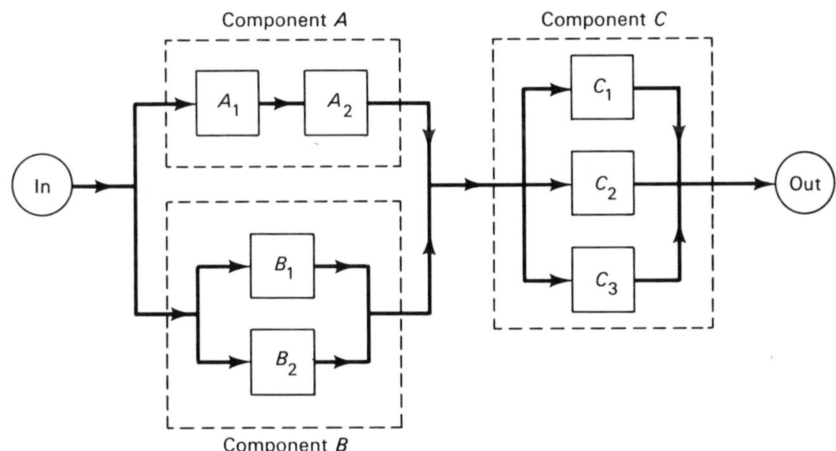

functions. The entire subsystem will function only if Component C functions, and if either Component A and/or Component B functions. Components A_1 and A_2 each has a reliability of .90. Components B_1 and B_2 each has a reliability of .80. Components C_1, C_2, and C_3 each has a reliability of .70. Assuming that each individual part operates independently of every other part, what is the reliability of the entire subsystem?

5.36 An electronic subsystem is composed of 3 components—A, B, and C—as shown in the diagram below. Component A will function only if both parts A_1 and A_2 function. Component B will function as long as at least 1 of its 3 parts—B_1, B_2, and B_3—functions. Component C will function only if C_1 and C_4 *both* function, *and* if either C_2 and/or C_3 functions. The entire subsystem will function only if Component C functions, and if either Component A and/or Component B functions.

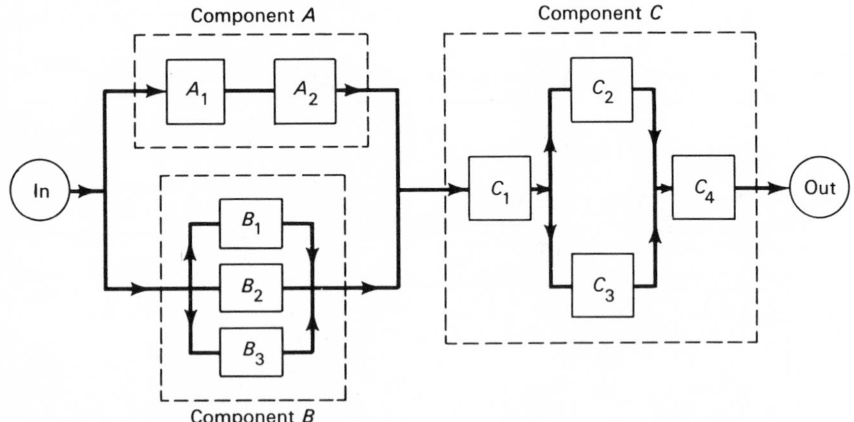

The reliability of the individual parts are as follows:

Part	*Reliability*	*Part*	*Reliability*	*Part*	*Reliability*
A_1	.90	B_2	.90	C_2	.80
A_2	.95	B_3	.90	C_3	.80
B_1	.90	C_1	.96	C_4	.98

Assuming that each individual part operates independently of every other part, what is the reliability of the entire subsystem?

5.37 How many cards must be dealt from the top of a well-shuffled, 52-card deck so that there is better than a 50-50 chance of obtaining:

(a) At least 1 heart?
(b) At least 1 king?

5.38 A box contains 4 black balls and 6 white balls. If 3 balls are drawn at random in succession without being replaced, what is the probability that:

(a) All 3 are white?
(b) Exactly 1 is black?
(c) Exactly 2 are black?
(d) Exactly 3 are black?
(e) At least 1 is black?

5.39 Consider a particular type of missile that has a .60 probability of hitting its target and a .40 probability of missing it. Let us assume that individual missile firings constitute independent trials of a random process.

(a) Determine the minimum number of missiles that must be fired at a target so that the probability of hitting the target is at least .98.

(b) In this situation, what is the meaning of the assumption of "independent trials?"

5.40 A selected group of 50 common stocks contains 25 industrials, 10 railroads, and 15 utilities. On the first of June, 20 of the industrials, 7 of the railroads, and 3 of the utilities rose; the rest fell. If a single stock picked at random from the group is discovered to have fallen on the first of June, what is the probability that it is a railroad issue?

5.41 A marketing research firm is interested in interviewing married couples about certain product preferences. An interviewer from this research firm arrives at a building, which has 3 apartments. He reads the names on the mail boxes and sees that 1 of the 3 apartments is inhabited by 2 women, another has 2 men living in it, and the third is occupied by a married couple. Proceeding into the building, the interviewer is surprised to find that there are no names or numbers on the doors of the 3 apartments. He therefore selects a door at random and rings the bell. A woman answers. Assuming that the woman who answered the bell is an occupant of the apartment, what is the probability that the interviewer has mistakenly chosen the apartment occupied by the 2 women rather than that of the married couple?

5.42 Table 5.13 presented joint probabilities of products of various grades receiving various inspector's classifications. Use the data in that table to answer the following questions:

(a) If the inspector classifies a unit as Grade B, what is the probability that the unit is actually Grade A? Grade B? Grade C?

(b) If the inspector classifies a unit as Grade C, what is the probability that the unit is actually Grade A? Grade B? Grade C?

5.43 A medical research group has developed a new diagnostic test for glombitis. The group's research data indicate the following: (1) Of people with glombitis, the test correctly gives a positive result 90 per cent of the time. (2) Of people free of glombitis, the test incorrectly gives a positive result 4 per cent of the time. From a large population of which only 2/10 of 1 per cent have glombitis, 1 person is selected at random, given the test, and the result is positive. What is the probability that the person actually has glombitis?

5.44 Management's prior probability assessment of demand for a newly developed product is shown below:

Demand	*Probability*
High	.60
Low	.10
Average	.30

A survey, taken to help estimate the true demand level for the product, indicates that demand will be average. The reliability of the survey procedure is such that it will indicate "average" demand 30 per cent of the time when it really will be high, 50 per cent of the time when it really will be low, and 90 per cent of the time when it really will be average. In the light of this information, what would be the revised probabilities of the 3 possible demand levels?

5.45 A cosmetics firm is considering whether or not to introduce its newly developed "Night in Eden" face cream into a particular marketing area. On the basis of the sales of other of the firm's products in this market, management estimates the following probabilities of demand during the first 6 months:

Demand Level	*Probability*
High	.30
Moderate	.50
Low	.20

To help reduce the amount of uncertainty in this decision, the firm's marketing research group conducts a survey in the proposed target area. The reliability of the survey procedure is summarized by the following table of conditional probabilities:

If Actual Demand Is	*Probability That Survey Prediction Will Be*		
	High	*Moderate*	*Low*
High	.60	.30	.10
Moderate	.10	.70	.20
Low	.10	.10	.80

(a) If the research group's report predicts that demand for "Night in Eden" will be "high," what is the probability that actual demand will be high? Moderate? Low?
(b) If the research group's report predicts that demand for "Night in Eden" will be "moderate," what is the probability that actual demand will be high? Moderate? Low?
(c) If the research group's report predicts that demand for "Night in Eden" will be "low," what is the probability that actual demand will be high? Moderate? Low?

5.46 In its long-range planning, Technox Associates, a major construction engineering firm in the Southwest, is analyzing the prospects for 2 different large-scale projects—an atomic power plant at Zorro Beach and a dam on the Osoblanco River. Since it is beyond the firm's capacity to undertake both of these projects, the planners must decide which has the greater potential. A study team is sent to the Zorro Beach area and reports that the major concern affecting the possible construction of the atomic plant is whether Juan Muir, an adamant conservationist, becomes the next majority leader in the state legislature. If Muir becomes majority leader, analysts estimate that the probability of state approval for the atomic plant is .20; but if Muir fails to become majority leader, they estimate a .70 probability of approval. Political experts estimate that there is a .40 probability of Muir's becoming the next majority leader. The study group then proceeds to the headquarters of the Osoblanco Water and Power District, where they find considerable uncertainty concerning the dam project. If the Zorro atomic plant is authorized, there appears to be only a .20 chance that the dam construction would be approved; but if the Zorro plant is not authorized, the probability is .90 that the dam project will be approved. From all of these subjective probability assessments, which project has the greater probability of being realized—the atomic plant or the dam?

Chapter 6
Random Variables

To simplify our discussion of some of the basic concepts presented in preceding chapters, many of our examples were concerned with random processes whose outcomes were described in *qualitative* terms such as "heads or tails," "successful or unsuccessful," "high or low," and "Grade A, Grade B, or Grade C." In applying probability theory to decision problems, however, it often is more practical to describe outcomes in *quantitative* terms, because numbers are more adaptable to mathematical manipulation than are qualitative expressions. Thus, if outcomes are described in qualitative terms, it frequently is desirable to convert such qualitative descriptions into quantitative terms. For example, in flipping a coin, we might adopt a "rule" that would assign a numerical value of 1 for the outcome heads and a value of 0 for the outcome tails.

Sometimes, even if the outcomes of a random process are naturally describable in numerical values, such values may not be of interest to a specific decision maker. For example, we have seen that the outcome of rolling a pair of dice may be described numerically as an ordered pair, such as (5,4). However, if a craps player's point is 9, he would not be particularly interested in which ordered pair he tosses, as long as the *sum* of the two values that constitute the pair is equal to 9.

This chapter is primarily concerned with describing the outcomes of random processes (whether the outcomes be qualitative or quantitative in nature) in *meaningful* numerical terms.

6.1 BASIC CONCEPT OF A FUNCTION

The reader may recall from his high school algebra that a *function* is a *rule* that associates with each element of one set, called the *domain*, a unique element of another set, called the *range*. This definition indicates that each function consists of three different parts: the rule, the domain, and the range.

As an illustration, consider the case in which the total manufacturing cost of a product depends on the quantity produced. Specifically, for the particular product concerned, the total manufacturing cost increases at a rate of \$8.00 for each additional unit produced if the total production does not exceed 100 units. In addition, there is a fixed cost of \$600. If we let q represent the production quantity and let c represent the total cost, then we may express c as a function of q as follows:

$$c = 600 + 8q$$

The domain of this function is the set of all possible values of q:

$$\{0,1, \ldots, 100\}$$

The rule of association is given by the mathematical expression $c = 600 + 8q$. In other words, for each of the 101 values of the variable q, a value for the variable c is determined by the rule "c equals 600 plus 8 times the value of q." This rule pairs each value of q with one and only one value of c, which is shown in the following table:

q	0	1	2	...	99	100
c	\$600	\$608	\$616	...	\$1,392	\$1,400

Thus, the range of this function is the set of all possible values of c:

$$\{\$600, \$608, \ldots, \$1{,}400\}$$

To denote a functional relationship between c and q, we may write

$$c = f(q)$$

where $f(q)$ is read "f of q." In this expression, the symbol f denotes some *function*, and $f(q)$ represents the *value* of the function at any specific value q. For instance, $c = f(2)$ means that c is equal to the value obtained when the function is evaluated at 2. That is, since $f(q) = 600 + 8q$, we obtain

$$c = f(2) = 600 + 8(2) = 616$$

This expression means that it will cost \$616 to produce two units.

Since the expression $c = 600 + 8q$ indicates that c is a function of q, the value of c *depends on* the value of q. Thus we regard c as the *dependent variable* and q as the *independent variable*. In these terms, the range of the function is the set of all possible values of the dependent variable, whereas the domain is the set of all possible values of the independent variable. The input to the function is some value of the independent variable, and the output of the function yields a corresponding value of the dependent variable.

The reader may find it helpful to think of a function in terms of a machine such as the one shown in Figure 6.1. In this figure, we consider the rule as the machine, the domain as the input to the machine, and the range as the output from the machine.

Figure 6.1

Illustration of a Function

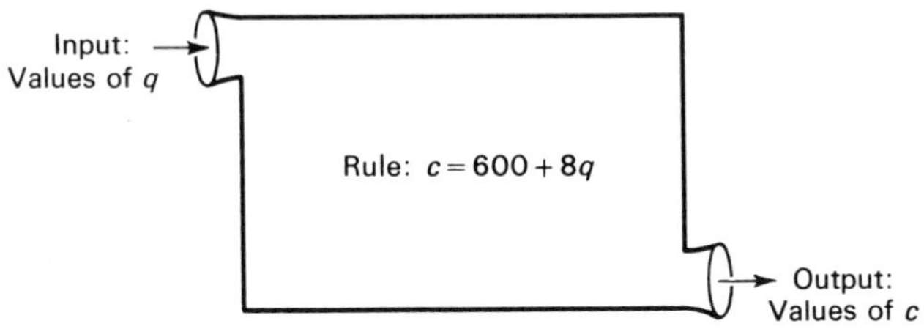

Notice that the domain and the range of the function in the above example each consists of a set of numerical values. Another type of mathematical function is one whose range is a set of numerical values but whose domain is a set of qualitative elements. Consider, for example, the common practice of assigning grades of A, B, C, D, and F in college courses. If a student wishes to compute his grade point average; he must convert his letter grades into numbers. Specifically, he may adopt the rule (function) that follows:

$$g = \begin{cases} 4 & \\ 3 & \\ 2 & \text{if} \\ 1 & \\ 0 & \end{cases} \begin{matrix} A \\ B \\ C \\ D \\ F \end{matrix}$$

In this function, the domain consists of a set of letter grades, $\{A,B,C,D,F\}$, and the range consists of a set of numerical values, $\{4,3,2,1,0\}$.

6.2 TRADITIONAL DEFINITION OF A RANDOM VARIABLE

The definition of a function presented in the preceding section can now be extended to situations involving random processes. Specifically, consider a function whose *range* is a set of *real numbers* and whose *domain* is the *sample space* of a random process. Then the function will assign a specific real number to each point in the sample space. The quantity that is represented by this type of functional relationship is called a *random variable*. This quantity is said to be random because its value is determined by the outcome of a random process. That is, the particular value yielded by the function cannot be known with certainty until the particular outcome of the random process has been observed. Insofar as possible, *we will use a capital letter to denote a random variable and the corresponding small letter to represent any specific value of the random variable.*

Example 1:

Suppose that an automatic machine produces widgets that may be classified as defective or nondefective. Consider the experiment of selecting a sample of two widgets at random from the output of this machine. If we denote a defective item by d and a nondefective item by d′, then the sample space for this experiment is:

$$\{(d',d'),\ (d',d),\ (d,d'),\ (d,d)\}$$

One way of converting these qualitatively described sample points into numerical values is to count the number of defective widgets in the sample. Clearly, this count is a random variable, since its value depends on the outcome of the experiment. If we let X denote this random variable, we can express the functional relationship between the count and the sample space as follows:

$$X = \begin{cases} 0 & \text{if} \quad (d',d') \\ 1 & \quad\quad (d',d) \quad or \quad (d,d') \\ 2 & \quad\quad (d,d) \end{cases}$$

Notice that in this function the range (that is, the set of all possible values of the random variable X) is {0,1,2}. It is important to remember that the function assigns one and only one value to each sample point, although the same value may be assigned to more than one sample point. This functional relationship is portrayed graphically in Figure 6.2.

Figure 6.2

Mapping a Sample Space into a Random Variable

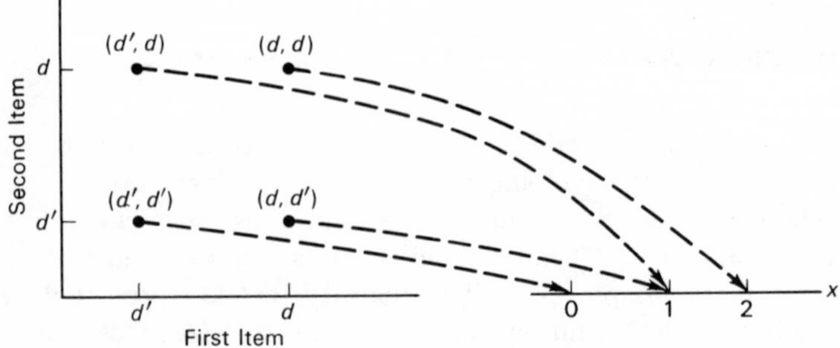

Example 2:

Let us consider again the sample space for the random process of rolling a pair of dice. Recall that each point in this sample space, which was presented in Figure 4.3, was defined in terms of an ordered pair of values. For instance, the ordered pair (1, 3) was used to denote the outcome that the red die comes up 1 and the green die comes up 3. Suppose that we now consider the rule of describing the outcome of a roll by adding the numbers of spots appearing on the upper faces of the two dice. If we let X represent this sum, then X is a random variable. All possible values of X, together with the sample points associated with each of these values, are shown in Table 6.1. Inspection of Table 6.1 indicates that each sample point is assigned a single value, although the same value may be assigned to several sample points. For instance, the sample point (1, 3) is assigned a single value of 4, although the same value of 4 is also assigned to sample points (3, 1) and (2, 2).

Table 6.1

Defining a Random Variable on a Sample Space

X: Sum of Spots	*Corresponding Sample Points*	*Number of Sample Points*
2	(1,1)	1
3	(1,2), (2,1)	2
4	(1,3), (2,2), (3,1)	3
5	(1,4), (2,3), (3,2), (4,1)	4
6	(1,5), (2,4), (3,3), (4,2), (5,1)	5
7	(1,6), (2,5), (3,4), (4,3), (5,2), (6,1)	6
8	(2,6), (3,5), (4,4), (5,3), (6,2)	5
9	(3,6), (4,5), (5,4), (6,3)	4
10	(4,6), (5,5), (6,4)	3
11	(5,6), (6,5)	2
12	(6,6)	1
		36

6.3 EXTENDED DEFINITION OF A RANDOM VARIABLE

When the term "random variable" was introduced into the vocabulary of probability theory, the discipline was limited essentially to the study of *objective* probabilities. Thus it was only natural that the definition of a random variable was originally formulated in terms of a clearly delineated sample space of a well-defined random process. During recent years, however, the study of *subjective* probability has received increasing attention and recognition, particularly as applied to business decision making. Since subjective probabilities are not necessarily related to well-defined random processes, it has become necessary to extend the traditional definition of a random variable to include situations in which existing uncertainties may not be describable in terms of random processes.

In the *extended sense*, a random variable is simply defined as *a quantity whose value is not known with certainty*. We can see that this definition applies to a random variable as traditionally defined as well as to any "uncertain quantity." This extended definition is illustrated by the following situations:

1. A crap shooter is about to roll the dice. He is uncertain about the sum that he will roll. Therefore, the sum is a random variable, which in this case fits the traditional definition as well as the extended definition.
2. You might have an argument with a friend concerning the number of games won by the New York Giants in the 1951 World Series. Although the outcome of the series is a matter of record, your vague memory makes the outcome uncertain to you. Therefore, the number of games won by the Giants is an uncertain quantity and may be considered as a random variable under the extended definition.
3. An investor is considering buying some stock that he plans to sell in three months. Due to various unpredictable factors, he is uncertain about what the price of the stock will be three months from now. Thus this price can be regarded as a random variable in the extended sense. In this case, the price of the stock is partly determined by a random process, but the mechanism of that process is not sufficiently understood to permit its logical analysis.

Example:

A specialty grocer stocks an exotic, highly perishable type of melon that he receives each morning by air shipment from a supplier in Pago Pago. He buys the melons at a cost of $2.00 each, and sells them for $5.00 each. For any given day, the grocer has found that the number of these melons that customers will request is unpredictable. On the basis of his past experience, the most typical daily demand is for six melons, which is the number he stocks each day. On some days, however, demand may be as low as zero, and on other days he may receive requests for considerably more than six melons. In other words, daily demand is an uncertain quantity.

In this example, we might define several random variables that could be of interest. We might, for instance, let D be a random variable representing the number of melons demanded daily. The possible values of D would be 0, 1, 2, . . . , with no definite upper limit, since any finite number of melons can be requested, regardless of the grocer's supply. We also might define a random variable S as the number of melons actually sold daily. As long as the value of D is no greater than 6, S will equal D; but S can never be larger than 6, which is the grocer's supply. This relationship may be expressed as follows:

$$S = \begin{cases} D & \text{if} \quad D = 0, 1, \ldots, 6 \\ 6 & \text{if} \quad D > 6 \end{cases}$$

Another random variable of interest would be the grocer's daily net gain, which might be denoted by G. Since the dealer's total cost (at \$2.00 per melon for six melons) is \$12.00, and his gross revenue from melon sales is \$5.00 multiplied by the number of melons sold, we might write:

$$G = \$5.00(S) - \$12.00$$

Notice here that we have expressed one random variable, G, as a function of another random variable S. We might alternatively express G as a function of D in the following manner:

$$G = \begin{cases} \$5.00(D) - \$12.00 & \text{if} \quad D = 0, 1, \ldots, 6 \\ \$18.00 & \text{if} \quad D > 6 \end{cases}$$

Observe that this expression clearly indicates that regardless of how great demand might be, maximum possible profit is limited to \$18.00 by virtue of the grocer's limited inventory.

6.4 DISCRETE AND CONTINUOUS RANDOM VARIABLES

In working with a random variable, it is important to recognize whether it is "discrete" or "continuous." A random variable is said to be *discrete* if it can take on only a countable number of values so that these values can be listed. In other words, the set of all possible values of a discrete random variable can be described by the listing method explained in Section 4.1. All the examples presented so far in this chapter are concerned with discrete random variables. For instance, in the dice example, the set of all possible values of the random variable X can be listed as follows:

$$\{2,3,4,5,6,7,8,9,10,11,12\}$$

This set contains eleven values and X can take on only these specific values. Thus X is a discrete random variable.

Although a discrete random variable can take on only a countable number of isolated values, a *continuous* random variable can take on any of an infinite number of values along a continuum between specific limits. As such, a continuous random variable has an infinite number of possible values. Therefore, the set of all possible values of a continuous random variable can be defined by the describing method but not by the listing method. For instance, a customer in a self-service supermarket fills a paper bag with a dozen oranges. Although he does not know the exact weight of the bag of oranges, he feels sure that it weighs at least four pounds but no more than seven pounds. Thus the weight of the bag of oranges is a random variable that may take on any real value between the limits of 4 and 7. If we denote this random variable by W, we may express the possible values of W in set notation as follows:

$$\{w|4 \leq w \leq 7\}$$

In this expression, notice that the lower-case w is used to denote the values of the random variable, which is designated by the upper-case W.

In making a continuous measurement—such as weighing an object on a scale or measuring its length with a yardstick—it is theoretically possible for the measurement to have any of an infinite number of values within a particular range. For example, even though a yardstick might be marked off in one-quarter inch units, it would be theoretically possible to mark it off in any size units regardless of how small. As a practical matter, however, there is a limit to the precision with which measurements can be obtained from such devices. For example, a strand of wire might be exactly 8.4726 centimeters in length, but if you were measuring the wire with an ordinary centimeter scale ruled off in one-tenth centimeter units, you would not be able to make a reading with the refinement required to give accuracy to the fourth decimal place. Most likely you would be satisfied to announce your measurement as "approximately" 8.5 centimeters. Thus, even if our supermarket customer were to place his bag of oranges on a scale, he would not be able to determine its *exact* weight because of the limited precision of the scale.

Example:

In an assembly plant, a particular quality control procedure requires that an inspector select and test a sample of ten items from each lot of steel springs received from a supplier. We might define the random variable D as the number of defectives obtained when a sample of ten items is inspected from a lot; this would be a discrete random variable that can assume only the isolated values 0, 1, . . . , 10. We might, however, define another random variable W as the combined weight in ounces of the ten items in the sample; this would be a continuous random variable, since the combined weight of the ten items could assume any of an infinite number of values along a continuous scale.

6.5 DEFINING EVENTS IN TERMS OF RANDOM VARIABLES

In Chapter Four we defined an event as any subset of a sample space. That is, an event corresponds to one or more sample points. In this chapter, we have seen how the concept of a random variable enables us to assign numerical values to sample points. Thus we are able to describe an event in terms of the numerical values assigned to the sample points of that event. In other words, events may be described as subsets of all possible values of a random variable.

Example 1:

In Section 6.2, we used the example of selecting a sample of two widgets from the output of an automatic machine. The sample space for this example is:

$$\{(d',d'),\ (d',d),\ (d,d'),\ (d,d)\}$$

For this sample space, it is possible to define sixteen events since the total number of subsets is equal to 2^4. Some of these events are shown below:

$$E_1 = \{(d',d')\} \qquad E_4 = \{(d',d),\ (d,d'),\ (d,d)\}$$
$$E_2 = \{(d',d),\ (d,d')\} \qquad E_5 = \{(d',d'),\ (d',d),\ (d,d')\}$$
$$E_3 = \{(d,d)\} \qquad E_6 = \{(d',d'),\ (d',d),\ (d,d'),\ (d,d)\}$$

Each of these events is shown graphically in Figure 6.3.

Figure 6.3

Sample Points Corresponding to Values of a Random Variable

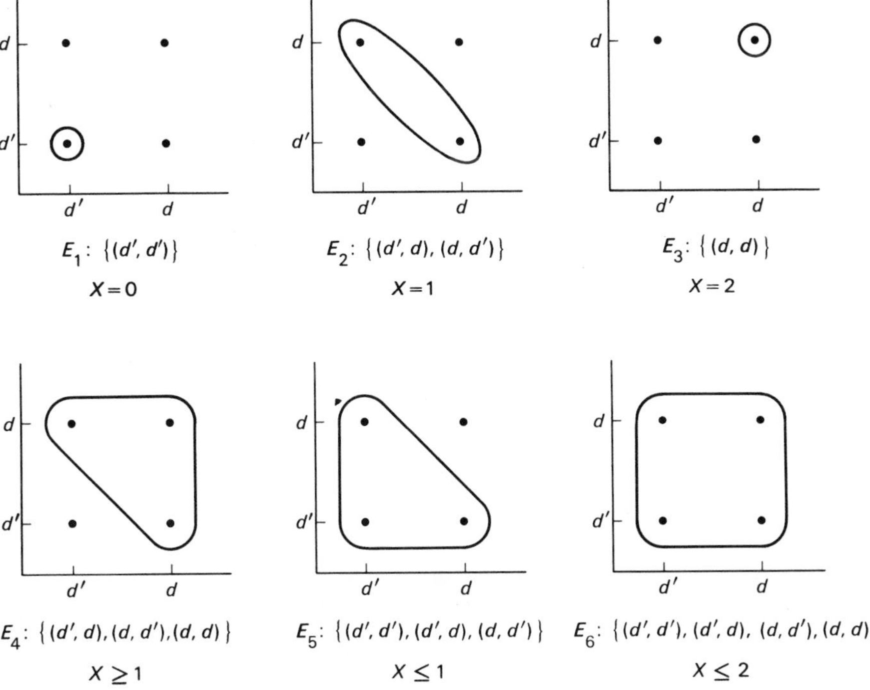

In this example, we defined the random variable X as the number of defective widgets in the sample. The set of possible values of X is $\{0,1,2\}$. Since we are able to list all possible values of the random variable, it is obvious that X is discrete. Thus we may redefine the above six events as subsets of the possible values of X. As shown in Figure 6.3, these redefined events are:

$$E_1\colon X = 0 \qquad E_4\colon X \geq 1$$
$$E_2\colon X = 1 \qquad E_5\colon X \leq 1$$
$$E_3\colon X = 2 \qquad E_6\colon X \leq 2$$

Example 2:

Let T be a random variable denoting the time (in hours) required to produce a delicate, hand-assembled medical instrument. The profit that the producer makes on a unit depends on the time required to produce it, as shown below:

Production Time	*Profit*
Less than 3 hours	*$300*
Between 3 and 8 hours	*$200*
Over 8 hours	*$100*

The producer has just received an order for one of these units. He feels sure that it will take at least one hour to produce the unit but no more than twelve hours. Therefore, we may describe the set of all possible values of T as follows:

$$\{t \mid 1 \leq t \leq 12\}$$

It should be obvious that T is a continuous random variable, since the production time for the unit may be any conceivable value between one and twelve hours. Because the profit to be made on the unit will depend on its production time, the producer might be particularly interested in the following three events:

$$E_1 = \{t \mid 1 \leq t < 3\}$$
$$E_2 = \{t \mid 3 \leq t \leq 8\}$$
$$E_3 = \{t \mid 8 < t \leq 12\}.$$

PROBLEMS

6.1 An American traveler, arriving at the Manila airport, wishes to exchange some of his U.S. dollars for Philippine pesos. He finds that 1 peso is equivalent to 24¢ in U.S. currency. Write a mathematical statement that expresses y (the number of pesos he receives) as a function of x (the number of dollars he exchanges).

6.2 In retailing, the markup on an item of merchandise is the difference between the cost, c, and the retail selling price, s. Two common methods of determining the amount of markup on an item are (i) to establish markup as a percentage of the cost; (ii) to establish markup as a percentage of selling price.

(a) A particular bookstore computes markup as 25 per cent of cost. Write a mathematical statement expressing selling price as a function of cost.

(b) A particular haberdasher computes markup as 40 per cent of selling price. Write a mathematical statement expressing selling price as a function of cost.

6.3 Consider the function

$$y = f(x) = 1{,}000 + 10x, \qquad x = 1, 2, 3, \ldots, 50$$

(a) Find $f(1)$, $f(5)$, and $f(10)$.

(b) Specify the domain and range of this function.

(c) Identify the dependent and independent variables.

6.4 Consider the function

$$y = f(x) = x^2 - 5x + 10, \qquad x = 0, 1, \ldots, 5$$

(a) What is the value of $f(2)$?

(b) Specify the domain and range of this function.

6.5 The table below defines a function that assigns a value of y to each value of x.

x	1	2	3	4	5
y	203	206	209	212	215

(a) Write a corresponding mathematical statement which expresses y as a function of x.

(b) Specify the domain and range of this function.

(c) What is the value of $f(4)$?

6.6 The total cost of producing a particular type of hand-assembled device is \$20.00 for materials plus \$5.00 per hour of labor. Thus, the cost of materials is a fixed cost, and the cost of labor is a variable cost that depends on the amount of labor required to assemble any single device.

(a) Let T be a random variable denoting the total cost of producing one of these devices. Let H be a random variable denoting the amount of labor (in hours) required to produce a device. Write a mathematical equation that expresses T as a function of H.

(b) Suppose that these devices are priced at \$50.00 each. Let P be a random variable denoting the manufacturer's profit on a device.

(i) Write a mathematical equation that expresses P as a function of T.

(ii) Write a mathematical equation that expresses P as a function of H.

6.7 A real estate developer has just won a cost-plus-incentive-fee contract to build a shopping center. The client wishes to have the center completed within a reasonably short period of time. Thus, the contract provides an incentive for completing the project at the earliest possible date. Specifically, the contract specifies the following formula for determining the incentive fee to be paid by the client as a function of the time taken to complete the project:

$$c = f(t) = \begin{cases} 500 - 20(t - 7)^2 & \text{if } t \geq 7 \\ 500 & \text{if } t < 7 \end{cases}$$

where c = incentive fee to be paid by the client (in thousands of dollars)
t = time taken to complete the projects (in months)

In determining the fee, the contract states that the value of t will be rounded to the nearest 1/10 of a month.

(a) Using the above function, determine the incentive fee to be paid by the client if it takes 10 months to complete the contract.
(b) Determine the break-even time beyond which the developer would suffer an an actual loss.
(c) If the developer wishes to receive an incentive fee of at least $455,000, what is the maximum time he may take to complete the project?
(d) The scheduled completion date for the project is 10 months. However, the developer indicates that such projects do not always go according to schedule. Thus he is uncertain about the actual amount of time required to complete the project. Under these conditions, can we consider the completion time as a random variable under the traditional definition or extended definition?
(e) Suppose we agree to consider the completion time as a random variable. Then would the incentive fee be a random variable?

6.8 Joe Schlutz has the beer and pretzel concession at the annual fraternity row block party to support Panhellinic activities. He is faced with the problem of how many cans of beer and how many bags of pretzels to stock. Joe is uncertain about how many people will attend the picnic, although he feels sure that there will be at least 100 people and no more than 200 people. On the basis of previous years' experience, each person who does attend will consume 4 cans of beer and 2 bags of pretzels. Joe will buy the beer for 10¢ a can and sell it for 25¢ a can. Also, he will buy the pretzels for 5¢ a bag and sell them for 15¢ a bag. To analyze his problem, Joe wishes to establish certain functional relationships. For this purpose, he has defined X as a random variable denoting the number of people who will attend the picnic. He has also identified the following random variables, each of which is a function of X:

Q_b = quantity (number of cans) of beer to be sold
Q_p = quantity (number of bags) of pretzels to be sold
R_b = revenue from sale of beer
R_p = revenue from sale of pretzels
R_t = total revenue
G_b = gross profit on beer
G_p = gross profit on pretzels
G_t = total gross profit

The two quantities that Joe must determine at the completion of his analysis are

s_b = number of cans of beer to stock
s_p = number of bags of pretzels to stock

These two quantities, rather than being random variables, are the *decision variables*. The decision problem consists of determining optimal values of these quantities. (In this exercise, however, we will not actually solve this problem, but merely examine functional relationships among the variables.)

(a) Express Q_b as a function of X and s_b.
(b) Express Q_p as a function of X and s_p.
(c) Express R_b as a function of Q_b.

(d) Express R_b as a function of X and s_b.
(e) Express R_p as a function of Q_p.
(f) Express R_p as a function of X and s_p.
(g) Express R_t as a function of R_b and R_p.
(h) Express R_t as a function of Q_b and Q_p.
(i) Express R_t as a function of X, s_b, and s_p.
(j) Express G_b as a function of R_b and s_b.
(k) Express G_p as a function of R_p and s_p.
(l) Express G_t as a function of G_b and G_p.
(m) Express G_t as a function of R_b, R_p, s_b, and s_p.

6.9 Consider the experiment of tossing a coin. Let H denote heads and T denote tails.

(a) Suppose that the coin is to be tossed twice. Specify the sample space for this experiment. If X is the random variable denoting the number of heads resulting from this experiment, draw a diagram (such as Figure 6.2) showing the relationship between this random variable and the sample space.

(b) Suppose that the coin is to be tossed 3 times. Specify the sample space for this experiment. If Y is used to denote the random variable representing the number of heads from this experiment, can you draw a diagram similar to the one above to show the relationship between this random variable and the sample space? What difficulty do you encounter in this diagraming? Can you resolve it by showing an alternate diagram?

6.10 Consider the following experiment: An urn contains 4 tags. One tag is marked with a "2", one with a "3", one with a "5" and one with a "9". One of these tags is to be drawn blindly from the urn, and its value recorded. Then, without replacing the first tag, a second tag is to be drawn and its number recorded.

(a) Each elementary outcome of this experiment may be represented by an ordered pair of values. For example, the ordered pair (5,3) indicates that the first tag was marked "5" and the second tag was marked "3". How many such elementary outcomes comprise the sample space of this experiment?

(b) Diagram the sample space of this experiment.

(c) Let S be a random variable denoting the sum of the values on the two tags that are drawn from the urn. Specify the set of all possible values of S.

(d) Prepare a table similar to Table 6.1, indicating the specific sample points that correspond to each of the possible values of S.

(e) List the sample points corresponding to each of the following events:

$$E_1: S \leq 7$$
$$E_2: 7 \leq S \leq 11$$
$$E_3: 7 < S \leq 11$$

Chapter 7
Probability Functions

In Chapter 6, we presented the concept of a random variable and discussed how to define events numerically in terms of values of random variables. In this chapter we introduce the concept of "probability functions" and consider the use of such functions to obtain the probabilities of numerically defined events.

7.1 BASIC CONCEPT OF A PROBABILITY FUNCTION

To introduce the reader to the basic concept of a probability function, let us consider again the example of rolling a pair of dice. In our discussion of this example in Section 6.2, we let X denote a random variable representing the sum appearing when the two dice are rolled. In Table 6.1, we saw that each value of the random variable was associated with a specific set of sample points. Each possible value of X, together with the corresponding number of sample points, is shown in the first two columns of Table 7.1.

Table 7.1

Probability Table for Dice Example

$X = x$	*Number of Sample Points*	$P(X = x)$
2	1	1/36
3	2	2/36
4	3	3/36
5	4	4/36
6	5	5/36
7	6	6/36
8	5	5/36
9	4	4/36
10	3	3/36
11	2	2/36
12	1	1/36
Total	36	36/36

Each of the thirty-six sample points represents an equally likely outcome of rolling the dice. Therefore, the probability of any particular value of X may be obtained by applying Formula (2.1). For instance, since there are six sample points associated with $X = 7$, the probability of rolling a sum of 7 is:

$$P(X = 7) = \frac{6}{36}$$

The probabilities of obtaining the various possible values of X are shown in the last column of Table 7.1.

In Table 7.1, we see that for each possible value of X in the first column there is an associated probability value shown in the third column. In other words, the third column can be considered as a function of the first column. Mathematically, this functional relationship may be expressed as follows:

$$f(x) = \begin{cases} \dfrac{x - 1}{36} & \text{if } x = 2, 3, 4, 5, 6 \\ \dfrac{13 - x}{36} & \text{if } x = 7, 8, 9, 10, 11, 12 \\ 0 & \text{otherwise} \end{cases}$$

Notice that the input of this function is any specific value, x, of the random variable, X, and the output is the probability that the random variable will take on that value. This is an example of one type of probability function.

More generally, consider a function whose domain is the set of all possible values of a random variable and whose range is the set of non-negative real numbers. If the function permits us to compute the probability for any event that is defined in terms of values of the random variable, then this function is called a *probability function*. Probabilities obtained from such a function must satisfy those requirements specified at the beginning of Section 5.1. A probability function is frequently called a *probability distribution*. Although there is a minor technical difference, we will use these two terms interchangeably.

Just as there are discrete and continuous random variables, so there are discrete and continuous probability functions. Because there are important differences in the mathematical properties and treatment of discrete and continuous probability functions, we must be careful to distinguish between these two types of functions. To emphasize this distinction, our discussion will proceed according to the diagram on page 104.

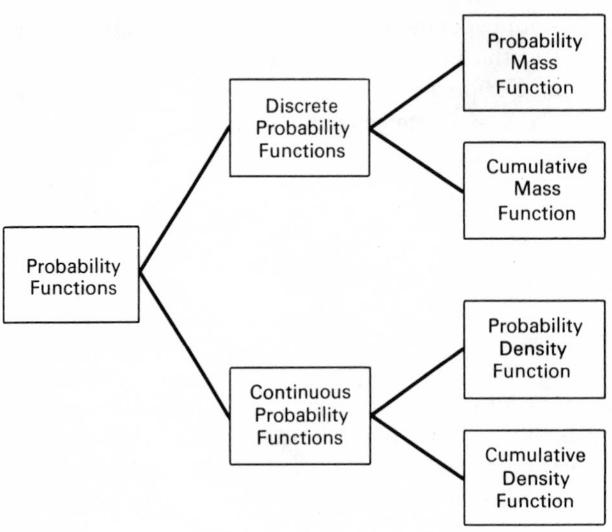

7.2 DISCRETE PROBABILITY FUNCTIONS

A probability function for a discrete random variable is called a *discrete probability function* since the domain of the function is discrete. In the following discussion of discrete probability functions, we will distinguish between the "probability mass function" and the "cumulative mass function."

7.2.1 Probability Mass Functions

A probability function that specifies the probability that any single value of a *discrete* random variable will occur is called a *probability mass function* (abbreviated as *p.m.f.*). That is, if $f(x)$ is the probability mass function of the random variable X, then $f(x) = P(X = x)$. As a probability mass function, $f(x)$ has the following properties:

1. $f(x) \geq 0$ for all values of X; that is, $f(x)$ cannot be negative for any value of X.
2. $\sum_{\text{all } x} f(x) = 1$; that is, the sum of the separate values of $f(x)$ over all x-values must equal 1.

Consider, for example, a highly fictionalized salesman who schedules four calls a day. He is certain of making at least one sale out of the four, and may make as many as four sales if he is artful and lucky. In fact, the probabilities of making various numbers of sales in a day are specified by the probability mass function:

$$f(x) = \begin{cases} (5 - x)/10 & \text{if } x = 1, 2, 3, 4 \\ 0 & \text{otherwise} \end{cases}$$

Notice that the p.m.f. specifies that $f(x) = 0$ for all values of X other than 1, 2, 3, and 4. This specification reflects that: (1) this particular salesman cannot make less than one sale in a day, nor can he make more than four sales; (2) the salesman cannot make a fractional number of sales, such as 2.7 sales, in a day.

Using the above probability function, we may obtain the probabilities shown in Table 7.2. In preparing such a probability table, it is customary to show only those x-values for which $f(x) > 0$. From the figures in this table, we see that the sum of the individual probabilities for all the separate values of X is equal to 1.

Table 7.2

Probability Mass Function of Salesman's Daily Sales

x *(Number of sales)*	*1*	*2*	*3*	*4*
$f(x)$	4/10	3/10	2/10	1/10

The probability mass function for our salesman example is shown graphically in Figure 7.1. The possible values of the random variable are repre-

Figure 7.1

Graphic Display of Probability Mass Function

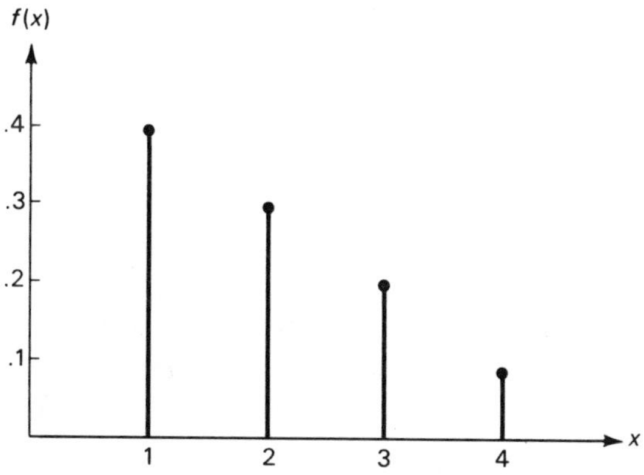

sented on the horizontal axis, and the associated probabilities are displayed on the vertical axis. Notice that the graph consists of unconnected straight lines rising from isolated points along the horizontal axis, which emphasizes the discrete nature of the distribution.

Example:

The credit manager of a large retailing firm has found that, as a long-run figure, 30 per cent of credit applicants are poor risks. Suppose that he selects from his files a random sample of five credit applications. Let R be a random variable denoting the number of poor risks in the sample. The credit manager determines that the probability mass function of R can be expressed in terms of the formula:

$$f(r) = \begin{cases} \dfrac{5!}{r!(5-r)!}(.3)^r(.7)^{5-r} & \text{if } r = 0, 1, 2, 3, 4, 5 \\ 0 & \text{otherwise} \end{cases}$$

After you have studied Chapter Nine, you will see why this formula might be regarded as a reasonable model in this situation. From application of the formula, we obtain the following probabilities:

$$f(0) = \frac{5!}{0!(5-0)!}(.3)^0(.7)^5 = .168$$

$$f(1) = \frac{5!}{1!(5-1)!}(.3)^1(.7)^4 = .361$$

$$f(2) = \frac{5!}{2!(5-2)!}(.3)^2(.7)^3 = .309$$

$$f(3) = \frac{5!}{3!(5-3)!}(.3)^3(.7)^2 = .132$$

$$f(4) = \frac{5!}{4!(5-4)!}(.3)^4(.7)^1 = .028$$

$$f(5) = \frac{5!}{5!(5-5)!}(.3)^5(.7)^0 = \underline{.002}$$

$$\Sigma f(r) = 1.000$$

Thus the probabilities are .168 that a random sample of five applications will contain no bad risks, .361 that it will contain one bad risk, .309 that it will contain two bad risks, .132 that it will contain three bad risks, .028 that it will contain four bad risks, and .002 that it will contain five bad risks. These six possibilities are mutually exclusive and collectively exhaustive, so that their probabilities add to 1.

7.2.2 Cumulative Mass Functions

If X is a discrete random variable with probability mass function $f(x)$, its *cumulative mass function* (abbreviated as c.m.f.), designated by $F(x)$, specifies the probability that an observed value of X will be no greater than x. That is, if $f(x)$ is a probability mass function, $F(x) = P(X \leq x)$. Consider again the salesman who makes four calls a day. The probability that he will make no more than two sales is equal to the sum of the probabilities that he will make one sale or two sales:

$$F(2) = f(1) + f(2) = \frac{4}{10} + \frac{3}{10} = \frac{7}{10}$$

Similarly, the probability that the salesman will make at most three sales is given by:

$$F(3) = f(1) + f(2) + f(3) = \frac{4}{10} + \frac{3}{10} + \frac{2}{10} = \frac{9}{10}$$

Since the maximum number of sales that this salesman can make in a day is four, it should be obvious that there should be a probability of 1 that his number of sales will be no greater than four. This is verified as follows:

$$F(4) = f(1) + f(2) + f(3) + f(4) = \frac{4}{10} + \frac{3}{10} + \frac{2}{10} + \frac{1}{10} = 1$$

In terms of a formula, we can express the cumulative mass function as:

$$F(x) = \begin{cases} 0 & \text{if } x < 1 \\ \sum_{t=1}^{x} (5 - t)/10 & \text{if } x = 1, 2, 3, 4 \\ 1 & \text{if } x > 4 \end{cases}$$

Since the above expression yields $P(X \leq x)$, which involves a summation of terms, the symbol x is used as the upper limit of the summation. To distinguish between this upper value of the summation and all values of the random variable that are included in the summation, it is necessary in the notation to substitute t for x in the expression for the mass function itself; that is, we write $(5 - t)/10$ rather than $(5 - x)/10$. When used in this manner, t often is called a "dummy variable." (Of course, any letter other than x may be used as a dummy variable for x). It should be emphasized that the dummy variable t simply is a substitution required for mathematical propriety and does not denote a different variable.

The cumulative probabilities obtained from the above c.m.f. are presented in Table 7.3. This cumulative function is displayed graphically in Figure 7.2.

Table 7.3

Cumulative Mass Function of Salesman's Daily Sales

x (*Number of sales*)	*1*	*2*	*3*	*4*
$F(x)$	4/10	7/10	9/10	1

Figure 7.2

Graphic Display of Cumulative Mass Function

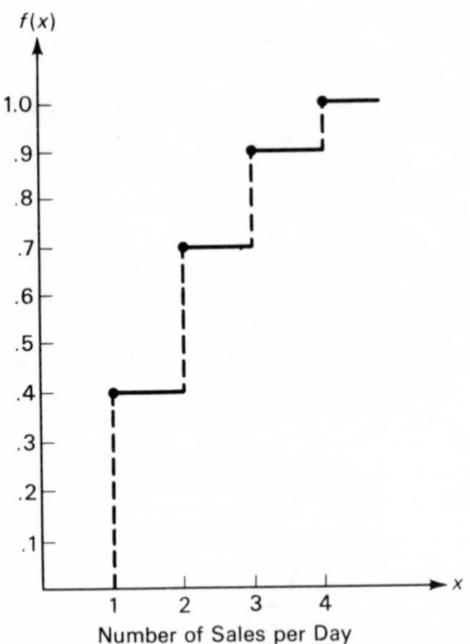

At this point we introduce a special notation that is used frequently in working with probability models. If X is a random variable, x_p is a value of X such that $P(X \leq x_p) = p$. That is, there is a probability equal to p that an observed value of the random variable X will be equal to or less than x_p. For our salesman example, in which X represents the number of his daily sales, $x_{.40} = 1$, $x_{.70} = 2$, and $x_{.90} = 3$. For a randomly selected day, the probability is .40 that he will make no more than one sale, .70 that he will make no more than two sales, and .90 that he will make no more than three sales. In this usage, $x_{.40}$ is called the 40th *percentile* of the probability distribution of the random variable X; similarly, $x_{.70}$ is the 70th percentile, and $x_{.90}$ is the 90th percentile.

Example:

In a particularly hazardous occupation, nationwide figures gathered over a period of several years indicate that, as a long-run average, there is one fatal accident per month in the occupation. Let X be a random variable representing the number of such fatal accidents in a month. The cumulative mass function for X is shown in Table 7.4.

Table 7.4

Cumulative Probability Distribution of Monthly Number of Fatal Accidents

x	$F(x)$
0	.36788
1	.73576
2	.91970
3	.98101
4	.99634
5	.99941
6	.99992
7	.99999
8	1.00000

From Table 7.4, we may answer the following questions:

1. *What is the probability that there will be no more than three fatal accidents in a given month?*

$$\text{From the table, } F(3) = .98101$$

2. *What is the probability that there will be at most two fatal accidents in a given month?*

$$\text{From the table, } F(2) = .91970$$

3. *What is the probability that there will be less than four fatal accidents in a given month?*

Less than four accidents means three or less accidents.
From the table, $F(3) = .98101$

4. *What is the probability that there will be more than four fatal accidents in a given month?*

   Since total probability equals 1,

$$P(X > 4) = 1 - P(X \leq 4) = 1 - .99634 = .00366$$

5. *What is the probability that there will be no less than two fatal accidents in a given month?*

$$P(X \geq 2) = 1 - P(X \leq 1) = 1 - .73576 = .26424$$

6. *What is the probability that there will be at least four fatal accidents in a given month?*

$$P(X \geq 4) = 1 - P(X \leq 3) = 1 - .98101 = .01899$$

7. *What is the probability that in a given month there will be exactly two fatal accidents?*

$$\begin{aligned} P(X = 2) &= P(X \leq 2) - P(X < 2) = P(X \leq 2) - P(X \leq 1) \\ &= .91970 - .73576 = .18394 \end{aligned}$$

8. *What is the probability that in a given month the number of fatal accidents will be between three and five, inclusive?*

$$\begin{aligned} P(3 \leq X \leq 5) &= P(X \leq 5) - P(X < 3) = P(X \leq 5) - P(X \leq 2) \\ &= .99941 - .91970 = .07971 \end{aligned}$$

7.3 CONTINUOUS PROBABILITY FUNCTIONS

A probability function for a continuous random variable is called a *continuous probability function* since the domain of the function is continuous. In discussing continuous probability functions, we will distinguish between the "probability density function" and the "cumulative density function."

7.3.1 Probability Density Functions

As we have seen, the probability mass function $f(x)$ of a discrete random variable X specifies probabilities for specific values of X. For a continuous random variable, the corresponding function $f(x)$ is called a *probability density function* (abbreviated as *p.d.f*). Unlike a probability mass function, however, a probability density function does not specify probabilities for specific individual values of the random variable.

What, then, is the meaning of the probability function $f(x)$ when X is continuous? To answer this question, we must first observe that it is meaningless to define events in terms of single values of a continuous random variable, because a continuous random variable has an infinite number of possible values. Thus, for a continuous random variable, an event must be defined in terms of an *interval* of values. For example, we might define some event E in terms of a continuous random variable X as follows:

$$E = \{x|3 \leq x \leq 5\}$$

In this expression, the event E is defined as the set of all real values in the interval from 3 to 5, inclusive.

In general, for a continuous random variable X, any event E is defined in terms of the values in an interval between two limits, a and b. Then the probability of the event is the probability that an observed value will fall within the specified interval. Symbolically,

$$P(E) = P(a \leq X \leq b)$$

If $f(x)$ is a probability density function, the above probability is obtained by integrating $f(x)$ from a to b. Specifically,

$$P(a \leq X \leq b) = \int_a^b f(x)\,dx \tag{7.1}$$

Graphically, we may view a p.d.f. as a mathematical function that describes a curve. For any given value x, $f(x)$ is simply the corresponding ordinate (height of the curve above the horizontal axis). That is, $f(x)$ does not represent a probability. Rather, the probability of an event E is represented by the *area* under the curve between the limits that define the event. Furthermore, for a continuous random variable, the probability of any specific single value is mathematically defined as zero.

As a probability density function, $f(x)$ has the following properties:

1. $f(x) \geq 0$ for all real values; that is, $f(x)$ cannot be negative for any value of X between $-\infty$ and ∞.
2. $\int_{-\infty}^{\infty} f(x)\,dx = 1$; that is, the total area under the curve described by $f(x)$ must equal 1.

Example:

Let X be a continuous random variable with the following probability density function:

$$f(x) = \begin{cases} 1/5 & \text{if } 2 \leq x \leq 7 \\ 0 & \text{otherwise} \end{cases}$$

Graphically, this function describes a rectangle, as shown in Figure 7.3. By inspecting the function, we see that $f(x)$ is not less than zero for any value of X between $-\infty$ and ∞. Also, we may verify that the total area under the curve is equal to one:

$$\int_{-\infty}^{\infty} f(x)\,dx = \int_{-\infty}^{2} (0)\,dx + \int_{2}^{7} (1/5)\,dx + \int_{7}^{\infty} (0)\,dx$$

$$= 0 + \int_{2}^{7} \frac{dx}{5} + 0 = \frac{x}{5}\Big]_{2}^{7} = \frac{7}{5} - \frac{2}{5} = 1$$

Suppose that we define the event E in terms of the interval $(3 \leq X \leq 5)$. Applying Formula (7.1), this probability is given by:

$$\int_{3}^{5} f(x)\,dx = \int_{3}^{5} \frac{1}{5}\,dx = \frac{x}{5}\Big]_{3}^{5} = \frac{5}{5} - \frac{3}{5} = .40$$

This probability is represented as an area by the shaded portion of Figure 7.3.

Figure 7.3

Probability Density Function

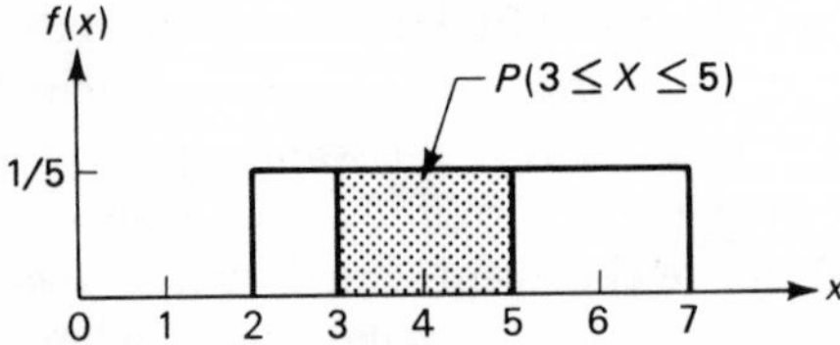

7.3.2 Cumulative Density Functions

Corresponding to the cumulative mass function of a discrete random variable, the *cumulative density function*[1] (abbreviated as c.d.f.), of a contin-

[1] Collectively, cumulative mass functions and cumulative density functions are referred to as *distribution functions.* The term "cumulative distribution function" often is used to denote a distribution function; however, the word "cumulative" is redundant in this context since a distribution function is cumulative by definition.

uous random variable specifies the probability that an observed value of X will be no greater than x. In general, if $f(x)$ is the probability density function of the random variable X, then the corresponding cumulative density function is given by:

$$F(x) = P(X \leq x) = \int_{-\infty}^{x} f(t)\,dt \tag{7.2}$$

In other words, for a continuous probability function, $P(X \leq x)$ is given by the definite integral of the density function evaluated between minus infinity and x. Notice that since x is the upper limit of the integration, the "dummy" expression $f(t)\,dt$ is used in place of $f(x)\,dx$ in the integrand. This procedure is analogous to the use of the dummy variable previously discussed in connection with the cumulative mass function for a discrete random variable in Section 7.2.2.

For a continuous random variable,

$$P(a \leq X \leq b) = P(X \leq b) - P(X \leq a)$$

since $P(X = a)$ is zero for any specific value a. We now see that $P(X \leq b) = F(b)$, and $P(X \leq a) = F(a)$. Hence, $P(a \leq X \leq b)$ may be expressed as:

$$P(a \leq X \leq b) = F(b) - F(a) \tag{7.3}$$

We have already seen from Formula (7.1) that if $f(x)$ is a p.d.f., then $P(a \leq X \leq b)$ may be computed by integrating the p.d.f. However, if the cumulative density function is known, it is unnecessary to integrate the p.d.f. in order to obtain this probability. Rather, as shown by Formula (7.3), this probability may be obtained simply by subtracting $F(a)$ from $F(b)$. This is the real advantage of using the cumulative density function.

Example:

Let us again consider the density function:

$$f(x) = \begin{cases} 1/5 & \text{if } 2 \leq x \leq 7 \\ 0 & \text{otherwise} \end{cases}$$

Over the interval $(-\infty < x < 2)$ *we have* $f(x) = 0$. *Since* X *is continuous,* $P(X = 2)$ *is zero. Thus,* $P(X < 2) = P(X \leq 2)$. *Hence, for* $x < 2$, *the cumulative density function is given by:*

$$F(x) = \int_{-\infty}^{x} f(t)\,dt = \int_{-\infty}^{x} 0\,dt = 0$$

Over the interval $(2 \leq x \leq 7)$, *we have* $f(x) = 1/5$. *The cumulative density up to the lower limit of this interval is* $F(2)$. *Therefore, for any value of* X *between 2 and 7,* $F(x)$ *is given by:*

$$F(x) = F(2) + \int_2^x f(t)\,dt = 0 + \int_2^x (1/5)\,dt = \frac{t}{5}\Big]_2^x = \frac{x-2}{5}$$

Over the interval $(7 < x < \infty)$, *we have* $f(x) = 0$. *The cumulative density up to the lower limit of this interval is* $F(7) = \dfrac{7-2}{5} = 1$. *Therefore, for any value of* X *greater than 7,* $F(x)$ *is given by:*

$$F(7) + \int_7^\infty f(t)\,dt = 1 + \int_7^\infty 0\,dt = 1 + 0 = 1$$

We now summarize this cumulative density function as follows:

$$F(x) = \begin{cases} 0 & \text{if } x < 2 \\ \dfrac{x-2}{5} & \text{if } 2 \leq x \leq 7 \\ 1 & \text{if } x > 7 \end{cases}$$

This cumulative density function is shown graphically in Figure 7.4. For any given x-value, the ordinate of Figure 7.4 is equal to the area to the left of that x-value in Figure 7.3. This is generally true of the relationship

Figure 7.4

Cumulative Density Function

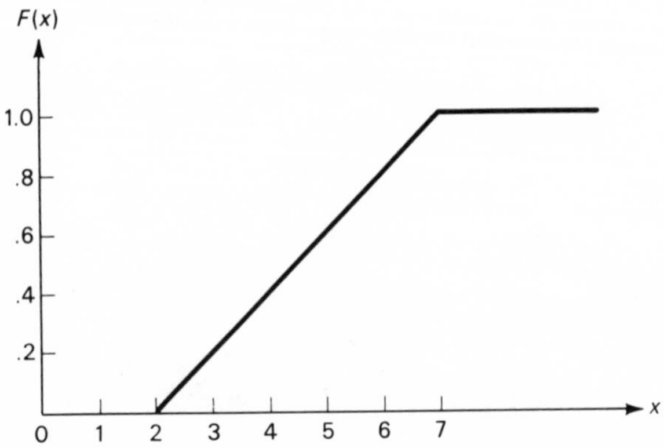

between any p.d.f. and its corresponding c.d.f.; i.e., for a specified p.d.f., the area under the curve to the left of a given x-value is equal to the corresponding $F(x)$ for that x-value.

Suppose that we wish to determine $P(2.2 \leq X \leq 3.4)$. Using Formula (7.1), we may compute this probability by integration as follows:

$$\int_{2.2}^{3.4} \frac{1}{5}\,dx = \frac{x}{5}\Big]_{2.2}^{3.4} = \frac{3.4}{5} - \frac{2.2}{5} = .24$$

Alternatively, we may determine this same probability by applying Formula (7.3) as follows:

$$F(2.2) = \frac{2.2 - 2}{5} = .04 \qquad F(3.4) = \frac{3.4 - 2}{5} = .28$$

$$P(2.2 \leq X \leq 3.4) = F(3.4) - F(2.2) = .28 - .04 = .24$$

7.3.3 Applications of Continuous Probability Functions

In our discussion of continuous probability functions, we introduced several new concepts that the reader may find somewhat difficult to digest completely on first encounter. The following cases are presented to help enrich the reader's comprehension of these concepts.

Case 1. Consider a highly sensitive recording apparatus employed in medical research. This device contains a particularly delicate component that has a useful life of only a few hours at most. Whenever this component fails, it is replaced; if it should last for as long as four hours, it is routinely replaced and discarded. Thus the maximum possible useful life of one of these components is four hours. If we let H designate a random variable representing the useful life (in hours) of such a component, experience has demonstrated that H has the probability density function:

$$f(h) = \begin{cases} h/8 & \text{if } 0 \leq h \leq 4 \\ 0 & \text{otherwise} \end{cases}$$

Applying Formula (7.2), the reader should be able to verify that the corresponding cumulative density function is:

$$F(h) = \begin{cases} 0 & \text{if } h < 0 \\ h^2/16 & \text{if } 0 \leq h \leq 4 \\ 1 & \text{if } h > 4 \end{cases}$$

Graphs of the p.d.f. and c.d.f. are shown in Figure 7.5.

Figure 7.5

Probability Functions for the Useful Life of a Delicate Component

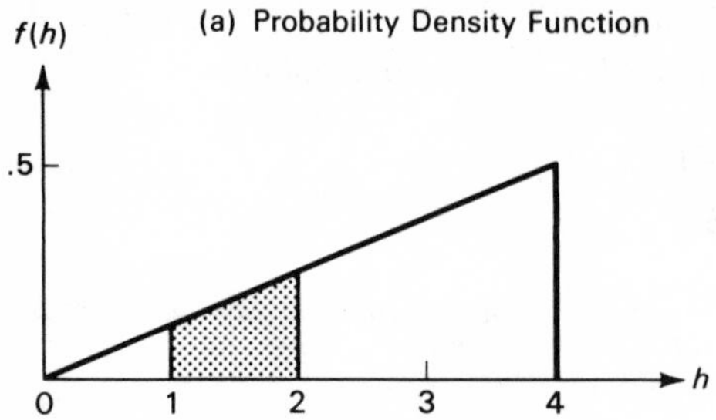

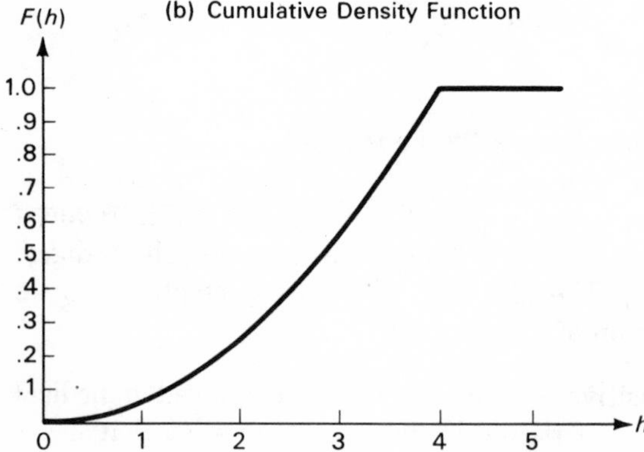

Suppose that we wish to determine the probability that one of these components will not have a useful life longer than one hour. In Figure 7.5a this probability, $P(H \leq 1)$, is represented by the area of the small triangle to the left of $h = 1$. From the cumulative density function, this area is given by:

$$P(H \leq 1) = F(1) = \frac{1}{16} = .0625$$

Likewise, the probability that a component will last at least two hours is given by:

$$P(H \geq 2) = 1 - F(2) = 1 - 4/16 = 1 - .250 = .750$$

Suppose that we want to know the probability that the useful life of one of these components will be between one and two hours—that is, not less than one hour and not more than two hours. In Figure 7.5a, this probability is represented by the area of the shaded portion under the curve. From the c.d.f., this probability is given by:

$$P(1 \leq H \leq 2) = F(2) - F(1) = .2500 - .0625 = .1875.$$

Case 2. A particular northern California community generally experiences occasional light snowfall during the winter. Let W be a random variable denoting the winter's snowfall (in feet). Suppose the probability density function of W is given by:

$$f(w) = \begin{cases} 4w - 3w^2 & \text{if } 0 \leq w \leq 1 \\ 0 & \text{otherwise} \end{cases}$$

Notice that this density function, although written in two parts, actually partitions the domain of the function into three subsets and could be written in the following more detailed form:

$$f(w) = \begin{cases} 0 & \text{if } w \in S_1 \quad \text{where } S_1 = \{w|w < 0\} \\ 4w - 3w^2 & \text{if } w \in S_2 \quad \text{where } S_2 = \{w|0 \leq w \leq 1\} \\ 0 & \text{if } w \in S_3 \quad \text{where } S_3 = \{w|w > 1\} \end{cases}$$

This probability density function is shown graphically in Figure 7.6a. It is worth noting that this function does indeed qualify as a probability density function in that:

1. From Figure 7.6a we can see that $f(w)$ is non-negative for all values of W from $-\infty$ to ∞.
2. The total area under the curve is equal to 1. That is:

$$\int_{-\infty}^{\infty} f(w)\,dw = \int_{-\infty}^{0} 0\,dw + \int_{0}^{1} (4w - 3w^2)\,dw + \int_{1}^{\infty} 0\,dw$$

$$= \int_{0}^{1} (4w - 3w^2)\,dw = \left[\frac{4w^2}{2} - \frac{3w^3}{3}\right]_0^1$$

$$= \left(\frac{4(1^2)}{2} - \frac{3(1^3)}{3}\right) - \left(\frac{4(0^2)}{2} - \frac{3(0^3)}{3}\right) = 1$$

To obtain the cumulative density function of w, we must derive a separate expression of $F(w)$ for each of the subsets implied by the statement of $f(w)$. We begin by cumulating from $-\infty$. Over the first interval, $S_1 = \{w|w < 0\}$, we have seen that $f(w) = 0$. Therefore, for $w < 0$:

$$F(w) = \int_{-\infty}^{w} f(t)\,dt = \int_{-\infty}^{w} 0\,dt = 0$$

Figure 7.6

Probability Functions for Snowfall

(a) Probability Density Function

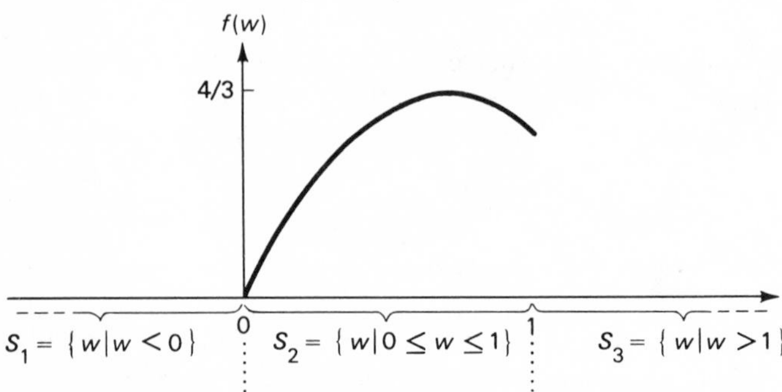

(b) Cumulative Density Function

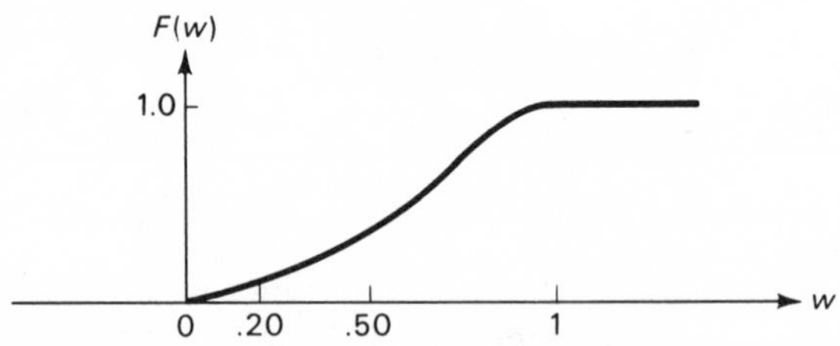

(c) A Shaded Area Showing a Probability

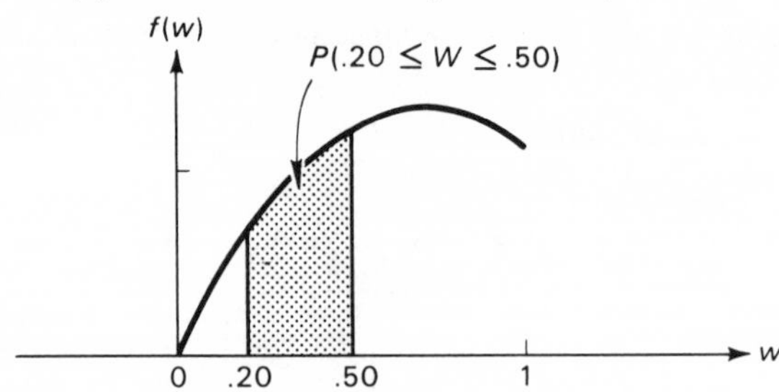

Over the next interval, $S_2 = \{w|0 \leq w \leq 1\}$, we have $f(w) = 4w - 3w^2$. The cumulative density up to the lower limit of this interval is $F(0) = 0$. Therefore, for any w-value contained in S_2, $F(w)$ is given by:

$$F(w) = F(0) + \int_0^w f(t)\,dt$$

$$= 0 + \int_0^w (4t - 3t^2)\,dt = \left[\frac{4t^2}{2} - \frac{3t^3}{3}\right]_0^w$$

$$= \left(\frac{4w^2}{2} - \frac{3w^3}{3}\right) - \left(\frac{4(0)^2}{2} - \frac{3(0)^3}{3}\right)$$
$$= 2w^2 - w^3$$

Over the final interval, $S_3 = \{w|w > 1\}$, the density is $f(w) = 0$. The cumulative density up to the lower limit of this interval is:

$$F(1) = 2(1^2) - 1^3 = 1$$

Therefore, for any given value of W contained in S_3, $F(w)$ is given by:

$$F(w) = F(1) + \int_1^w f(t)\,dt = 1 + \int_1^w 0\,dt = 1 + 0 = 1$$

The cumulative density function is summarized as follows:

$$F(w) = \begin{cases} 0 & \text{if } w < 0 \\ 2w^2 - w^3 & \text{if } 0 \le w \le 1 \\ 1 & \text{if } w > 1 \end{cases}$$

This function is shown graphically in Figure 7.6b.

Suppose that we wish to determine the probability that the community's winter snowfall in a given year will be between .20 and .50 feet. By applying Formula (7.3), we have:

$$P(.20 \le W \le .50) = F(.50) - F(.20)$$

From the cumulative density function, we see that:

$$F(.50) = 2(.50)^2 - (.50)^3 = .375$$

and

$$F(.20) = 2(.20)^2 - (.20)^3 = .072$$

Therefore, $P(.20 \le W \le .50) = .375 - .072 = .303$. This is shown by the shaded area of Figure 7.6c.

Case 3. Let Q be a random variable denoting the monthly quantity demanded (in tons) for a particular chemical product. The probability density function for this demand is:

$$f(q) = \begin{cases} q - 1 & \text{if } 1 \le q < 2 \\ 3 - q & \text{if } 2 \le q \le 3 \\ 0 & \text{otherwise} \end{cases}$$

This density function, which is illustrated in Figure 7.7a, partitions the domain of the function into four subsets:

$$S_1 = \{q | q < 1\}$$
$$S_2 = \{q | 1 \leq q < 2\}$$
$$S_3 = \{q | 2 \leq q \leq 3\}$$
$$S_4 = \{q | q > 3\}$$

Figure 7.7

Probability Functions for Chemical Product Demand

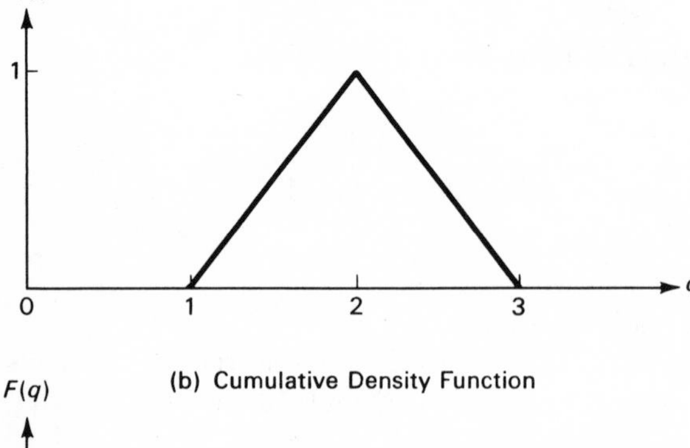

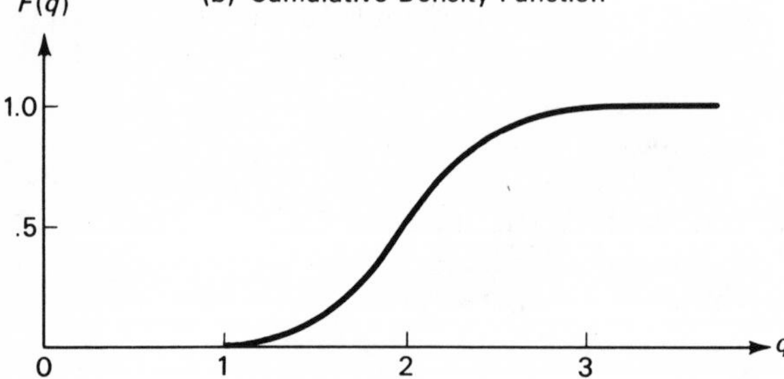

For any q-value contained in the first interval, $S_1 = \{q | q < 1\}$, the cumulative density is given by:

$$F(q) = \int_{-\infty}^{q} 0 \, dt = 0$$

Over the next interval, $S_2 = \{q | 1 \leq q < 2\}$, the density function is $f(q) = q - 1$. The cumulative density up to the lower limit of this interval is $F(1) = 0$. Therefore, for any q-value contained in S_2, $F(q)$ is given by:

$$F(q) = F(1) + \int_{1}^{q} f(t) \, dt = 0 + \int_{1}^{q} (t - 1) \, dt$$

$$= \left[\frac{t^2}{2} - t\right]_1^q = \left(\frac{q^2}{2} - q\right) - \left(\frac{1}{2} - 1\right)$$
$$= \frac{q^2}{2} - q + \frac{1}{2}$$
$$= \frac{q^2 + 1}{2} - q$$

Over the next interval, $S_3 = \{q|2 \leq q \leq 3\}$, the density function is $f(q) = 3 - q$. The cumulative density up to the lower limit of this interval is:

$$F(2) = \frac{(2)^2 + 1}{2} - 2 = \frac{1}{2}$$

Therefore, for any q-value contained in S_3, $F(q)$ is given by:

$$F(q) = F(2) + \int_2^q f(t)\,dt = \frac{1}{2} + \int_2^q (3 - t)\,dt$$
$$= \frac{1}{2} + \left[3t - \frac{t^2}{2}\right]_2^q = \frac{1}{2} + \left(3q - \frac{q^2}{2}\right) - \left(3(2) - \frac{(2)^2}{2}\right)$$
$$= 3q - \frac{q^2 + 7}{2}$$

Over the final interval, $S_4 = \{q|q > 3\}$, the density function is $f(q) = 0$. The cumulative density up to the lower limit of this interval is:

$$F(3) = 3(3) - \frac{(3)^2 + 7}{2} = 1$$

Therefore, for any q-value contained in S_4, $F(q)$ is given by:

$$F(q) = F(3) + \int_3^q f(t)\,dt = 1 + \int_3^q 0\,dt = 1 + 0 = 1$$

The cumulative density function is summarized as follows:

$$F(q) = \begin{cases} 0 & \text{if } q < 1 \\ \dfrac{q^2 + 1}{2} - q & \text{if } 1 \leq q < 2 \\ 3q - \dfrac{q^2 + 7}{2} & \text{if } 2 \leq q \leq 3 \\ 1 & \text{if } q > 3 \end{cases}$$

This function is shown graphically in Figure 7.7b. If we wish to know the probability that demand for this product will be between 1.2 tons and 2.4 tons during a given month, we have:

$$P(1.2 \le Q \le 2.4) = \int_{1.2}^{2.4} f(q)\,dq = F(2.4) - F(1.2)$$

$$= \left[3(2.4) - \frac{(2.4)^2 + 7}{2}\right] - \left[\frac{(1.2)^2 + 1}{2} - 1.2\right]$$

$$= .80$$

Case 4. Let X be a random variable denoting the continuous burning life (in hours) of a particular type of flood lamp. The probability density function of X is:

$$f(x) = \begin{cases} 0 & \text{if } x < 10 \\ \dfrac{10}{x^2} & \text{if } x \ge 10 \end{cases}$$

This density function, shown in Figure 7.8a, partitions the domain of the function into two parts:

$$S_1 = \{x | x < 10\}$$
$$S_2 = \{x | x \ge 10\}$$

The cumulative density function for any x-value contained in S_1 is given by:

$$F(x) = \int_{-\infty}^{x} f(t)\,dt = \int_{-\infty}^{x} 0\,dt = 0$$

For any x-value contained in $S_2 = \{x | x \ge 10\}$, the density function is $f(x) = \dfrac{10}{x^2}\cdot$ Therefore, the cumulative density function for values of X contained in S_2 is:

$$F(10) + \int_{10}^{x} f(t)\,dt = 0 + \int_{10}^{x} \frac{10}{t^2}\,dt = -\frac{10}{t}\Big]_{10}^{x}$$

$$= \left(-\frac{10}{x}\right) - \left(-\frac{10}{10}\right)$$

$$= 1 - \frac{10}{x}$$

The complete cumulative density function, illustrated in Figure 7.8b, may be written:

$$F(x) = \begin{cases} 0 & \text{if } x < 10 \\ 1 - \dfrac{10}{x} & \text{if } x \geq 10 \end{cases}$$

If we wish to know the probability that a given floodlamp of this type will burn continuously for more than twenty hours, we have:

$$P(X > 20) = \int_{20}^{\infty} \frac{10}{x^2}\,dx = 1 - F(20) = 1 - \left(1 - \frac{10}{20}\right) = .50$$

Figure 7.8

Probability Functions for Flood Lamp Lives

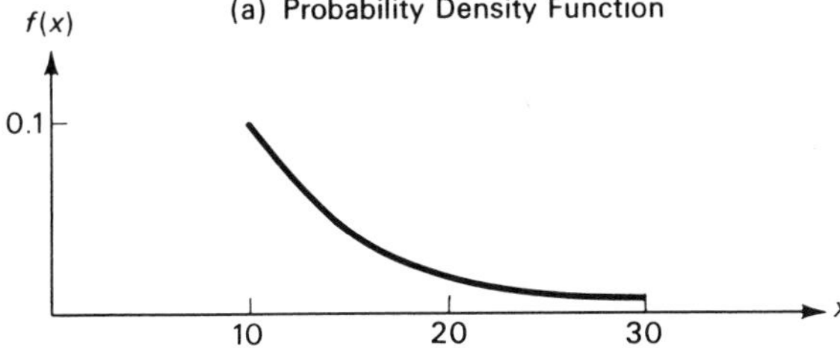

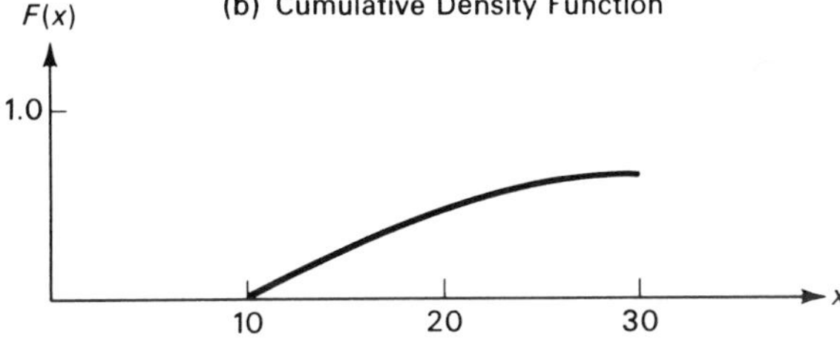

Case 5. Let T be a random variable denoting the time (in minutes) an industrial worker must wait for service at a supply room. The probability density function of T is given by:

$$f(t) = \begin{cases} 0 & \text{if } t < 0 \\ 2e^{-2t} & \text{if } t \geq 0 \end{cases}$$

This density function, which is shown in Figure 7.9a, partitions the domain of the function into two parts:

$$S_1 = \{t|t < 0\}$$
$$S_2 = \{t|t \geq 0\}$$

To obtain the cumulative density function for any t-value contained in S_1, we have:

$$F(t) = \int_{-\infty}^{t} 0\,dy = 0$$

Notice that, in the above expression, y is used to denote the dummy variable for t. For any t-value contained in $S_2 = \{t|t \geq 0\}$, the probability density function is $f(t) = 2e^{-2t}$. The cumulative density function for any t-value contained in S_2 is therefore given by:

$$\begin{aligned} F(t) &= F(0) + \int_0^t f(y)\,dy = 0 + \int_0^t 2e^{-2y}\,dy \\ &= \left.\frac{2e^{-2y}}{-2}\right]_0^t = \frac{2e^{-2t}}{-2} - \left(\frac{2e^{-2(0)}}{-2}\right) \\ &= -e^{-2t} - (-1) = 1 - e^{-2t} \end{aligned}$$

The complete cumulative density function, illustrated in Figure 7.9b, may be written:

$$F(t) = \begin{cases} 0 & \text{if } t < 0 \\ 1 - e^{-2t} & \text{if } t \geq 0 \end{cases}$$

To determine the probability that a worker will have to wait between one and two minutes for service at the supply room, we compute:

$$\begin{aligned} P(1 \leq T \leq 2) &= \int_1^2 2e^{-2t}\,dt = F(2) - F(1) \\ &= (1 - e^{-2(2)}) - (1 - e^{-2(1)}) = e^{-2} - e^{-4} \\ &= .1353 - .0183 = .1170 \end{aligned}$$

In the above solution the values of e^{-2} and e^{-4} were obtained from Appendix Table C.

Figure 7.9

Probability Functions for Waiting Time

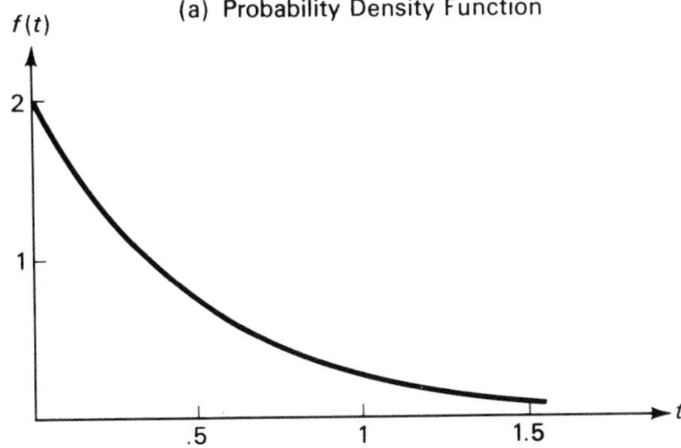

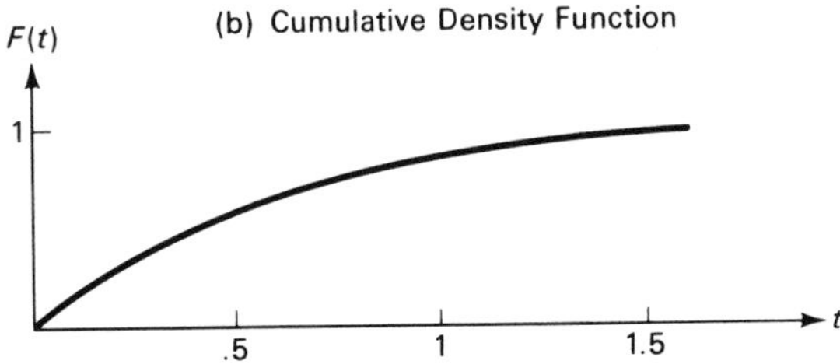

PROBLEMS

7.1 A discrete random variable X has the probability mass function:

$$f(x) = \begin{cases} \dfrac{x+2}{25} & \text{if } x = 1, 2, 3, 4, 5 \\ 0 & \text{otherwise} \end{cases}$$

(a) Present the mass function in the form of a probability table.
(b) Determine the cumulative mass function.
(c) Present the cumulative mass function in the form of a table.

7.2 A discrete random variable X has the probability mass function:

$$f(x) = \begin{cases} \dfrac{x+1}{50} & \text{if } x = 2, 3, 4, 5, 6 \\ \dfrac{11-x}{20} & \text{if } x = 7, 8, 9, 10 \\ 0 & \text{otherwise} \end{cases}$$

(a) Present the probability mass function in the form of a table.
(b) Determine the cumulative mass function.
(c) Present the cumulative mass function in the form of a table.

7.3 A discrete random variable X has the probability mass function:

$$f(x) = \begin{cases} \dfrac{x^2+4}{50} & \text{if } x = 0, 1, 2, 3, 4 \\ 0 & \text{otherwise} \end{cases}$$

Determine the following probabilities:
(a) $P(X = 2)$; (b) $P(X < 2)$; (c) $P(X \leq 2)$;
(d) $P(X < 5)$; (e) $P(X \leq 3)$.

7.4 A discrete random variable X has the cumulative mass function shown in the following table:

x:	2	3	4	5	6	7	8	9
$F(x)$:	.018	.099	.324	.607	.852	.971	.994	1.000

Determine the following probabilities:
(a) $P(X = 5)$;
(b) $P(X > 6)$;
(c) $P(X \geq 6)$;
(d) $P(X < 7)$;
(e) $P(X \leq 5)$;
(f) $P(4 \leq X \leq 7)$;
(g) $P(5 < X \leq 8)$.

7.5 The probability mass function of discrete random variable Z is:

$$f(z) = \begin{cases} \dfrac{1}{z!\, e} & \text{if } z = 0, 1, 2, \ldots \\ 0 & \text{otherwise} \end{cases}$$

(a) Determine the cumulative mass function.
(b) Determine $P(2 \leq Z \leq 4)$.

7.6 A box of a dozen eggs contains 7 good eggs and 5 bad eggs. A sample of 4 eggs is to be selected at random from the box. Let G be a random variable denoting the number of good eggs in the sample. Under these conditions, the probability mass function of G can be shown to be:

$$f(g) = \begin{cases} \dfrac{\binom{7}{g}\binom{5}{4-g}}{\binom{12}{4}} & \text{if } g = 0, 1, 2, 3, 4 \\ 0 & \text{otherwise} \end{cases}$$

What is the probability that the sample will contain:
(a) No good eggs?
(b) Exactly 1 good egg?

(c) Exactly 2 good eggs?
(d) Less than 2 good eggs?
(e) No more than 2 good eggs?
(f) More than 2 good eggs?
(g) No less than 2 good eggs?

7.7 The probability mass function of a discrete random variable R is:

$$f(r) = \begin{cases} \binom{4}{r} .2^r(.8)^{4-r} & \text{if } r = 0, 1, 2, 3, 4 \\ 0 & \text{otherwise} \end{cases}$$

(a) Determine $P(R = 1)$.
(b) Determine $P(R \leq 2)$.

7.8 The probability mass function of a discrete random variable X is:

$$f(x) = \begin{cases} kx & \text{if } x = 0, 1, 2, 3, 4 \\ 0 & \text{otherwise} \end{cases}$$

Determine the value of k.

7.9 Let D represent the number of defects produced in an hour's run by a particular automatic machine. The probability mass function of D is given by:

$$f(d) = \begin{cases} .10 & \text{if } d = 0 \\ kd & \text{if } d = 1, 2, 3 \\ k(6 - d) & \text{if } d = 4, 5 \\ 0 & \text{otherwise} \end{cases}$$

(a) Determine the value of k.
(b) What is the probability that, in an hour's run, the machine will produce at least 4 defects, given that it does produce at least 1 defect?

7.10 The probability mass function of a discrete random variable X is:

$$f(x) = \begin{cases} kx & \text{if } x = \frac{1}{5}, \frac{2}{5}, \frac{3}{5}, \frac{4}{5} \\ 0 & \text{otherwise} \end{cases}$$

(a) Determine the value of k.
(b) What is the probability that X is at most 3/5?

7.11 The probability mass function of a discrete random variable X is:

$$f(x) = \begin{cases} kx^{x-2} & \text{if } x = 3, 4, 5, 6 \\ 0 & \text{otherwise} \end{cases}$$

Determine the value of k.

7.12 Let K be a random variable denoting the number of customers waiting for service in a garden supply shop at any particular moment on a Monday morning. The probability function of K is given by:

$$f(k) = \begin{cases} .5^{(k+1)} & \text{if } k = 0, 1, 2, \ldots \\ 0 & \text{otherwise} \end{cases}$$

(a) Is $f(k)$ a p.m.f. or a p.d.f.? Explain.
(b) Determine the cumulative probability function corresponding to $f(k)$.
(c) At exactly 10:15 on a particular Monday morning, what is the probability that:
 (i) There will be no customers waiting?
 (ii) There will be exactly 1 customer waiting?
 (iii) There will be exactly 2 customers waiting?
 (iv) There will be exactly 3 customers waiting?
 (v) There will be no more than 1 customer waiting?
 (vi) There will be at most 2 customers waiting?
 (vii) There will be at least 2 customers waiting?
 (viii) There will be more than 2 customers waiting?
 (ix) There will be at least 1 but no more than 3 customers waiting?
 (x) There will be at least 1 but less than 3 customers waiting?

7.13 Let X represent the demand for a particular product. The *cumulative mass function* of x is given by:

$$F(x) = \begin{cases} 0 & \text{if } x < 1 \\ x^2/25 & \text{if } x = 1, 2, 3, 4, 5 \\ 1 & \text{if } x > 5 \end{cases}$$

Of the 4 functions that appear below, 1 and only 1 is the *probability mass function* corresponding to the above cumulative mass function. Your task is to (i) identify this corresponding probability mass function, and (ii) explain why each of the other functions is not the appropriate one.

(a) $$f(x) = \begin{cases} 2x/25 & \text{if } 0 \le x \le 5 \\ 0 & \text{otherwise} \end{cases}$$

(b) $$f(x) = \begin{cases} 2x/25 & \text{if } x = 1, 2, 3, 4, 5 \\ 0 & \text{otherwise} \end{cases}$$

(c) $$f(x) = \begin{cases} (2x - 1)/25 & \text{if } x = 1, 2, 3, 4, 5 \\ 0 & \text{otherwise} \end{cases}$$

(d) $$f(x) = \begin{cases} (11 - 2x)/25 & \text{if } x = 1, 2, 3, 4, 5 \\ 0 & \text{otherwise} \end{cases}$$

7.14 The probability density function of a continuous random variable X is:

$$f(x) = \begin{cases} x/2 & \text{if } 0 < x < 2 \\ 0 & \text{otherwise} \end{cases}$$

(a) Determine the cumulative density function of X.
(b) Determine the following probabilities:
 (i) $P(0 < X < 1)$
 (ii) $P(.5 < X < 1.5)$

7.15 The probability density function of a continuous random variable G is:

$$f(g) = \begin{cases} 2g & \text{if } 0 \le g \le 1 \\ 0 & \text{otherwise} \end{cases}$$

(a) Verify that $f(g)$ is a p.d.f.
(b) Graph $f(g)$.

(c) Determine the cumulative density function corresponding to $f(g)$.
(d) Determine the following probabilities:
  (i) $P(G \leq .50)$
  (ii) $P(G > .60)$
  (iii) $P(.40 < G < .70)$

7.16 The probability density function of a continuous random variable Y is:

$$f(y) = \begin{cases} 3y^2 & \text{if } 0 \leq y \leq 1 \\ 0 & \text{otherwise} \end{cases}$$

(a) Graph the probability density function of Y.
(b) Determine the corresponding cumulative density function.
(c) Graph the cumulative density function in (b).
(d) From the graph obtained in (c), determine the following probabilities:
  (i) $P(Y \leq .60)$
  (ii) $P(Y \leq .80)$
(e) By direct integration, confirm the probabilities obtained in (d).

7.17 The probability density function of a continuous random variable W is:

$$f(w) = \begin{cases} w/8 & \text{if } 0 \leq w \leq 4 \\ 0 & \text{otherwise} \end{cases}$$

(a) Determine the corresponding cumulative density function.
(b) From the function obtained in (a), determine the following probabilities:
  (i) $P(1 < W < 2)$
  (ii) $P(W < 3)$
  (iii) $P(W > 2.5)$
(c) By directly integrating the probability density function of W, confirm each of the probabilities obtained in (b).

7.18 The probability density function of a continuous random variable X is:

$$f(x) = \begin{cases} 1/6 & \text{if } 2 \leq x \leq 8 \\ 0 & \text{otherwise} \end{cases}$$

(a) Determine the cumulative density function of X.
(b) Determine the following probabilities:
  (i) $P(2 < X < 5)$
  (ii) $P(5 < X < 6)$

7.19 The probability density function of a continuous random variable H is:

$$f(h) = \begin{cases} 6h(1-h) & \text{if } 0 \leq h \leq 1 \\ 0 & \text{otherwise} \end{cases}$$

(a) Verify that $f(h)$ is a p.d.f.
(b) Graph the probability density function of H.
(c) Determine the corresponding cumulative density function.
(d) Determine the following probabilities:
  (i) $P(H \leq .20)$
  (ii) $P(H > .85)$
  (iii) $P(.25 \leq H \leq .75)$

7.20 The probability density function of a continuous random variable X is:

$$f(x) = \begin{cases} x^2/9 & \text{if } 0 \leq x \leq 3 \\ 0 & \text{otherwise} \end{cases}$$

(a) Graph the probability density function of X.
(b) Determine the corresponding cumulative density function.
(c) Graph the cumulative density function in (b).
(d) From the graph obtained in (c), determine the following probabilities:
   (i) $P(X \leq 2.0)$
   (ii) $P(X \geq 2.4)$
   (iii) $P(2.0 \leq X < 2.4)$

7.21 Let H represent the burning life (in hours) of a particular type of photographic floodlamp. Experience has shown that the burning life of this type of bulb is a random variable which follows the density function

$$f(h) = \begin{cases} .3(h - h^2 + 2) & \text{if } 0 \leq h \leq 2 \\ 0 & \text{otherwise} \end{cases}$$

(a) Determine the cumulative density function corresponding to $f(h)$.
(b) Determine the probability that one of these lamps will last no longer than 1 hour.
(c) Determine the probability that one of these lamps will last at least 1.5 hours.

7.22 The probability density function of a continuous random variable H is:

$$f(h) = \begin{cases} 3h^2/26 & \text{if } 1 \leq h \leq 3 \\ 0 & \text{otherwise} \end{cases}$$

(a) Graph the probability density function of H.
(b) Determine the corresponding cumulative density function.
(c) Graph the cumulative density function in (b).
(d) From the graph obtained in (c), determine the following probabilities:
   (i) $P(H < 2)$
   (ii) $P(H > 1.5)$
   (iii) $P(2.1 < H < 2.5)$
(e) By directly integrating the probability density function of H, confirm each of the probabilities obtained in (d).

7.23 The probability density function of a continuous random variable M is:

$$f(m) = \begin{cases} 2(m - 1) & \text{if } 1 \leq m \leq 2 \\ 0 & \text{otherwise} \end{cases}$$

(a) Determine the cumulative density function corresponding to $f(m)$.
(b) From the function obtained in (a), determine the following probabilities:
   (i) $P(M < .10)$
   (ii) $P(M \geq .40)$
   (iii) $P(1.5 < M \leq 1.8)$
(c) By directly integrating the probability density function of M verify the probabilities obtained in (b).

7.24 The probability density function of a continuous random variable S is:

$$f(s) = \begin{cases} 2s - 4 & \text{if } 2 \leq s \leq 3 \\ 0 & \text{otherwise} \end{cases}$$

(a) Graph the probability density function of S.
(b) Determine the corresponding cumulative density function.
(c) Graph the cumulative density function in (b).

(d) From the graph obtained in (c), determine the following probabilities:
  (i) $P(S < 2.2)$
  (ii) $P(S > 2.5)$
  (iii) $P(2.4 < S < 2.6)$

(e) By directly integrating the probability density function of S, confirm each of the probabilities obtained in (d).

7.25 The probability density function of a continuous random variable V is:

$$f(v) = \begin{cases} v - 2 & \text{if } 2 \leq v < 3 \\ 4 - v & \text{if } 3 \leq v \leq 4 \\ 0 & \text{otherwise} \end{cases}$$

(a) Graph the probability density function of V.
(b) Determine the corresponding cumulative density function.
(c) Graph the cumulative density function in (b).
(d) Determine each of the following probabilities by using the c.d.f. obtained in (b):
  (i) $P(V \leq 2.5)$
  (ii) $P(V \leq 3.5)$
  (iii) $P(V > 3.2)$
  (iv) $P(2.4 < V < 3.6)$

7.26 The probability density function of a continuous random variable X is:

$$f(x) = \begin{cases} 1 + x & \text{if } -1 \leq x < 0 \\ 1 - x & \text{if } 0 \leq x \leq 1 \\ 0 & \text{otherwise} \end{cases}$$

(a) Graph the probability density function of X.
(b) Determine the corresponding cumulative density function.
(c) Graph the cumulative density function in (b).
(d) Determine each of the following probabilities by using the c.d.f. obtained in (b):
  (i) $P(X \leq -.50)$
  (ii) $P(X \leq .50)$
  (iii) $P(-.70 < X < .40)$

7.27 The probability density function of a continuous random variable X is:

$$f(x) = \begin{cases} 3x^2/2 & \text{if } 0 \leq x < 1 \\ 1/x^2 & \text{if } 1 \leq x \leq 2 \\ 0 & \text{otherwise} \end{cases}$$

(a) Graph the probability density function of X.
(b) Determine the corresponding cumulative density function.
(c) Graph the cumulative density function in (b).
(d) Determine the following probabilities:
  (i) $P(X < 1)$
  (ii) $P(X < 1.5)$
  (iii) $P(X > .50)$
  (iv) $P(.20 < X < .80)$
  (v) $P(1.2 < X < 1.4)$
  (vi) $P(.30 < X < 1.8)$

7.28 Let X be a random variable denoting the time (in weeks) required to complete a small contract. The probability density function of x is given by:

$$f(x) = \begin{cases} \dfrac{x}{16} - \dfrac{1}{8} & \text{for } 2 \le x \le 6 \\ \dfrac{5}{8} - \dfrac{x}{16} & \text{for } 6 < x \le 10 \\ 0 & \text{otherwise} \end{cases}$$

(a) Graph the above probability density function. Label the axes and the important points.
(b) Determine the corresponding cumulative density function.
(c) The profit of the contract depends on the time required to complete it, as shown by the following function:

$$\text{Profit (in dollars)} = 100 - 10X$$

Determine the probability that the profit will be smaller than \$60.

7.29 Let X be a continuous random variable with the probability density function:

$$f(x) = \begin{cases} x/8 & \text{of } 0 \le x \le 4 \\ 0 & \text{otherwise} \end{cases}$$

Let the events A_1, A_2, and A_3 be defined by:

$$A_1 = \{x| -\infty < x \le 2\}$$
$$A_2 = \{x|1 < x \le 3\}$$
$$A_3 = \{x|2 \le x < \infty\}$$

(a) Determine the cumulative density function corresponding to $f(x)$.
(b) Determine the following probabilities:
  (i) $P(A_1)$;
  (ii) $P(A_1 \cap A_2)$;
  (iii) $P(A_2 \cap A_3)$;
  (iv) $P(A_2 \cup A_3)$;
  (v) $P(A_2|A_3)$;
(c) Are A_2 and A_3 independent events? Explain.

7.30 Of the 4 functions that appear below, 1 and only 1 meets all the requirements of a probability density function, whereas each of the remaining 3 functions fails (for one reason or another) to qualify as a probability density function. Your task is to (i) identify the probability density function, and (ii) explain why each of the other functions is not a probability density function.

(a) $$f(x) = \begin{cases} 3x^2 & \text{if } 0 \le x \le 2 \\ 0 & \text{otherwise} \end{cases}$$

(b) $$f(x) = \begin{cases} \dfrac{4x}{3} - 1 & \text{if } 1 \le x \le 2 \\ 0 & \text{otherwise} \end{cases}$$

(c) $$f(x) = \begin{cases} \dfrac{3x}{2} - 1 & \text{if } 0 \le x \le 2 \\ 0 & \text{otherwise} \end{cases}$$

(d)
$$f(x) = \begin{cases} x/10 & \text{if } x = 1, 2, 3, 4 \\ 0 & \text{otherwise} \end{cases}$$

7.31 The probability density function of a random variable X is:

$$f(x) = \begin{cases} 2e^{-2x} & \text{if } x \geq 0 \\ 0 & \text{if } x < 0 \end{cases}$$

(a) Determine the cumulative density function of X.
(b) Determine the following probabilities:
  (i) $P(1 < X < 2)$
  (ii) $P(.2 < X < 1.6)$

7.32 The probability density function of a random variable X is:

$$f(x) = \begin{cases} 6.05\, e^{-x} & \text{if } x \geq 1.8 \\ 0 & \text{otherwise} \end{cases}$$

(a) Demonstrate that this is a probability density function.
(b) Determine $P(1.9 < X < 2.1)$.

7.33 Let X be a random variable with probability density function:

$$f(x) = \begin{cases} e^{-x} & \text{if } 0 \leq x < \infty \\ 0 & \text{otherwise} \end{cases}$$

Two independent observations x_1 and x_2, are made. What is the probability that the maximum of these 2 values will be no greater than 3?

7.34 Let X represent the life (in hours) of a particular type of delicate component. The probability density function of X is:

$$f(x) = \begin{cases} 4e^{-4x} & \text{if } x \geq 0 \\ 0 & \text{otherwise} \end{cases}$$

What is the probability that the life of 1 of these components will be:
(a) Less than .25 hour?
(b) More than .50 hour?
(c) Between .20 hour and .30 hour?

7.35 The cumulative density function of a continuous random variable X is:

$$F(x) = \begin{cases} e^{x} & \text{if } x \leq 0 \\ 1 & \text{if } x > 0 \end{cases}$$

(a) Determine the probability density function of X.
(b) Demonstrate that $f(x)$ is a p.d.f.
(c) Determine $P(-2 < X < -1)$.

7.36 The probability density function of a random variable X is:

$$f(x) = \begin{cases} ke^{-x} & \text{if } x \geq 5.3 \\ 0 & \text{otherwise} \end{cases}$$

Determine the value of k.

7.37 The probability density function of a continuous random variable X is:

$$f(x) = \begin{cases} \frac{1}{2} e^{x} & \text{if } x < 0 \\ \frac{1}{2} e^{-x} & \text{if } x \geq 0 \\ 0 & \text{otherwise} \end{cases}$$

(a) Graph the probability density function of X.
(b) Determine the corresponding cumulative density function.
(c) Graph the cumulative density function in (b).
(d) Determine the following probabilities:
  (i) $P(X < -.40)$
  (ii) $P(X < .50)$
  (iii) $P(-.20 < X < .30)$

Chapter 8

Central Tendency and Variability

In Chapter Six we examined the basic concept of a random variable, and in Chapter Seven we explored methods of using probability functions to compute the probabilities of events that are defined in terms of values of random variables. This chapter is concerned with ways of describing the "behavior" of a random variable in terms of certain characteristics of its probability distribution. Specifically, this chapter will consider various measures of "central tendency" and "variability."

8.1 MEASURES OF CENTRAL TENDENCY: AVERAGES

Without ever taking a course in probability or statistics, most people have an intuitive understanding of the general concept of an "average" as a sort of "central" value. An average is frequently taken to be more or less "representative" of a set of individual values. Summarizing a collection of values in the form of an average simplifies our thinking about the set, particularly if the set is large.

The term "average" has a much broader meaning to the statistician than to the layman. Depending on the situation, there are actually several different kinds of averages from which to choose. The averages that we will consider in this chapter are the median, mode, and arithmetic mean.[1]

8.1.1 The Median

Loosely speaking, the *median* of a probability distribution is the value that divides the distribution in half, in the sense that the chances are 50-50 that the random variable will have a value either above or below the median. As we will see, it is not always possible to specify a value for the median such that the distribution will be split *exactly* in half by this value. Therefore, the median of the probability distribution of a random variable X is more formally defined as the smallest x-value for which $F(x)$ is at least .50.

[1] Other averages include the geometric mean, harmonic mean, and quadratic mean.

The Median of a Discrete Distribution. To illustrate the procedure for determining the median of a discrete distribution, consider a random variable X that has the following probability mass function:

$$f(x) = \begin{cases} \dfrac{.4 + x}{20} & \text{if } x = 0, 1, 2, 3, 4 \\ \dfrac{7.5 - x}{10} & \text{if } x = 5, 6 \\ 0 & \text{otherwise} \end{cases}$$

To determine the median for such a function, we first tabulate the values for $f(x)$ and $F(x)$ as shown in Table 8.1. This table shows that there is no x-value for which $F(x)$ is exactly .50. In other words, none of the possible x-values will split the distribution exactly in half. Therefore, to determine the value of the median in such a case, we must locate the smallest x-value for which $F(x)$ is greater than .50. From Table 8.1, we can see that $F(3) = .38$ and $F(4) = .60$. Thus, since 4 is the smallest x-value for which $F(x) \geq .50$, the median of the distribution is equal to 4.

Table 8.1

Tabulation of a Discrete Probability Distribution

x	$f(x)$	$F(x)$
0	.02	.02
1	.07	.09
2	.12	.21
3	.17	.38
4	.22	.60
5	.25	.85
6	.15	1.00

The Median of a Continuous Distribution. We have observed that, for a discrete distribution, it is usually not possible to find a value that splits the distribution exactly in half. It is generally possible, however, to determine such a value for a continuous distribution, except under rare theoretical conditions that we may ignore for practical purposes.[2] Thus, in the case of a

[2] For instance, consider the p.d.f.

$$f(x) = \begin{cases} .25(2 - x) & \text{if } 0 \leq x \leq 2 \\ x - 3 & \text{if } 3 \leq x \leq 4 \\ 0 & \text{otherwise} \end{cases}$$

As the reader may verify, $F(x) = .50$ for all values in the interval $2 \leq x < 3$. Thus, any x-value in this interval will satisfy the definition of a median.

continuous random variable X, we simply say that the median is the x-value for which $F(x) = .50$. Graphically, the median of a continuous distribution will divide the distribution into two equal parts in the sense that the area under the curve will be exactly .50 on either side of the median. For a continuous random variable X, the median is computed by solving the following equation for the value of x_{md}:

$$\int_{-\infty}^{x_{md}} f(x)\,dx = .50 \tag{8.1}$$

To illustrate this procedure, consider again Case 1 presented in Section 7.3.3. In that case, the random variable H represented the useful life (in hours) of a component. The probability density function for that random variable was given as:

$$f(h) = \begin{cases} \dfrac{h}{8} & \text{if } 0 \le h \le 4 \\ 0 & \text{otherwise} \end{cases}$$

Applying Formula (8.1) to this function, we have:

$$\int_{-\infty}^{h_{md}} f(h)\,dh = .50$$

$$\int_{-\infty}^{0} 0\,dh + \int_{0}^{h_{md}} \frac{h}{8}\,dh = .50$$

$$\left.\frac{h^2}{16}\right]_0^{h_{md}} = .50$$

$$h_{md}^2 = (.50)(16) = 8$$

Taking the positive square root (since h_{md} must lie between 0 and 4), we have:

$$h_{md} = 2.828$$

Thus the median life for this type of component is 2.828 hours.

8.1.2 The Mode

Another measure that is sometimes used to describe the central tendency of a probability distribution is the *mode*. The mode of the probability distribution of a random variable may be roughly defined as the value of the random variable that is most likely to occur. Thus, for the probability mass function in Figure 8.1a, it should be clear that the mode is equal to 6, since the random variable X has the highest probability for that value. Similarly, from Figure 8.1b, we can see that the distribution has a mode at 4. We can also see that the value of 10 has a greater probability than the values to either side of it. Such a value is also considered to be a mode. However, the rough definition given above fails to take into account this type of situation. Therefore, a mode of a probability distribution is more formally defined as a value of the random variable at which the probability function reaches a *local maximum*. In other words, a mode is an x-value for which $f(x)$ is greater than for any other x-value in its *immediate neighborhood*. Graphically, a mode occurs at any x-value at which the graph of the distribution reaches a peak. This is illustrated by the various parts of Figure 8.1.

Figure 8.1

The Modes of Probability Distributions

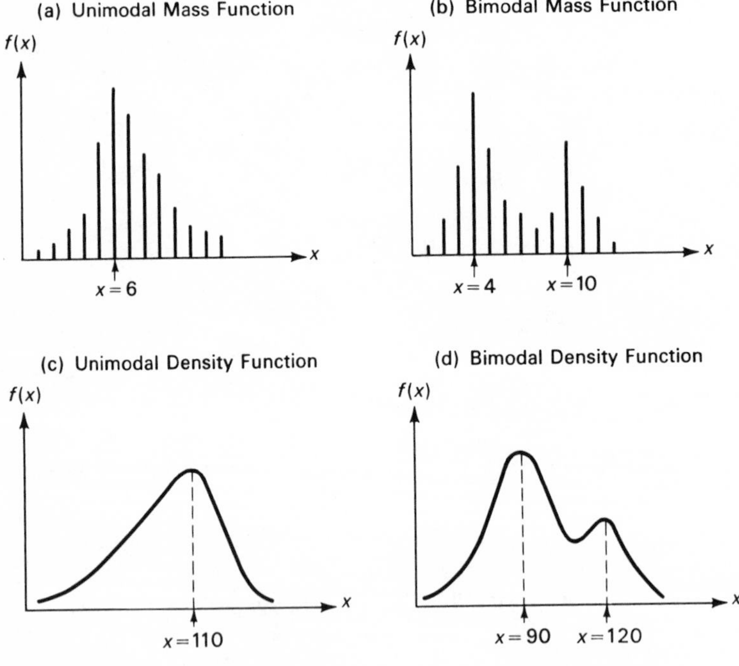

A distribution that has a single mode is called *unimodal*, whereas a distribution that has two modes is called *bimodal*. Of course, a distribution may have even more than two modes. Therefore, the use of the mode as a measure of central tendency will be meaningful only if the distribution is unimodal.

Determining the value of a mode for a discrete random variable is a simple matter of identifying any x-value whose probability is greater than the probabilities of the x-values on either side of that value. This may be accomplished by inspecting either a graph or a table of the mass function (such as Figures 8.1a and 8.1b, or Table 8.1).

For a continuous random variable, the *approximate* value of a mode may be obtained by inspecting a carefully drawn graph of the density function. However, to obtain the *exact* value of a mode, we usually apply differential calculus. Specifically, if the probability density function $f(x)$ is twice differentiable, a mode is any x-value that satisfies the following two expressions:

$$\begin{cases} f'(x) = 0 \\ f''(x) < 0 \end{cases} \tag{8.2}$$

where $f'(x)$ is the first derivative of $f(x)$, and $f''(x)$ is the second derivative of $f(x)$.

To illustrate this application of differential calculus, consider again Case 2 presented in Section 7.3.3. In that case, the random variable W represented winter snowfall (in feet). The probability density function for that random variable was:

$$f(w) = \begin{cases} 4w - 3w^2 & \text{if } 0 \leq w \leq 1 \\ 0 & \text{otherwise} \end{cases}$$

The first derivative of this function is:

$$f'(w) = 4 - 6w$$

Setting this expression equal to zero, we obtain the following solution for w:

$$4 - 6w = 0$$
$$w = 2/3$$

The second derivative is:

$$f''(w) = -6$$

Therefore, since $f''(w) < 0$, the density function has a mode at $w = 2/3$. This distribution is unimodal since the solution for $f'(w) = 0$ is unique.

8.1.3 The Arithmetic Mean

The *arithmetic mean*, often referred to simply as the *mean*, is the measure of central tendency that corresponds to the layman's usual idea of the average. That is, for a *finite set* of values, the mean is the sum of the values of all the elements in the set divided by the total number of elements. To illustrate this idea, consider the board of directors of a manufacturing firm. This board consists of eight members whose ages are:

$$40, 40, 40, 44, 46, 49, 49, 52$$

The arithmetic mean of these ages is computed as follows:

$$\frac{40 + 40 + 40 + 44 + 46 + 49 + 49 + 52}{8} = \frac{360}{8} = 45$$

Notice in this example that some of the elements have identical values. Three men are 40, two men are 49, and each one of the remaining men has a different age. Thus, an alternative way of calculating the mean is:

$$\frac{40(3) + 44(1) + 46(1) + 49(2) + 52(1)}{8} = \frac{360}{8} = 45$$

By using simple algebra, we may re-write the above expression as:

$$40\left(\frac{3}{8}\right) + 44\left(\frac{1}{8}\right) + 46\left(\frac{1}{8}\right) + 49\left(\frac{2}{8}\right) + 52\left(\frac{1}{8}\right) = 45$$

In this last expression, each of the fractions contained in parentheses represents the proportion of the number of men of a specific age to the total number of men of all ages. That is, of the total board members, there are 3/8 with an age of 40, 1/8 with an age of 44, and so on. Each of these proportions may be regarded as a relative frequency. Thus, one way to compute the arithmetic mean is to "weight" each value by the corresponding relative frequency. In this sense, the arithmetic mean may be regarded as a kind of weighted average.

To gain a clearer understanding of the arithmetic mean as a sort of weighted average, consider the following set of nine values:

$$\{11,12,12,13,13,13,14,14,15\}$$

As the reader may verify, the mean of this set is 13. Suppose that we build a physical model of this set by arranging blocks on a plank as shown in Figure 8.2a. In this figure, each block represents an individual element of the set. At equal intervals along the plank, values are marked corresponding to the values of the elements. If we wished to place a fulcrum under the plank so that the plank would balance, the fulcrum would be placed at such a point that the sum of the forces on the left side of the fulcrum is equal to the sum of the forces on the right side of the fulcrum. This point, which is the *center of gravity* of the pile of blocks, corresponds to the arithmetic mean. Thus the fulcrum is placed directly beneath the value 13, which is the mean of the set.

Figure 8.2

Demonstration of the Arithmetic Mean as the Center of Gravity

(a) Symmetrical Pattern

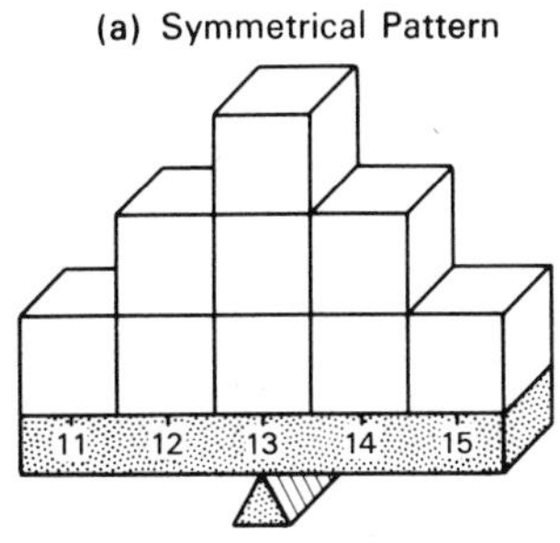

(b) Asymmetrical Pattern

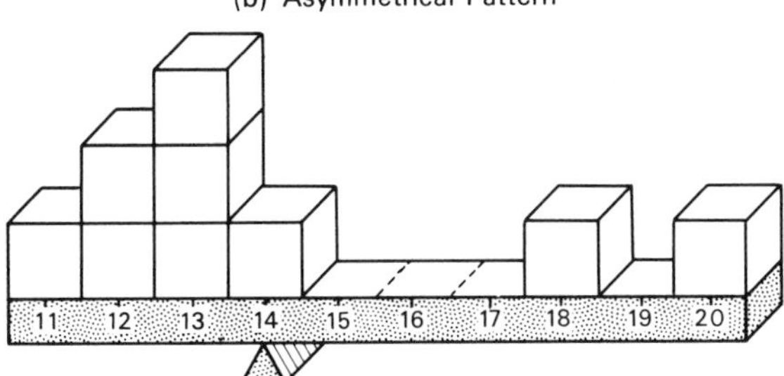

The distribution of values illustrated in Figure 8.2a is *symmetrical* around the value 13. In other words, the blocks form a symmetrical pattern with the value 13 at the center. In contrast, Figure 8.2b represents an *asymmetrical* pattern, which is formed from the following set of values:

$$\{11,12,12,13,13,13,14,18,20\}$$

If the fulcrum were placed under the value 13, the plank would tip to the right. In order for the plank to balance, the fulcrum must be placed directly beneath the value corresponding to the mean, which, as the reader is urged to verify, is equal to 14.

The Mean of a Discrete Distribution. As developed above, the concept of the mean of a finite set may be readily extended to a discrete probability distribution. Specifically, if a discrete random variable X has a probability mass function $f(x)$, then its mean μ is given by

$$\mu = \sum_{\text{all } x} x\, f(x) \tag{8.3}$$

In words, the mean of a discrete random variable is computed by multiplying each of the possible values of the random variable by the corresponding probability and summing the products. Since we may regard the values of $f(x)$ as relative frequencies, we can interpret the mean of a discrete random variable as the weighted average of the possible x-values, where each x-value is weighted by its relative frequency.

The mean of a probability distribution is commonly referred to as the *expected value* (or *expectation*) of the random variable. For a random variable X, the expected value is often symbolized as $E(X)$, which is an alternative symbol to μ.

To illustrate the computation of the expected value of a discrete random variable, consider a random variable X that has the following probability mass function:

$$f(x) = \begin{cases} \dfrac{.4 + x}{20} & \text{if } x = 0, 1, 2, 3, 4 \\ \dfrac{7.5 - x}{20} & \text{if } x = 5, 6 \\ 0 & \text{otherwise} \end{cases}$$

This function was presented previously in the discussion of the median of a discrete random variable, and the values for $f(x)$ were shown in Table 8.1. Using these values, we may compute the mean of the distribution as follows:

$$\begin{aligned} \mu = E(X) &= 0(.02) + 1(.07) + 2(.12) + 3(.17) + 4(.22) + 5(.25) + 6(.15) \\ &= 3.85 \end{aligned}$$

Using a computation table, these same calculations may be performed as shown in Table 8.2.

Table 8.2

Computation of Expected Value of a Discrete Random Variable

x	$f(x)$	$x\,f(x)$
0	.02	0
1	.07	.07
2	.12	.24
3	.17	.51
4	.22	.88
5	.25	1.25
6	.15	.90
		$\Sigma = 3.85 = \mu = E(X)$

The concept of expected value was first used in connection with games of chance. For example, the type of roulette wheel used in Las Vegas casinos has thirty-eight different "equally likely" values on which a person may bet. Of these values eighteen are red, eighteen are black, and two are green. In one type of play, a person can bet on either red or black but not on green. If a person bets on red, and the outcome of the spin is either black or green, he loses the amount of his bet; if the outcome is red, he retains the amount of his wager and wins an additional amount equal to his bet. Suppose that a person sits all evening at the table and makes a \$1.00 bet on red each time the wheel is spun. On any spin, he has an 18/38 probability of making a \$1.00 profit and a 20/38 probability of a \$1.00 loss. Then his expectation is:

$$\frac{18}{38}(\$1.00) + \frac{20}{38}(-\$1.00) = -\$.053$$

That is, in the long run, the player can expect an average loss close to 5 cents per play. Notice, in our present frame of reference, that we used the word "expect" in a very special sense. The player does not actually expect a \$.053 loss *on any given play*; in fact, such a loss is possible only as an average. Thus, in one sense, the expected value is the average value that one would expect to obtain *in the long run*.

On any given play, our roulette player can either win \$1.00 or lose \$1.00. The random variable in this case has only two possible values (+\$1.00 and −\$1.00). The expected value is computed simply by weighting each of these two values by its corresponding probability. Thus, in another sense, the expected value may be regarded as the weighted average of the two values.

The Mean of a Continuous Distribution. The formula for the expected value of a continuous random variable is analogous to that for a discrete random variable, with the operation of integration replacing the operation of summation. That is, if X is a continuous random variable with probability density function $f(x)$, the expected value of X is given by:

$$\mu = E(X) = \int_{-\infty}^{\infty} x\, f(x)\, dx \tag{8.4}$$

The application of Formula (8.4), is illustrated by the following examples.

Example 1:

In Case 1 of Section 7.3.3, we considered a component whose useful life (in hours) was described by the probability density function:

$$f(h) = \begin{cases} \dfrac{h}{8} & \text{if } 0 \le h \le 4 \\ 0 & \text{otherwise} \end{cases}$$

Then the mean life of this type of component is computed as follows:

$$E(H) = \int_{-\infty}^{\infty} h\,f(h)\,dh = \int_{-\infty}^{0} h\,0\,dh + \int_{0}^{4} h(h/8)\,dh + \int_{4}^{\infty} h\,0\,dh$$

$$= \int_{0}^{4} h\,\frac{h}{8}\,dh = \int_{0}^{4} \frac{h^2}{8}\,dh$$

$$= \frac{h^3}{24}\Big]_0^4 = \frac{64}{24} - \frac{0}{24} = \frac{8}{3} = 2.67$$

Example 2:

In Case 3 of Section 7.3.3, Q was a random variable denoting monthly quantity demanded (in tons) for a chemical product, with the p.d.f.:

$$f(q) = \begin{cases} q - 1 & \text{if } 1 \leq q < 2 \\ 3 - q & \text{if } 2 \leq q \leq 3 \\ 0 & \text{otherwise} \end{cases}$$

Then, expected monthly demand is given by:

$$E(Q) = \int_{-\infty}^{\infty} q\,f(q)\,dq$$

$$= \int_{-\infty}^{1} q\,0\,dq + \int_{1}^{2} q(q-1)\,dq + \int_{2}^{3} q(3-q)\,dq + \int_{3}^{\infty} q\,0\,dq$$

$$= \int_{1}^{2} (q^2 - q)\,dq + \int_{2}^{3} (3q - q^2)\,dq$$

$$= \left[\frac{q^3}{3} - \frac{q^2}{2}\right]_1^2 + \left[\frac{3q^2}{2} - \frac{q^3}{3}\right]_2^3$$

$$= \left[\left(\frac{8}{3} - \frac{4}{2}\right) - \left(\frac{1}{3} - \frac{1}{2}\right)\right] + \left[\left(\frac{27}{2} - \frac{27}{3}\right) - \left(\frac{12}{2} - \frac{8}{3}\right)\right]$$

$$= 2$$

That is, long-run average monthly demand is two tons.

8.1.4 Expected Value of a Function of a Random Variable

We have seen how to compute the expected value of any random variable X whose probability function $f(x)$ is known. In applying probability theory to decision problems, however, we often encounter a random variable, say Y, whose probability function is not readily available. Nevertheless, if a relationship between Y and X can be specified by a mathematical function, $Y = g(X)$, then it becomes possible to compute the expected value of Y. Specifically, if X is a discrete random variable with mass function $f(x)$, and if $g(X)$ is some other function of X, then the expected value of $g(X)$ is given by:

$$E[g(X)] = \sum_{\text{all } x} g(x)f(x) \tag{8.5}$$

If X is a continuous random variable with density function $f(x)$, and if $g(X)$ is some other function of X, then the expected value of $g(X)$ is given by:

$$E[g(X)] = \int_{-\infty}^{\infty} g(x)f(x)\,dx \tag{8.6}$$

Example 1:

Let X be a random variable denoting a job shop's monthly demand for an expensive hand-assembled device. Experience has shown that demand behaves according to the mass function:

$$f(x) = \begin{cases} \dfrac{x-2}{10} & \text{if } x = 3, 4, 5, 6 \\ 0 & \text{otherwise} \end{cases}$$

The shop's monthly profit, as a function of demand, is given by:

$$Y = g(X) = \$1{,}000(X^2 - X + 6)$$

For these two functions, we may prepare the following table:

x	$f(x)$	$g(x)$
3	.10	$12,000
4	.20	$18,000
5	.30	$26,000
6	.40	$36,000

Then, applying Formula (8.5) to obtain expected monthly profit, we have:

$$\begin{aligned} E(Y) &= E[g(X)] \\ &= 12{,}000(.10) + 18{,}000(.20) + 26{,}000(.30) + 36{,}000(.40) \\ &= \$27{,}000 \end{aligned}$$

Example 2:

Let X be a random variable denoting the time (in hours) required to produce a hand-assembled article. The probability density function of X is given by:

$$f(x) = \begin{cases} .40(x+1) & \text{if } 1 \le x \le 2 \\ 0 & \text{otherwise} \end{cases}$$

The profit (in dollars) that the producer makes on an article is given by:

$$Y = g(X) = 3 - X^2$$

To obtain the expected profit per article, we apply Formula (8.6), which yields:

$$\begin{aligned} E(Y) = E[g(X)] &= \int_1^2 (3 - x^2)(.40)(x+1)\,dx \\ &= .40 \int_1^2 (3 + 3x - x^2 - x^3)\,dx \\ &= .40\left[3x + \frac{3x^2}{2} - \frac{x^3}{3} - \frac{x^4}{4}\right]_1^2 \\ &= .40\left(5\frac{1}{3} - 3\frac{11}{12}\right) \\ &= .57 \text{ dollars (approximately)} \end{aligned}$$

8.1.5 Mathematical Properties of Expectations

In working with expectations, it is helpful to be aware of certain mathematical properties that may be used to simplify algebraic manipulations. Several of these properties are presented below.

1. For any constant a,

$$E(a) = a \tag{8.7}$$

If a random variable can assume only a single value a, the expected value of this random variable is equal to a. For example, suppose that a magician has a deck of fifty-two identical cards, each card being the ten of hearts. Then any time a person draws a card at random from this deck, the value of that card must be 10. Obviously then, the expected value must also be 10.

2. For any constant a and any random variable X,

$$E(a + X) = a + E(X) \tag{8.8}$$

This simply says that the expectation of the sum of a constant and a random variable is equal to the sum of the constant and the expected value of the random variable. To illustrate this relationship, consider the random variable X whose probability mass function is shown in Table 8.3.

Table 8.3

Probability Mass Function of X

x	1	4	6
$f(x)$	.3	.2	.5

The expected value of the random variable X is:

$$\begin{aligned} E(X) &= 1(.3) + 4(.2) + 6(.5) \\ &= .3 + .8 + 3 \\ &= 4.1 \end{aligned}$$

Suppose that we define a new variable $Y = (5 + X)$. Recognizing that Y is a function of X, we may obtain the expected value of Y by applying Formula (8.5) as follows:

$$\begin{aligned} E(5 + X) &= (5 + 1)(.3) + (5 + 4)(.2) + (5 + 6)(.5) \\ &= 9.1 \end{aligned}$$

This same result may be obtained more readily by applying Formula (8.8):

$$E(5 + X) = 5 + E(X) = 5 + 4.1 = 9.1$$

3. For any constant b and any random variable X,

$$E(bX) = bE(X) \tag{8.9}$$

If we multiply a random variable X by a constant b, then the expectation of this product will be equal to the constant b times the expected value of the random variable X. Consider again the random variable X in Table

8.3. If we define a new random variable as $Y = 7X$, the expected value of Y may be computed by applying Formula (8.5):

$$\begin{aligned} E(7X) &= 7(1)(.3) + 7(4)(.2) + 7(6)(.5) \\ &= 7(.3) + 28(.2) + 42(.5) \\ &= 28.7 \end{aligned}$$

We may obtain the same result more easily by applying Formula (8.9) as follows:

$$E(7X) = 7E(X) = 7(4.1) = 28.7$$

4. By combining Formulas (8.8) and (8.9), we obtain:

$$E(a + bX) = a + bE(X) \tag{8.10}$$

If we multiply X by some constant b and then add another constant a, the expected value of this quantity will be equal to the sum of the constant a and the product of b times $E(X)$. For example, suppose that we wish to find the expected value of the quantity $(5 + 7X)$. Using the probability mass function in Table 8.3 and applying Formula (8.5), we compute:

$$\begin{aligned} E(5 + 7X) &= (5 + 7)(.3) + (5 + 28)(.2) + (5 + 42)(.5) \\ &= 12(.3) + 33(.2) + 47(.5) \\ &= 33.7 \end{aligned}$$

By direct application of Formula (8.10), we may obtain the same result, as shown below:

$$\begin{aligned} E(5 + 7X) &= 5 + 7E(X) = 5 + 7(4.1) \\ &= 5 + 28.7 = 33.7 \end{aligned}$$

To simplify our discussion of mathematical properties of expectations, the above demonstrations of Formulas (8.7) through (8.10) were limited to discrete random variables. It should be pointed out, however, that all the four properties shown by these formulas are valid regardless of whether the random variable is discrete or continuous. For discrete random variables, these properties may be proved mathematically by using Formulas (8.3) and (8.5). For continuous random variables, the proof may be obtained by using Formulas (8.4) and (8.6).

8.2 MEASURES OF VARIABILITY

We have seen that the mean is a measure of central tendency that may be more or less representative of all the values in a distribution. The extent to which the mean is representative of a random variable depends on how closely the distribution is concentrated around the mean. In other words, the wider the spread of the distribution, the less representative the mean becomes. This spread of a distribution is usually referred to as *variability* or *dispersion*. Thus, in order to describe a probability distribution, it is helpful to have some measure of variability in addition to a measure of central tendency. The need for a measure of variability is particularly apparent if we realize that two distinctly different probability distributions may have the same mean. One way to distinguish between two such distributions is to compare their variabilities. As an illustration, consider two random variables X and Y whose probability mass functions are, respectively:

$$f(x) = \begin{cases} (5 - x)/10 & \text{if } x = 1, 2, 3, 4 \\ 0 & \text{otherwise} \end{cases}$$

$$f(y) = \begin{cases} 3/10 & \text{if } y = 0 \\ y/10 & \text{if } y = 1, 2, 3 \\ 1/20 & \text{if } y = 5, 7 \\ 0 & \text{otherwise} \end{cases}$$

As the reader may verify by using Formula (8.3), both $E(X)$ and $E(Y)$ are equal to 2. That is, the two random variables have identical means. However, as we can see from Figure 8.3, the distribution of Y has a wider spread than

Figure 8.3

Probability Distributions of Two Random Variables With Identical Means but Different Dispersions

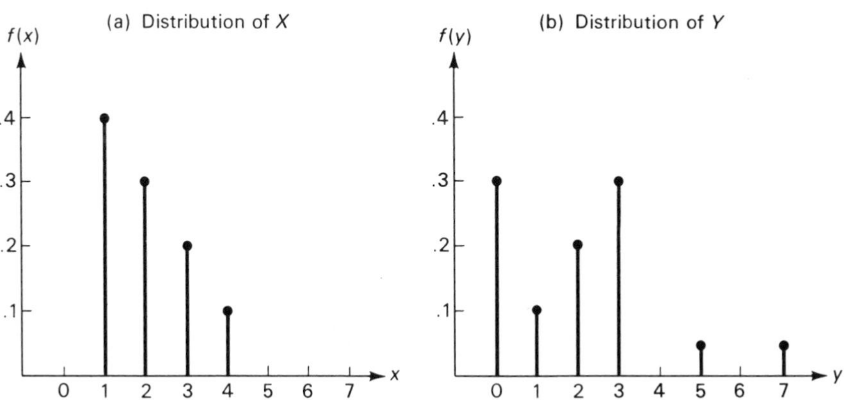

the distribution of X. Thus, if we were to measure the dispersions of these two distributions, the measure for the distribution of Y should be greater than that for the distribution of X.

8.2.1 Expected Deviation

The most commonly used measures of dispersion are based on the concept of "deviations" from the mean. For any possible value x of a random variable X, the *deviation* is defined as the *algebraic* difference between that value and the mean of the distribution. Symbolically, such a deviation may be expressed as $(x - \mu)$. Thus, if a value is less than the mean, its deviation is negative, and if the value is greater than the mean, its deviation is positive.

For any possible value in a distribution, there is a corresponding deviation from the mean. Thus, there are as many deviations as there are possible values of the random variable. Now, the question is: How may these deviations be employed to obtain a single measure for representing the variability of the entire distribution?

Recognizing that the quantity $(X - \mu)$ is a function of the random variable X, the reader might be inclined to suggest that we simply compute the expected value of this quantity. Unfortunately, this would be futile since, for any random variable X, it is always true that:

$$E(X - \mu) = 0 \tag{8.11}$$

To show that the expression $E(X - \mu)$ must always be zero, we first observe that μ is a constant. Then applying Formula (8.8), we obtain:

$$E(X - \mu) = E(X) - \mu$$

Since $E(X)$ and μ are merely alternative symbols for the mean, we may rewrite the above expression as:

$$E(X - \mu) = \mu - \mu$$

Then it is obvious that $E(X - \mu)$ must always be zero. This conforms with the idea that the mean is the center of gravity. That is, the negative forces on one side of the mean counterbalance the positive forces on the other side of the mean.

As an illustration of Formula (8.11), consider the distribution shown previously in Figure 8.3a. The computation of $E(X - \mu)$ is shown in Table 8.4. As we can see from the last column in this table, this expectation is zero, which agrees with Formula (8.11).

Table 8.4

Calculation of Expected Algebraic Deviation

x	$f(x)$	$xf(x)$	$(x - \mu)$	$(x - \mu)f(x)$
1	.4	.4	$1 - 2 = -1$	$-.4$
2	.3	.6	$2 - 2 = 0$	0
3	.2	.6	$3 - 2 = +1$	$+.2$
4	.1	.4	$4 - 2 = +2$	$+.2$
Sum		$2.0 = \mu$		$0 = E(X - \mu)$

We have seen that $E(X - \mu)$ is useless as a measure of the dispersion of a distribution since it is always equal to zero. This is due to the counterbalancing of positive and negative deviations. One way to avoid this difficulty is to disregard the algebraic signs of the deviations. In other words, rather than working with *algebraic* deviations, we could work with the *absolute* values of the deviations. This implies that, when measuring dispersion, we are primarily interested in the *magnitudes* of the deviations rather than their *signs*. That is, we are concerned with *how far* the values of the random variable depart from the mean, but not with the *direction* in which they depart. Thus, we may obtain a meaningful measure of dispersion by computing the *expected absolute deviation*, $E(|X - \mu|)$, instead of the useless $E(X - \mu)$. Specifically, for a discrete random variable X, the expected absolute deviation is given by:

$$A.D. = E(|X - \mu|) = \sum_{\text{all } x} |x - \mu| \, f(x) \tag{8.12}$$

Similarly, for a continuous random variable X, the expected absolute deviation is given by:

$$\begin{aligned} A.D. = E(|X - \mu|) &= \int_{-\infty}^{\infty} |x - \mu| \, f(x) \, dx \\ &= \int_{-\infty}^{\mu} (\mu - x) \, f(x) \, dx + \int_{\mu}^{\infty} (x - \mu) \, f(x) \, dx \end{aligned} \tag{8.13}$$

Example 1:

Consider again the distribution shown in Table 8.4, in which $E(X - \mu)$ was computed. In contrast, by applying Formula (8.12), the expected absolute deviation, $E(|X - \mu|)$, is calculated in Table 8.5. As we can see from this table, the expression is computed by weighting each possible absolute deviation, $|x - \mu|$, by the corresponding $f(x)$, and summing the products. The result is equal to .8. As a measure of dispersion, this figure indicates that, "on the average," the values of the random variable deviate .8 from the mean.

Table 8.5

Calculation of Expected Absolute Deviation

x	$f(x)$	$x\,f(x)$	$\lvert x - \mu \rvert$	$\lvert x - \mu \rvert\, f(x)$
1	.4	.4	$\lvert 1 - 2 \rvert = 1$	.4
2	.3	.6	$\lvert 2 - 2 \rvert = 0$	0
3	.2	.6	$\lvert 3 - 2 \rvert = 1$	.2
4	.1	.4	$\lvert 4 - 2 \rvert = 2$	.2
Sum		$2.0 = \mu$		$.8 = E(\lvert X - \mu \rvert)$

Example 2:

In our previous discussion of the mean of a continuous random variable, we considered the random variable H whose probability density function was:

$$f(h) = \begin{cases} \dfrac{h}{8} & \text{if } 0 \le h \le 4 \\ 0 & \text{otherwise} \end{cases}$$

The reader may recall that the mean for this distribution is equal to 8/3. Applying Formula (8.13), we may compute the expected absolute deviation as follows:

$$\begin{aligned} A.D. = E(|X - \mu|) &= \int_0^{8/3} (8/3 - h)\,\frac{h}{8}\,dh + \int_{8/3}^{4} (h - 8/3)\,\frac{h}{8}\,dh \\ &= \left[\frac{h^2}{6} - \frac{h^3}{24}\right]_0^{8/3} + \left[\frac{h^3}{24} - \frac{h^2}{6}\right]_{8/3}^{4} \\ &= .7901 \end{aligned}$$

8.2.2 Variance and Standard Deviation

Although the expected absolute deviation of a distribution is relatively easy to compute and understand, it seldom is employed in practice because its "poor" mathematical properties limit its use for more extensive analysis. Two other measures, which are much more commonly used to describe variability around the mean, are the *variance* and *standard deviation*. Both of these measures, which are closely related, work with squared deviations. Squaring not only makes all deviations positive but, more important, also provides mathematical properties that are more desirable for many types of further analysis.

Definition of the Variance. Unlike the expected absolute deviation (which is computed from the *absolute* deviations from the mean), the variance is computed by using the *squared* deviations from the mean. Just as the symbols μ and $E(X)$ are used alternately to represent the mean of a random variable X, the symbols σ^2 (sigma squared) and $V(X)$ are used interchangeably to denote the variance. Mathematically, the variance is defined as:

$$\sigma^2 = V(X) = E[(X - \mu)^2] \tag{8.14}$$

To gain a clear understanding of this definition, let us compute the variance of the random variable whose probability distribution was shown previously in Figure 8.3a. The calculation of the variance is summarized in Table 8.6. As we can see from the last column of this table, the variance is computed by weighting each possible squared deviation $(x - \mu)^2$ by the corresponding $f(x)$, and then summing the products. This sum is equal to 1, which is the variance of the random variable X. Thus, "on the average," the squared deviation between the values of the random variable and the mean is 1. Since the variance represents the average of the squared deviations, the variance is often referred to as the *mean-square-deviation*.

Table 8.6

Calculation of the Variance of a Discrete Random Variable

x	$f(x)$	$x\,f(x)$	$(x - \mu)$	$(x - \mu)^2$	$(x - \mu)^2 f(x)$
1	.4	.4	$(1 - 2) = -1$	1	.4
2	.3	.6	$(2 - 2) = 0$	0	0
3	.2	.6	$(3 - 2) = 1$	1	.2
4	.1	.4	$(4 - 2) = 2$	4	.4
Sum	1.0	2.0			1.0

$\mu = E(X) = \Sigma\, x\,f(x) = 2.0$

$\sigma^2 = V(X) = E[(X - \mu)^2] = \Sigma(x - \mu)^2 f(x) = 1.0$

Following the same procedure as shown in Table 8.6, we may compute the variance of the random variable Y in Figure 8.3b. As the reader is urged to verify, the variance of Y is equal to 3.3. Thus, for the two distributions in Figure 8.3, the variance of the random variable Y is much larger than the variance of the random variable X. This is in agreement with our intuitive observation that the distribution of Y has a wider spread than the distribution of X.

Computation of the Variance. We have used Formula (8.14) to compute variances in order to reinforce the meaning of the variance as an average squared deviation. In practice, however, this formula is seldom used, because it is often cumbersome to work with squared deviations from the mean. Fortunately, it can be proved that Formula (8.14) is identical to:

$$\sigma^2 = V(X) = E(X^2) - \mu^2 \tag{8.15}$$

In this formula, $E(X^2)$ is the mean of the squared values of X, and μ^2 is the square of the mean. Thus, the variance is sometimes defined as the *mean square* minus the *squared mean.*

For purposes of computation, Formula (8.15) is generally preferable to Formula (8.14), since Formula (8.15) requires less computational work. To apply Formula (8.15), we need only compute the quantities $E(X^2)$ and μ^2. We have already seen how to compute μ for either a discrete or continuous random variable, and μ^2 is merely the square of μ. For a discrete random variable, $E(X^2)$ is given by:

$$E(X^2) = \sum_{\text{all x}} x^2 f(x) \tag{8.16}$$

For a continuous random variable, $E(X^2)$ is given by:

$$E(X^2) = \int_{-\infty}^{\infty} x^2 f(x)\, dx \tag{8.17}$$

To illustrate the computation of the variance using Formula (8.15) together with Formula (8.16), consider again the distribution shown in Figure 8.3a. Applying these formulas, the variance is computed in Table 8.7. The reader may recall that, with greater computational effort, the same variance was obtained with Formula (8.14), as shown in Table 8.6.

Table 8.7

Alternative Way of Calculating the Variance of a Discrete Random Variable

x	$f(x)$	$x\,f(x)$	x^2	$x^2 f(x)$
1	.4	.4	1	.4
2	.3	.6	4	1.2
3	.2	.6	9	1.8
4	.1	.4	16	1.6
Sum		$2.0 = \mu$		$5.0 = E(X^2)$

$$\mu = E(X) = \Sigma\, x\,f(x) = 2.0$$

$$E(X^2) = \Sigma\, x^2 f(x) = 5.0$$

$$\sigma^2 = V(X) = E(X^2) - \mu^2 = 5 - 2^2 = 1$$

Example 1:

On a particular milkman's route, the number of quarts of milk consumed daily by a household varies from one to four quarts. Let Q be a random variable denoting the number of quarts consumed daily by a household. Analysis of the milkman's records shows that the probability mass function of Q is given by:

$$f(q) = \begin{cases} q^2/30 & \text{if } q = 1, 2, 3, 4 \\ 0 & \text{otherwise} \end{cases}$$

From the mass function, we obtain the following probabilities:

q	$f(q)$
1	.033
2	.133
3	.300
4	.533

Applying Formula (8.3), the mean of Q is:

$$\mu = E(Q) = .033(1) + .133(2) + .300(3) + .533(4) = 3.33$$

Applying Formula (8.16), we obtain the mean square:

$$E(Q^2) = .033(1^2) + .133(2^2) + .300(3^2) + .533(4^2) = 11.79$$

Finally, using Formula (8.15) to compute the variance, we have:

$$\sigma^2 = V(Q) = E(Q^2) - \mu^2 = 11.79 - 3.33^2 = .70$$

Thus, the variance of the milk consumption is .70 square quarts.

Example 2:

Consider once more the component whose useful life in hours was described by the density function

$$f(h) = \begin{cases} h/8 & \text{if } 0 \le h \le 4 \\ 0 & \text{otherwise} \end{cases}$$

We found earlier that the mean life was $E(H) = \mu = 2.67$ hours. Applying Formula (8.17), the mean square is:

$$E(H^2) = \int_0^4 h^2 f(h)\, dh = \int_0^4 h^2 \frac{h}{8}\, dh$$
$$= \int_0^4 \frac{h^3}{8}\, dh = \frac{h^4}{32}\Big]_0^4 = 8$$

Then, from Formula (8.15), the variance of H is:

$$\sigma^2 = V(H) = E(H^2) - \mu^2 = 8 - (2.67^2) = .87$$

Thus, the variance of the lives of this particular type of component is .87 square hours.

Mathematical Properties of the Variance. In introducing the concept of the variance, we mentioned that the variance possesses certain mathematical properties that are particularly desirable for further analysis. We now present several of these properties as follows:

1. For any constant a,

$$V(a) = 0 \tag{8.18}$$

This simply says that the variance of any constant is zero. We may recall from Formula (8.7) that if a random variable can assume only a single value a, the expected value of the random variable is equal to a itself. Thus, it is obvious that the deviation of the value of the random variable from the mean must be zero. Therefore, the variance, which is the expectation of the squared deviation, must also be zero. This is reasonable since, if a random variable can take on only a single value, this random variable will have no variability.

2. For any random variable X, the variance is always non-negative. Algebraically,

$$V(X) \geq 0 \tag{8.19}$$

This property follows from the definition of the variance given in Formula (8.14). Generally $V(X)$ is greater than zero; $V(X)$ is equal to zero only in the special case in which the random variable can assume only a single value.

3. For any constant a and any random variable X,

$$V(a + X) = V(X) \tag{8.20}$$

In words, the variance of the sum of a constant and a random variable is simply equal to the variance of the random variable. This shows that adding a constant to every possible value of a random variable has no effect on the variance. That is, adding a constant merely shifts the location of the distribution without changing the dispersion of the distribution.

4. For any constant b and any random variable X,

$$V(bX) = b^2V(X) \tag{8.21}$$

If we multiply a random variable X by some constant b, then the variance of this product will be equal to b^2 times the variance of X. In other words, multiplying each possible value of a random variable by a constant multiplies the variance by the square of the constant.

5. By combining Formulas (8.20) and (8.21), we obtain:

$$V(a + bX) = b^2V(X) \tag{8.22}$$

If we multiply X by some constant b and then add another constant a, the variance of this quantity will be merely b^2 times $V(X)$. This reinforces the idea that adding a constant has no effect on the variance.

6. For any constant c and any random variable X whose mean is μ, it is always true that:

$$E[(X - \mu)^2] \leq E[(X - c)^2] \tag{8.23}$$

To clarify this statement, suppose that we choose some constant c and compute the expected squared deviation from c:

$$E[(X - c)^2]$$

Then, it can be shown mathematically that:

$$E[(X - c)^2] = E[(X - \mu)^2] + (\mu - c)^2$$

In the above expression, the first term on the right is simply the variance. Since the second term is a squared number, it is always non-negative. Thus, $E[(X - c)^2]$ must be greater than or equal to the variance. It can be equal to the variance only if $c = \mu$. In other words, the expected squared deviation from the mean (i.e., the variance) is less than the expected squared deviation from any other value. Hence, the expected squared deviation will have its minimum value when calculated from the mean. This characteristic is known as the "least squares" property of the mean.

The Standard Deviation. Due to its desirable mathematical properties, the variance is used extensively in statistical *analysis*. Difficulty arises, however, in *interpreting* the variance because, when we square the deviations, we also square the units of measurement in which those deviations are expressed. For instance, in our previous examples of the computation of the variance, the random variable Q was expressed in "quarts," whereas the variance was expressed in "square quarts." Similarly, the random variable H was expressed in "hours," whereas the variance was expressed in "square hours."

The difficulty arising from expressing the variance in square units of measurement may be resolved by taking the positive square root. This square root operation permits us to re-convert the square units of measurement into original units of measurement. The positive square root of the variance is known as the *standard deviation*, which is commonly denoted by σ. Thus,

$$\sigma = \sqrt{\sigma^2} = \sqrt{V(X)} \tag{8.24}$$

For our random variable Q, the standard deviation is:

$$\sigma = \sqrt{.70 \text{ square quarts}} = .84 \text{ quarts}$$

For the random variable H, the standard deviation is:

$$\sigma = \sqrt{.87 \text{ square hours}} = .93 \text{ hours}$$

8.3 CHEBYSHEV'S INEQUALITY

In this chapter, we have discussed various ways of measuring the central tendency and the variability of a probability distribution. Loosely speaking, when the standard deviation is relatively small, the possible values of the random variable are closely concentrated around the mean. When the standard deviation is relatively large, the possible values of the random variable are widely dispersed from the mean.

To be more specific, let us refer again to the two random variables X and Y shown in Figure 8.3. From our discussion of measures of dispersion, we have observed that:

1. X and Y have identical means—that is, $E(X) = E(Y) = 2$.
2. X has a smaller dispersion than Y. In terms of the variances, $V(X) = 1$, whereas $V(Y) = 3.3$. Therefore, $\sigma_X = 1$, whereas $\sigma_Y = 1.8$.

Now suppose that we wish to determine the probability that the random variable X will lie within one unit of its mean. Since $E(X)$ is equal to 2, this is equivalent to calculating $P(1 \leq X \leq 3)$. From the probability mass function of X, we obtain:

$$P(1 \leq X \leq 3) = \sum_{x=1}^{3} f(x) = .9$$

Similarly, we can determine the probability that the random variable Y will lie within one unit of its mean. Since $E(Y)$ also equals 2, this is equivalent to computing $P(1 \leq Y \leq 3)$. From the probability mass function of Y, we have:

$$P(1 \leq Y \leq 3) = \sum_{y=1}^{3} f(y) = .6$$

As we should have anticipated, $P(1 \leq X \leq 3)$ is *greater* than $P(1 \leq Y \leq 3)$, since X has a *smaller* standard deviation than Y. In other words, the probability that the random variable will lie *within* the interval between 1 and 3, inclusive, is greater for X than for Y. Therefore, the probability that the random variable will lie *outside* this interval (i.e., in the "tails" of the distribution) will be smaller for X than for Y.

In the above discussion, we were able to determine exact probabilities because we knew the probability functions of both X and Y. Suppose, however, that we knew the mean and standard deviation of some random variable but did not know its probability distribution. Then, could we make similar statements regarding the probabilities that a randomly observed value will lie within certain distances from the mean? The answer is that we can set limits on such probabilities with the aid of a mathematical theorem known as *Chebyshev's Inequality*. In one of its forms, Chebyshev's Inequality may be stated that, for any $k \geq 1$,

$$P(|X - \mu| \leq k\sigma) \geq 1 - \frac{1}{k^2} \tag{8.25}$$

Although this inequality may not look so easy at first glance, it simply says that the probability that a randomly observed value of a random variable will be within $k\sigma$ from the mean is at least $1 - 1/k^2$. In an alternate form, Chebyshev's Inequality may be stated as:

$$P(|X - \mu| > k\sigma) \leq \frac{1}{k^2} \tag{8.26}$$

In this form the inequality says that the probability that a randomly observed value of a random variable will be more than $k\sigma$ from the mean in either direction can be no greater than $1/k^2$. Stating the theorem in other words, for any given random variable X with mean μ and variance σ^2:

1. The *minimum* probability that the value of a random observation of X will lie *within* the range $\mu \pm k\sigma$ is $1 - 1/k^2$.

2. The *maximum* probability that the value of a random observation of X will lie *outside* the range $\mu \pm k\sigma$ is $1/k^2$.

Example:

A statistical abstract reports that, for the several thousand companies incorporated within a particular state, mean net income in 1971 was \$400,000, with a standard deviation of \$100,000. The abstract fails to indicate, however, how these corporate earnings are distributed. If one of these companies is selected at random, what is the minimum *probability that its 1971 net earnings were between \$200,000 and \$600,000? What is the* maximum *probability that its 1971 net earnings were either below \$100,000 or above \$700,000?*

If X represents net earnings of a corporation, μ = \$400,000 and σ = \$100,000. Thus, \$200,000 is 2σ below μ, and \$600,000 is 2σ above μ. From Formula (8.25), the probability that the net earnings of a randomly selected company will lie within this range is at least $(1 - 1/4) = .75$. Similarly, \$100,000 is 3σ below μ, and \$700,000 is 3σ above μ. From Formula (8.26), the probability that the net earnings of a randomly selected company will lie outside this range is at most $1/9 = .11$.

PROBLEMS

8.1 The probability mass function of a discrete random variable X is:

$$f(x) = \begin{cases} \dfrac{x^2 + 4}{50} & \text{if } x = 0, 1, 2, 3, 4 \\ 0 & \text{otherwise} \end{cases}$$

Determine the mean, median, and mode of X.

8.2 The probability mass function of a discrete random variable X is:

$$f(x) = \begin{cases} \dfrac{x + 2}{25} & \text{if } x = 1, 2, 3, 4, 5 \\ 0 & \text{otherwise} \end{cases}$$

Determine the mean, median, and mode of X.

8.3 The probability mass function of a discrete random variable X is:

$$f(x) = \begin{cases} \dfrac{x+1}{50} & \text{if } x = 2, 3, 4, 5, 6 \\ \dfrac{11-x}{20} & \text{if } x = 7, 8, 9, 10 \\ 0 & \text{otherwise} \end{cases}$$

Determine the mean, median, and mode of X.

8.4 According to the *Commissioners 1941 Standard Ordinary Mortality Table*, the probability that a person age 25 will live through the year is .997. An insurance company offers a 1-year term policy at age 25 in an amount of \$1,000 for a premium of \$8.00. Assuming this mortality experience fits the expected company experience, what should be the expected gain on a \$1,000 policy, not considering selling and administrative expenses?

8.5 The mortality tables used by a particular life insurance company show that there is a .00243 probability that a 20-year-old person will die within 1 year.
(a) What would be the company's expected payment on a \$10,000 one-year term policy issued to a 20-year-old person?
(b) How large a premium should the company charge per thousand dollars of insurance in order to gross \$5,000 above expected payments from sales on 1-year term policies to 20-year-olds totaling \$1,000,000 in face value?

8.6 The probability density function of a continuous random variable X is:

$$\begin{cases} x/2 & \text{if } 0 \le x \le 2 \\ 0 & \text{otherwise} \end{cases}$$

Determine the mean and median of X.

8.7 The probability density function of a continuous random variable X is:

$$f(x) = \begin{cases} 1/6 & \text{if } 2 \le x \le 8 \\ 0 & \text{otherwise} \end{cases}$$

Determine the mean and median of X.

8.8 The probability density function of a continuous random variable T is:

$$f(t) = \begin{cases} 3t(2-2t) & \text{if } 0 \le t \le 1 \\ 0 & \text{otherwise} \end{cases}$$

Determine the mean and mode of T.

8.9 The probability density function of a continuous random variable X is:

$$f(x) = \begin{cases} \dfrac{3x^2}{8} & \text{if } 0 \le x \le 2 \\ 0 & \text{otherwise} \end{cases}$$

Find the median value of X.

8.10 The probability density function of a continuous random variable G is:

$$f(g) = \begin{cases} 10g - 3g^2 - 7 & \text{if } 1 \le g \le 2 \\ 0 & \text{otherwise} \end{cases}$$

What is the mode of the distribution of G?

8.11 Let X be a random variable denoting the life span (in months) of a particular mechanical part. The probability density function of X is given by:

$$f(x) = \begin{cases} 0.4 & \text{if } 0 \leq x < 2 \\ a + bx & \text{if } 2 \leq x \leq 3 \\ 0 & \text{otherwise} \end{cases}$$

(a) Assuming that $f(3) = 0$, determine the values of a and b in the above function.
(b) Determine the probability that the life span of the part is at most 2.5 months.
(c) Determine the expected life span of the part.

8.12 The probability density function of a random variable W is:

$$f(w) = \begin{cases} 2e^{-2w} & \text{if } w \geq 0 \\ 0 & \text{if } w < 0 \end{cases}$$

Determine the expectation and median of W.

8.13 Let T be a random variable denoting the time (in minutes) that a telephone solicitor spends on a call. The probability density function of T is:

$$f(t) = \begin{cases} \dfrac{e^{-t/4}}{4} & \text{if } t \geq 0 \\ 0 & \text{otherwise} \end{cases}$$

(a) What is the mean length of a call by this solicitor?
(b) What is the probability that this solicitor will spend at least 2 minutes on a call?
(c) If the solicitor has already spent 2 minutes on a call, what is the probability that he will spend at least another 2 minutes on that call?

8.14 The random variable X represents the operating life (in days) of a particular type of satellite component. The probability density function of X is given by:

$$f(x) = \begin{cases} 0 & \text{if } x < 60 \\ \dfrac{60}{x^2} & \text{if } x \geq 60 \end{cases}$$

(a) What is the expected life of this type of component?
(b) If a satellite system contains 3 of these components, and if they operate independently, what is the probability that all 3 of these components will have operating lives that exceed 100 days?

8.15 Let X be a random variable denoting the number of automobile accidents on a certain stretch of highway in a 24-hour period. The probability mass function of X is given by:

$$f(x) = \begin{cases} .15 & \text{if } x = 0 \\ .05x^2 & \text{if } x = 1, 2 \\ .10(6 - x) & \text{if } x = 3, 4, 5 \\ 0 & \text{otherwise} \end{cases}$$

The daily cost of towing away the wrecks is given by:

$$C = \$10(10X - X^2)$$

(a) Find the probability that, during any given 24-hour period, there will be:
(i) More than 3 accidents on this stretch of highway.
(ii) At least 1 but less than 4 accidents on this stretch of highway.
(b) What is the expected daily cost of towing away the wrecks?

8.16 Let X represent a retailer's weekly demand for a low-volume, high-quality item. Experience has shown that the demand for this item is a random variable that follows the probability mass function:

$$f(x) = \begin{cases} .10(14 - x) & \text{if } x = 10, 11, 12, 13 \\ 0 & \text{otherwise} \end{cases}$$

The retailer's weekly profit on this item, as a function of demand, is given by:

$$Y = g(X) = \$100(X^2 - X)$$

(a) Determine the probability that the weekly demand is at most 11 units.
(b) Determine the expected weekly demand.
(c) Determine the expected weekly profit.

8.17 Let X be a random variable denoting the number of defective components in a manufacturing sub-assembly. The probability mass function of X is given by:

$$f(x) = \begin{cases} .15 & \text{if } x = 0 \\ .11x & \text{if } x = 1 \text{ or } 2 \\ .02x^2 & \text{if } x = 3 \text{ or } 4 \\ .02 & \text{if } x = 5 \end{cases}$$

The cost, in hundreds of dollars, of replacing the defective components is given by:

$$C = X^2 - 2X + 2$$

(a) What is the mean number of defective components in a sub-assembly?
(b) What is the expected cost of replacing the defective components?

8.18 Let X be a random variable representing a firm's monthly demand for an expensive hand-assembled device. Experience has shown that demand behaves according to the mass function:

$$f(x) = \begin{cases} \dfrac{x - 2}{10} & \text{if } x = 3, 4, 5, 6 \\ 0 & \text{otherwise} \end{cases}$$

(a) Determine expected monthly demand.
(b) Determine expected monthly profit if the firm's monthly profit, as a function of demand, is given by:

$$Y = \$1{,}000(X^2 - X + 6)$$

8.19 Let X be a random variable with the following probability density function:

$$f(x) = \begin{cases} x/8 & \text{if } 3 \le x \le 5 \\ 0 & \text{otherwise} \end{cases}$$

(a) Determine the expectation of X.
(b) If $Y = 4(\sqrt{X})$, determine the expectation of Y.

8.20 Let X be a random variable denoting monthly demand (in tons) for a chemical product. The probability density function of X is given by:

$$f(x) = \begin{cases} .02x & \text{if } 0 \leq x \leq 10 \\ 0 & \text{otherwise} \end{cases}$$

Assume monthly profit (in hundreds of dollars), as a function of demand, is given by:

$$Y = g(X) = X^2 + X - 20.$$

What is the expected monthly profit?

8.21 Let T be a random variable denoting the time (in hours) required to produce a hand-assembled article. The probability density function of t is given by:

$$f(t) = \begin{cases} 2t - 4 & \text{if } 2 \leq t \leq 3 \\ 0 & \text{otherwise} \end{cases}$$

The profit (in dollars) that the manufacturer makes on an article is given by:

$$Y = g(T) = 8 - T$$

(a) Determine the expected time to produce an article.
(b) Determine the expected profit per article.
(c) Determine the probability that the profit on an individual article will be greater than \$5.50.

8.22 The profit realized from the yearly yield of a large alfalfa ranch is a function of annual rainfall. For the region in which the ranch is located, the probability density function for annual rainfall (in inches) R is given by:

$$f(r) = \begin{cases} \dfrac{e^{-r/5}}{5} & \text{if } r \geq 0 \\ 0 & \text{otherwise} \end{cases}$$

Profit (in thousands of dollars) is given by:

$$Y = g(R) = 20(1 - 3e^{-R})$$

(a) What is the average yearly rainfall in this region?
(b) What is the ranch's expected annual profit?
(c) What amount of annual rainfall is needed for the ranch to break even?

8.23 Let X be a random variable denoting the time (in hours) required to produce a delicate device. The probability density function of X is given by:

$$f(x) = \begin{cases} x/50 & \text{if } 0 \leq x \leq 10 \\ 0 & \text{otherwise} \end{cases}$$

The profit that the manufacturer makes on a unit depends on the time required to produce it, as shown by the following table:

Production Time	*Profit*
Less than 3 hours	\$300
Between 3 and 8 hours	\$200
Over 8 hours	\$100

(a) What is the expected number of hours required to produce a device?
(b) What is the manufacturer's expected profit per unit?

8.24 Determine the expected absolute deviation, variance, and standard deviation of the random variable in Problem 8.1.

8.25 With reference to the random variable described in Problem 8.2:

(a) Determine the expected absolute deviation, variance, and standard deviation.

(b) Demonstrate that the expected algebraic deviation is equal to zero.

8.26 For the random variable in Problem 8.3, determine the expected absolute deviation, variance, and standard deviation.

8.27 Let X be a random variable representing a worker's production of an expensive hand-assembled device. Experience has shown that X has the following probability mass function:

$$f(x) = \begin{cases} \dfrac{x-2}{15} & \text{if } x = 3, 4, 5, 6, 7 \\ 0 & \text{otherwise} \end{cases}$$

(a) Present the probability mass function and the cumulative mass function in the form of a table.

(b) Calculate the expected production quantity, $E(X)$.

(c) Calculate the variance and standard deviation of X.

8.28 For the random variable in Problem 8.6, determine the expected absolute deviation, variance, and standard deviation.

8.29 For the random variable in Problem 8.7, determine the variance and standard deviation.

8.30 For the random variable in Problem 8.8, determine the variance and standard deviation.

8.31 For the random variable in Problem 8.12, determine the expected absolute deviation, variance, and standard deviation.

8.32 The probability density function of a random variable S is:

$$f(s) = \begin{cases} 3s^2 & 0 \le s \le 1 \\ 0 & \text{otherwise} \end{cases}$$

(a) Determine the expected absolute deviation, variance, and standard deviation of S.

(b) Demonstrate that the expected algebraic deviation of S is equal to zero.

8.33 Let X be a continuous random variable with the probability density function:

$$f(x) = \begin{cases} x/8 & \text{if } 0 \le x \le 4 \\ 0 & \text{otherwise} \end{cases}$$

(a) Determine and state completely the corresponding cumulative density function.

(b) Determine the expected value of X.

(c) Determine the variance and standard deviation of X.

(d) Using the cumulative density function, determine $P(2 < X < 3)$.

8.34 Let X be a random variable denoting the life span (in hours) of a very delicate type of laboratory instrument. The distribution of X is described by the following density function:

$$f(x) = \begin{cases} 1/4 & \text{if } .5 \le x \le 4.5 \\ 0 & \text{otherwise} \end{cases}$$

(a) Graph the density function.
(b) Determine the mean and variance of X.

8.35 Use Formula (8.14) to compute the variance of the random variable Y whose probability distribution is shown in Figure 8.3b.

8.36 An electric utility company has 200,000 domestic subscribers. For the month of October, the mean electric bill for these 200,000 subscribers was $\mu = \$8.50$, with a standard deviation $\sigma = \$1.25$. No additional information is available. What is the minimum number of these 200,000 bills that would lie between \$6.00 and \$11.00?

8.37 A census of the households in a large city showed that the mean annual household income was $\mu = \$8{,}500$, with a standard deviation of $\sigma = \$2{,}000$. No additional information concerning the income distribution is available.
(a) What is the maximum probability that the annual income of a randomly selected household would be either less than \$3,500 or greater than \$13,500?
(b) Determine a range of household incomes (with μ lying at the center of the range) that will guarantee that at least 75 per cent of the households will have incomes within that range.

8.38 A seamstress hand-makes jabots that she sells to a ready-to-wear manufacturer in one-dozen lots. Since she sometimes wastes material when she works, her profit on a lot is a function of the total amount of material consumed in producing the lot. Her brother-in-law, who is her accountant, has determined that her profit (in dollars) for a lot is given by:

$$Z = 200 - 12\,Y$$

where Y is a random variable denoting the amount of material (in yards) consumed in producing the lot. The seamstress' next-door neighbor, who is a statistician, has determined that the probability density function for Y may be given as:

$$f(y) = \begin{cases} (y - 12)/4 & \text{if } 12 \le y < 14 \\ (16 - y)/4 & \text{if } 14 \le y \le 16 \\ 0 & \text{otherwise} \end{cases}$$

(a) Determine the mean and variance of Y.
(b) Using Formula (8.6), determine expected profit per lot.
(c) Apply Formula (8.10) to determine the expected profit, then compare this result with that in (b).
(d) Determine the variance of profit per lot.

8.39 A pharmaceutical firm has developed 2 different processes for extracting gamma globulin from whole blood. Let X be the yield (in cubic centimeters) of this substance from 1 pint of whole blood. With Process A, the probability density function of X is:

$$f(x) = \begin{cases} 1/10 & 40 \le x \le 50 \\ 0 & \text{otherwise} \end{cases}$$

With Process B, the probability density function of X is:

$$f(x) = \begin{cases} (70 - x)/400 & \text{if } 40 \le x \le 60 \\ 0 & \text{otherwise} \end{cases}$$

(a) For which process is mean yield the greater?
(b) Which process has the greater probability of a yield that is less than 45 cc?

(c) In terms of standard deviations, which process is the more variable?
(d) Which process has the greater probability of a yield that is greater than 46 cc?

8.40 Show that Formulas (8.8) through (8.10) hold for either a discrete or a continuous random variable.

8.41 Prove that Formula (8.15) is identical to Formula (8.14).
Hint: Expand $(X - \mu)^2$ and then use the mathematical properties in Section 8.1.5.

8.42 Show that Formula (8.19) holds for either a discrete or a continuous random variable.

8.43 Use the mathematical properties of expectations (Section 8.1.5) to prove Formulas (8.20) through (8.22).

8.44 For any constant c and any random variable X with mean μ and variance σ^2, show that:

$$E[(X - c)^2] = \sigma^2 + (\mu - c)^2$$

Hint: $E[(X - c)^2]$ can be written as: $E[(X - \mu + \mu - c)^2]$.

Chapter 9

Commonly Used Discrete Probability Models

In our discussion of random variables and probability functions in preceding chapters, we dealt with a variety of specific probability functions. These functions were presented to illustrate the basic concepts and methods introduced in those chapters. A key question now arises: Is it necessary, each time we encounter a new problem situation, to originate a special, tailor-made mass or density function for the particular situation at hand? As might be hoped, the answer to this question is, "No"!

In applying probability theory to decision problems, we often find that many problems share certain essential characteristics, even though each appears superficially to be different. By specifying a set of essential characteristics that are common to many problems, it may be possible to formulate a general probability function that is applicable to all problems having these common characteristics. Since such a function is general in its applicability, it is often called a *probability model.* In other words, a probability model is a general function that serves as a pattern for analyzing a "family" of probability problems. Over the years, many probability functions have been developed and found useful as models in analyzing statistical decision problems.

From earlier chapters, the reader may recall that, in statistical decision making, sampling from a population is an important device for obtaining information to reduce uncertainty. As the reader will see shortly, there are certain probability models that are particularly useful in applying the sampling process to obtain such information. In order to apply probability models to the sampling process, it is essential to use *probability sampling*.

Probability sampling is an approach whereby, before the sample is selected, each possible sample that might be drawn has a determinable probability of being the sample that actually will be selected. For example, if we were concerned with a population consisting of 500 households, and if we decided to draw a sample of 50 households from this population, there would be a total of $\binom{500}{50}$ different possible samples that might be selected. If we adopt some well-defined sampling procedure that permits us to specify the probability that each of these possible samples actually will be the one selected, then our sample will be a probability sample.

The most common type of probability sample is a *simple random sample.* A simple random sample is a sample that is selected in such a way that each of the possible samples that might be drawn has an equal chance of being the sample that actually is drawn. In practice, this is achieved by adopting some sampling procedure such that, at any point in the process of drawing cases from the population into the sample, each element that is available to be drawn has an equal probability of being the next one drawn. In applying probability models to sampling, we shall limit our discussion to situations that employ simple random sampling.

In this chapter we will examine several commonly used probability models for discrete random variables. In the next chapter, we will turn our attention to continuous probability models.

9.1 THE HYPERGEOMETRIC MODEL

Many decision problems are concerned with *finite* populations that are *dichotomous*—that is, populations consisting of a definite number of units or elements that may be divided into *two mutually exclusive categories*, such as favorable and unfavorable, defective and non-defective, successes and failures, profits and losses, and so on. Examples of finite dichotomous populations are:

1. A carton of one dozen eggs, of which ten are good and two are rotten.
2. A bin of 1,000 bolts, of which 800 have right-hand threads and 200 have left-hand threads.
3. A city of 10,000 households, of which 4,000 subscribe to a daily newspaper and 6,000 do not subscribe to a daily newspaper.
4. A company of 2,000 employees, of which 500 are white-collar and 1,500 are blue-collar.
5. A trade association of 600 merchants, of whom 500 favor and 100 do not favor publication of a new trade journal.

In dealing with such populations, we often do not know the actual number of elements in each category, so that uncertainty exists concerning the composition of the population. One way to reduce this uncertainty is to take a sample from the population. Such a sampling generally would be conducted *without replacement*—that is, once an element has been observed, it is removed from the population so that it will not be available to be observed a second time. In dealing with this type of sampling situation, a frequently used probability model is the *hypergeometric* distribution.

In actual practice, the reason for sampling is that we do not know the nature of the population. However, to understand how the hypergeometric model may be applied to sampling from a dichotomous finite population, it is helpful at this point to assume that we do know the composition of the population from which we are sampling.

Suppose that we have a box containing ten transistors, of which four are defective and six are good. We may regard this set of ten transistors as a finite, dichotomous population. Suppose that we take a random sample of three transistors from the box without replacing them. Let X represent the number of defective transistors in the sample. Then X is a random variable with the possible values $X = 0, 1, 2, 3$. Under these conditions, we might ask: "What is $P(X = 0)$? $P(X = 1)$? $P(X = 2)$? $P(X = 3)$?"

The probability that there will be zero defectives in the sample is the probability that there will be three good items in the sample. Let d represent a defective item and g represent a good item. If the sample items are selected one at a time in sequence, the probability that $X = 0$ is the probability of obtaining the sequence *ggg*. Similarly, the probability that $X = 1$ is the probability of obtaining any of the sequences *dgg*, *gdg*, or *ggd*. Likewise, the probability that $X = 2$ is the probability of obtaining any of the sequences *ddg*, *dgd*, or *gdd*. Finally, the probability that $X = 3$ is the probability of obtaining the sequence *ddd*. Since sampling is being done without replacement, the successive observations are *not* independent. That is, at any point in the sequence, the probability of selecting a defective item depends on the outcome of the selections at previous points in the sequence. For example, at the moment of selecting the first item, the probability of selecting a defective is 4/10. Once the first item is selected, only nine items remain in the population. If the first item selected were defective, the probability that the second item selected would be defective is 3/9; but if the first item were good, the probability that the second item would be defective is 4/9. Similarly, if the

Table 9.1

Probability Distribution of Number of Defective Transistors in a Sample Drawn from a Finite Population

Number of Defectives in the Sample (x)	*Corresponding Sequences*	*Probabilities*		
0	*ggg*	(6/10)(5/9)(4/8) = 120/720		= .167
1	*dgg*	(4/10)(6/9)(5/8) = 120/720	= 360/720	= .500
	gdg	(6/10)(4/9)(5/8) = 120/720		
	ggd	(6/10)(5/9)(4/8) = 120/720		
2	*ddg*	(4/10)(3/9)(6/8) = 72/720	= 216/720	= .300
	dgd	(4/10)(6/9)(3/8) = 72/720		
	gdd	(6/10)(4/9)(3/8) = 72/720		
3	*ddd*	(4/10)(3/9)(2/8) = 24/720		= .033
				1.000

first two items were both defective, the probability that the third item also would be defective is 2/8. Then the joint probability that all three items will be defective is:

$$P(X = 3) = (4/10)(3/9)(2/8) = 24/720 = .033.$$

Thus, the probabilities that we seek may be viewed as the products of sequences of conditional probabilities. Using this approach, computation of the probabilities for all the possible values of X is shown in Table 9.1. This table represents the probability distribution for the number of defective transistors in the random sample of three transistors.

9.1.1 The Hypergeometric Mass Function

Obviously, with larger populations and sample sizes, the approach that we have just taken would be impractical because of the tedious listing of sequences that would be involved. What is needed, then, is a probability model from which the desired probabilities may be determined in a more direct fashion. The model that meets our need is the hypergeometric distribution. To formalize our presentation of this model, consider a sampling situation in which:

1. The population being sampled is
   (a) Finite, in the sense that it consists of a fixed number of elements
   (b) Dichotomous, in the sense that it is composed of two categories, such as "successes" and "failures."
2. A simple random sample is selected without replacement from the population.
3. The random variable of interest is the number of "successes" observed in the sample.

Suppose that we let:

N = total number of elements in a finite dichotomous population
n = size of random sample selected from the population
D = number of "successes" in the population
X = number of "successes" that might be observed in the sample

Then the probability mass function of the random variable X is:

$$f_h(x|N,n,D) = \begin{cases} \dfrac{\dbinom{D}{x}\dbinom{N-D}{n-x}}{\dbinom{N}{n}} & \text{if } x = 0, 1, \ldots, k \\ 0 & \text{otherwise} \end{cases} \tag{9.1}$$

In this formula, k is equal to either n or D, whichever is smaller. N, n, and D are referred to as the *parameters* of the model. As a probability mass function, $f_h(x|N,n,D)$ yields $P(X = x|N,n,D)$.

To illustrate Formula (9.1), consider again our transistor example. For this example, $N = 10$ (the total number of elements in the population), $D = 4$ (the number of defective elements in the population), $n = 3$ (the total number of elements in the sample), and X is a random variable denoting the number of defective elements in the sample. Then, using Formula (9.1), we may compute $P(X = 2)$:

$$f_h(2|10,3,4) = \frac{\binom{4}{2}\binom{6}{1}}{\binom{10}{3}} = \frac{\frac{4!}{2!\,2!}\,\frac{6!}{1!\,5!}}{\frac{10!}{3!\,7!}}$$

Simplifying this expression, we obtain $f_h(2) = .300$, which is the same value of $f(2)$ that we originally obtained in Table 9.1. The reader is urged to apply this same procedure to verify the other values in Table 9.1.

In computing hypergeometric probabilities, calculation of factorials can become cumbersome. However, the use of tables of factorials, such as Appendix Table A, can lighten the computational burden. Of course, if parameter values become extremely large, beyond the range of tabulated factorials, hypergeometric calculations become the province of the computer. To relieve the practitioner from the necessity of making any calculations, tabulated values of the hypergeometric function are available.[1]

9.1.2 The Hypergeometric Cumulative Mass Function

The hypergeometric cumulative mass function is given by:

$$F_h(x|N,n,D) = \sum_{t=0}^{x} \frac{\binom{D}{t}\binom{N-D}{n-t}}{\binom{N}{n}} \tag{9.2}$$

Formula (9.2) simply specifies a summation of hypergeometric probabilities. In this formula, t is a dummy variable representing possible values of X,

[1] G. J. Lieberman and D. B. Owen, *Tables of the Hypergeometric Probability Distribution*, Stanford University Press, 1961. These provide individual and cumulative probabilities for $N = 2, 3, 4, \ldots, 50, 60, 70, 80, 90, 100$, as well as for selected values of N between 100 and 1,000. An appendix provides fifteen-place logarithms for $1!, 2!, \ldots, 2000!$.

and x represents the upper limit of the summation. Applying this formula to our transistor example, we have:

$$P(X \leq 0) = F_h(0) = \sum_{t=0}^{0} \frac{\binom{4}{t}\binom{6}{3-t}}{\binom{10}{3}}$$

$$= f_h(0) = .167$$

$$P(X \leq 1) = F_h(1) = \sum_{t=0}^{1} \frac{\binom{4}{t}\binom{6}{3-t}}{\binom{10}{3}}$$

$$= f_h(0) + f_h(1) = .167 + .500 = .667$$

$$P(X \leq 2) = F_h(2) = \sum_{t=0}^{2} \frac{\binom{4}{t}\binom{6}{3-t}}{\binom{10}{3}}$$

$$= f_h(0) + f_h(1) + f_h(2) = .167 + .500 + .300 = .967$$

$$P(X \leq 3) = F_h(3) = \sum_{t=0}^{3} \frac{\binom{4}{t}\binom{6}{3-t}}{\binom{10}{3}}$$

$$= f_h(0) + f_h(1) + f_h(2) + f_h(3) = .167 + .500 + .300 + .033$$

$$= 1.00$$

The above cumulative probabilities obtained from Formula (9.2), together with the individual probabilities obtained previously from Formula (9.1), are displayed in Table 9.2.

Table 9.2

Hypergeometric Probability Distribution and Cumulative Distribution ($N = 10$, $n = 3$, $D = 4$)

Number of Defectives in the Sample (x)	*Individual Probability* $f_h(x)$	*Cumulative Probability* $F_h(x)$
0	.167	.167
1	.500	.667
2	.300	.967
3	.033	1.000

To understand the probability mass function and the cumulative mass function in Table 9.2, consider a population of ten elements of which four are defective and six are not. Suppose that three elements are drawn from this population without being replaced. Then, using Table 9.2, we may answer the following questions:

1. What is the probability that the sample will contain exactly two defectives?

$$P(X = 2) = f_h(2) = .300$$

2. What is the probability that the sample will contain one defective or less?

$$P(X \leq 1) = F_h(1) = .667$$

3. What is the probability that the sample will contain less than three defectives?

$$P(X < 3) = P(X \leq 2) = F_h(2) = .967$$

4. What is the probability that the sample will contain more than two defectives?

$$P(X > 2) = 1 - P(X \leq 2) = 1 - F_h(2) = 1 - .967 = .033$$

5. What is the probability that the sample will contain at least two defectives?

$$\begin{aligned} P(X \geq 2) &= 1 - P(X < 2) = 1 - P(X \leq 1) \\ &= 1 - F_h(1) = 1 - .667 = .333 \end{aligned}$$

9.1.3 Mean and Variance of the Hypergeometric Distribution

From the figures in Table 9.2, we may compute the mean and standard deviation of our hypergeometric distribution. To compute the mean of the distribution, we apply Formula (8.3) to obtain:

$$E(X) = 0(.167) + 1(.500) + 2(.300) + 3(.033) = 1.2$$

Similarly, for the variance of the distribution, we apply Formula (8.15) to obtain:

$$V(X) = [(.167)(0^2) + (.500)(1^2) + (.300)(2^2) + (.033)(3^2)] - (1.2)^2 = .56$$

Thus the standard deviation of the distribution is $\sqrt{.56} = .75$ defective transistors per sample.

By substituting the hypergeometric mass function for $f(x)$ in Formula (8.3), it can be shown that the mean of the hypergeometric distribution is:

$$\mu = E(X) = n\left(\frac{D}{N}\right) \tag{9.3}$$

This makes it possible to compute the mean of a hypergeometric distribution directly from the parameters of the distribution, without having to make the laborious calculations required when Formula (8.3) is applied directly. Applying Formula (9.3) to the transistor example, we obtain $\mu = E(X) = 3\left(\frac{4}{10}\right) = 1.2$, which confirms the figure obtained previously through Formula (8.3).

Similarly, by using Formula (8.15), the variance of the hypergeometric distribution may be shown to be:

$$\sigma^2 = V(X) = n\left(\frac{D}{N}\right)\left(\frac{N-D}{N}\right)\left(\frac{N-n}{N-1}\right) \tag{9.4}$$

For the transistor example, this formula yields:

$$\sigma^2 = 3\left(\frac{4}{10}\right)\left(\frac{10-4}{10}\right)\left(\frac{10-3}{10-1}\right) = .56$$

which agrees with the figure obtained by using Formula (8.15) directly.

9.2 MODELS BASED ON THE BERNOULLI PROCESS

Our discussion of the hypergeometric model was concerned essentially with dichotomous, *finite* populations. Across the spectrum of business activities, however, we encounter types of decision problems that are concerned with dichotomous populations that, rather than being finite, are either *infinite* or so large that they may be considered effectively infinite.

Consider a dichotomous population composed of elements comprising two mutually exclusive categories that we will identify as "successes" or "failures." Assume that this population is essentially infinite or inexhaustible. Let us propose an experiment in which a random sample of n elements will be drawn from the population, and each element will be observed to determine whether it is a success or a failure. Each of these separate draws will be called a "trial" of the experiment. We may refer to the observation resulting from any single trial—i.e., whether the observed element is a success or a failure—as the "outcome" of the trial. If we let p represent the probability that an observation on a single trial will be a success, then $(1 - p)$ represents the probability that an observation will be a failure. The experiment will be conducted in such a manner that the successive trials are independent—that is, the results of previous trials have no effect on the probability that the next trial will result in a success or a failure. Furthermore, we will require that the value of p remain constant for all trials.

The reader will recognize that our proposed experiment specifies a random process conducted under very particular conditions. A random sampling experiment conducted under such conditions is an example of a *Bernoulli process* (named after Jacques Bernoulli, 1654-1705). A Bernoulli process is a random process in which:

1. Each trial results in one of just two possible outcomes, such as "success" or "failure," "yes" or "no," "good" or "defective," "hit" or "miss," "pass" or "fail," and so on.
2. Each trial is independent of all other trials. For example, on any given trial, the probability of a success would be the same regardless of whether the previous trial resulted in a success or a failure.
3. The probability, p, of the occurrence of an outcome, such as a success, remains constant from trial to trial.

Taking a random sample of n observations from a dichotomous population that is infinite, or effectively infinite, is tantamount to observing n trials of a Bernoulli process.

In practice, many common sampling procedures may be conceptualized essentially as Bernoulli processes. For example, in developing a new missile system, a manufacturer might be interested in test-firing a sample of missiles to estimate the probability that a firing will result in a "hit" on the target under particular conditions. In such a case he would be concerned with an infinite, dichotomous population composed of all possible similar firings that could be classified as hits and misses. If the test firings are conducted in such a way that the probability of a hit is the same for each firing, and if the firings may be considered to represent independent trials, then these trials may be regarded as Bernoulli trials.

As another example, consider a particular state that has several million registered voters. A political analyst might wish, on the basis of interviewing a random sample of 500 of these voters, to estimate the proportions of the total who "favor" or "disfavor" a referendum to be submitted at an upcoming election. In this instance, although the population is technically finite, it is so large (particularly in relation to the sample size) that it may be considered effectively infinite for all practical purposes. We may regard this sampling experiment as a Bernoulli process if we are willing to assume that (1) during the course of conducting the survey, the proportion of voters who favor the referendum remains essentially stable, and (2) the survey is conducted in such a manner that the results of the individual interviews may be regarded as essentially independent.

9.2.1 The Binomial Model

Suppose that we were to conduct a Bernoulli sampling experiment consisting of n trials. We might then focus our attention on the number, X, of successes that might be observed in the n trials. Then X would be a discrete random variable with the possible values $0, 1, 2, \ldots, n$. For example, suppose that a manufacturer produces items by means of an automatic production process that operates as a Bernoulli process with a .20 defective rate. That is, at any given moment, there is a constant probability of .20 that the next item that is produced will be defective. Furthermore, the defective and good items occur in random sequence such that the probability that the next item will be defective is independent of any preceding sequence of good and defective items. At any given moment we might propose to inspect a sample consisting of the next four items produced by the process, and to count the number of defective items in the sample of four items. In this case, if X represents the number of defective items to be found in the sample, the possible values of X would be 0, 1, 2, 3, 4. Under these conditions we might ask, "What is $P(X = 0)$? $P(X = 1)$? $P(X = 2)$? $P(X = 3)$? $P(X = 4)$?"

If we let d represent a defective item and g represent a good item, the probability that $X = 0$ is the probability of obtaining the sequence *gggg*. If the probability of obtaining a d at any point in the sequence is .20, then the probability of obtaining a g at any point in the sequence is .80. Furthermore, the successive events in the sequence are independent. Thus, the probability of the sequence *gggg* is $(.80(.80)(.80)(.80) = (.80)^4 = .4096$, which is $P(X = 0)$.

Table 9.3

Probability Distribution of Number of Defectives In a Bernoulli Sample ($n = 4, p = .20$)

Number of Defectives in the Sample (x)	*Possible Sequences*	*Probabilities*
0	*gggg*	$(.80)^4 = .4096$
1	*dggg, gdgg, ggdg, gggd*	$4(.20)(.80)^3 = .4096$
2	*ddgg, dgdg, dggd, gddg, gdgd, ggdd*	$6(.20)^2(.80)^2 = .1536$
3	*dddg, ddgd, dgdd, gddd*	$4(.20)^3(.80) = .0256$
4	*dddd*	$(.20)^4 = .0016$
		1.0000

The probability that $X = 1$ is the probability of obtaining any of the sequences *dggg, gdgg, ggdg, gggd*. Because of the independence of the elements in the sequence, each of these sequences (composed of one d and three g's in some order) has the same probability; namely $(.20)(.80)^3 = .1024$. Since there are four possible sequences, each with a probability of .1024, we have $4(.1024) = .4096$, which is $P(X = 1)$. Probabilities for the other possible values of X are obtained in a similar manner. The computations are summarized in Table 9.3.

The Binomial Mass Function. By abstracting the procedure employed in Table 9.3, we may arrive at a probability function that will serve as a general model for specifying the probability for the number of successes in a given number of Bernoulli trials. For this purpose, let:

p = the probability of a success on a single Bernoulli trial
n = the number of Bernoulli trials
X = the number of successes in the n trials

Then, the probability mass function of the random variable X is:

$$f_b(x|n,p) = \begin{cases} \binom{n}{x} p^x(1-p)^{n-x} & \text{if } x = 0, 1, \ldots, n \\ 0 & \text{otherwise} \end{cases} \tag{9.5}$$

where $0 < p < 1$ and $n = 1, 2, \ldots$.

This formula for the probability distribution of the number of successes in a series of Bernoulli trials is called the *binomial* probability distribution.

Formula (9.5) gives the probability of obtaining x successes and $(n - x)$ failures in n Bernoulli trials. To understand this formula, we may first observe that the term $p^x(1 - p)^{n-x}$ represents the probability of any single sequence of x successes and $(n - x)$ failures. Thus, to obtain the total probability of x successes and $(n - x)$ failures for all possible sequences, we must multiply the probability of a single sequence by the total number of equiprobable sequences. The reader may recognize that the number of possible sequences is equal to the number of distinct permutations of two kinds of objects when there are x objects of one kind and $(n - x)$ objects of the other kind. From Formula (3.4), this number of permutations is equal to $\frac{n!}{x!(n-x)!}$. As pointed out in Section 3.4, this expression is identical to the number of combinations of x objects selected from a set of n objects. Due to this mathematical identity, it is common practice, for notational convenience, to use the term $\binom{n}{x}$ to specify the operation $\frac{n!}{x!(n-x)!}$ in the binomial distribution. The term $\binom{n}{x}$ sometimes is referred to as the *binomial coefficient.* Binomial coefficients for selected values of n and x are tabulated in Appendix Table B.

Applying the binomial model to our example of sampling four items from a Bernoulli process with a .20 defective rate, we would let $p = .20$ and $n = 4$. The random variable X (number of defectives in the sample) has the possible values: 0, 1, 2, 3, 4. As given by Formula (9.5), the probabilities associated with each of these possible values are:

$$P(X = 0) = f_b(0|4,.20) = \binom{4}{0}(.20)^0(.80)^4 = (1)(1)(.4096) = .4096$$

$$P(X = 1) = f_b(1|4,.20) = \binom{4}{1}(.20)^1(.80)^3 = (4)(.20)(.512) = .4096$$

$$P(X = 2) = f_b(2|4,.20) = \binom{4}{2}(.20)^2(.80)^2 = (6)(.04)(.64) = .1536$$

$$P(X = 3) = f_b(3|4,.20) = \binom{4}{3}(.20)^3(.80)^1 = (4)(.008)(.80) = .0256$$

$$P(X = 4) = f_b(4|4,.20) = \binom{4}{4}(.20)^4(.80)^0 = (1)(.0016)(1) = .0016$$

These are the same results we obtained in Table 9.3.

The Binomial Cumulative Mass Function. The binomial cumulative mass function may be stated by:

$$F_b(x|n,p) = \sum_{t=0}^{x} \binom{n}{t} p^t(1 - p)^{n-t} \tag{9.6}$$

Formula (9.6) simply specifies a summation of individual binomial probabilities. In this summation, the dummy variable t represents possible values of X, taken in ascending order, and x represents the upper limit of the summation. Applying this formula to our example of a Bernoulli sample, with $n = 4$ and $p = .20$, we obtain:

$$P(X \le 0) = F_b(0) = \sum_{t=0}^{0} \binom{4}{t} (.20)^t(.80)^{4-t}$$

$$= f_b(0) = .4096$$

$$P(X \le 1) = F_b(1) = \sum_{t=0}^{1} \binom{4}{t} (.20)^t(.80)^{4-t}$$

$$= f_b(0) + f_b(1) = .4096 + .4096 = .8192$$

$$P(X \le 2) = F_b(2) = \sum_{t=0}^{2} \binom{4}{t} (.20)^t(.80)^{4-t}$$

$$= f_b(0) + f_b(1) + f_b(2) = .4096 + .4096 + .1536 = .9728$$

$$P(X \le 3) = F_b(3) = \sum_{t=0}^{3} \binom{4}{t} (.20)^t(.80)^{4-t}$$

$$= f_b(0) + f_b(1) + f_b(2) + f_b(3)$$

$$= .4096 + .4096 + .1536 + .0256 = .9984$$

$$P(X \le 4) = F_b(4) = \sum_{t=0}^{4} \binom{4}{t} (.20)^t(.80)^{4-t}$$

$$= f_b(0) + f_b(1) + f_b(2) + f_b(3) + f_b(4)$$

$$= .4096 + .4096 + .1536 + .0256 + .0016 = 1.000$$

Table (9.4) summarizes the results obtained from Formula (9.5) and Formula (9.6).

Table 9.4

Binomial Probability Distribution and Cumulative Distribution for Number of Defectives in a Bernoulli Sample ($n = 4$, $p = .20$)

Number of Defectives in the Sample (x)	*Individual Probability* $f_b(x)$	*Cumulative Probability* $F_b(x)$
0	.4096	.4096
1	.4096	.8192
2	.1536	.9728
3	.0256	.9984
4	.0016	1.0000

Assume that, with respect to quality of output (good or defective), a particular production process produces items like a Bernoulli process with 20 per cent defectives. Suppose that observations are made of a sample consisting of four consecutive items produced by the process. Then we may use Table 9.4 to answer the following questions:

1. What is the probability that the sample will contain exactly two defectives?

$$P(X = 2) = f_b(2) = .1536$$

2. What is the probability that the sample will contain one defective or less?

$$P(X \le 1) = F_b(1) = .8192$$

3. What is the probability that the sample will contain less than four defectives?

$$P(X < 4) = P(X \leq 3) = F_b(3) = .9984$$

4. What is the probability that the sample will contain more than two defectives?

$$P(X > 2) = 1 - P(X \leq 2) = 1 - F_b(2) = 1 - .9728 = .0272$$

5. What is the probability that the sample will contain at least two defectives?

$$\begin{aligned} P(X \geq 2) &= 1 - P(X < 2) = 1 - P(X \leq 1) \\ &= 1 - F_b(1) \\ &= 1 - .8192 = .1808. \end{aligned}$$

The Use of Binomial Tables. Although Formulas (9.5) and (9.6) provide mathematical definitions of the binomial mass function and binomial cumulative mass function, they are seldom required for actual computational purposes, since binomial tables are readily available.[2] Appendix Tables D and E contain values for the binomial mass function and cumulative mass function. The following examples will illustrate the use of these tables.

Example 1:

Suppose that one month before a senatorial election in a large state, 50 per cent of the voters are still undecided concerning which candidate they favor. From the roster of registered voters, a pollster selects a random sample of fifty voters to be interviewed.

1. *What is the probability that, of the fifty voters in the sample, exactly twenty will be undecided? For all practical purposes, we are dealing with an effectively infinite, dichotomous population composed of "decided" and "undecided" voters. Thus, drawing a random sample of fifty cases from this population may be conceived essentially as a Bernoulli process. Therefore, the desired probability,* $P(X = 20|n = 50,\ p = .50)$, *is given by the binomial expression:*

$$f_b(20|50, .50) = \binom{50}{20} (.50)^{20}(.50)^{30}$$

[2] Two of the most extensive volumes of binomial tables are:

1. National Bureau of Standards, *Tables of the Binomial Probability Distribution.* Washington, D.C., U.S. Government Printing Office, 1949 (corrected reprint, 1952). These tables provide binomial probabilities for $n = 2, 3, \ldots, 49$ and for $p = .01, .02, \ldots, .50$.
2. Harry G. Romig, *50-100 Binomial Tables.* New York, John Wiley and Sons, 1953. These tables provide binomial probabilities for $n = 50, 51, \ldots, 100$, and for p ranging from .001 to .90.

Rather than carrying out the computations specified by the above expression, we may obtain the desired probability directly from Appendix Table D. Entering this table with $n = 50$, $p = .50$, *and* $x = 20$, *we find the answer equal to .0419.*

2. *What is the probability that the sample of fifty voters will contain twenty or less undecided? In this case, the desired probability,* $P(X \leq 20|n = 50, p = .50)$, *is given by the binomial cumulative expression:*

$$F_b(20|50, .50) = \sum_{t=0}^{20} \binom{50}{t} (.50)^t(.50)^{50-t}.$$

We may avoid the computations indicated in the above expression by using Appendix Table E. Entering this table with $n = 50$, $p = .50$, *and* $x = 20$, *we find* $F_b(20) = .1013$.

3. *What is the probability that the sample of fifty voters will contain at least twenty-five undecided? In this situation, the probability we seek is given by the expression:*

$$P(X \geq 25|n = 50, p = .50) = 1 - F_b(24|50, .50)$$

From Appendix Table E, we find $F_b(24) = .4439$. *Therefore, our answer is* $1 - F_b(24) = 1 - .4439 = .5561$.

Example 2:

Assume that a production line operates as a Bernoulli process with a 10 per cent defective rate.

1. *If a sample of twenty items were taken from the production output, what is the probability that the sample would contain between three and five defectives, inclusive? Symbolically, the probability we desire is given by the expression:*

$$P(3 \leq X \leq 5|n = 20, p = .10) = F_b(5|20, .10) - F_b(2|20, .10)$$

With $n = 20$ *and* $p = .10$, *Appendix Table E indicates that* $F_b(5) = .9887$ *and* $F_b(2) = .6769$. *Then our answer is:*

$$F_b(5) - F_b(2) = .9887 - .6769 = .3118$$

2. *If a sample of ten items were taken from the process output, what is the probability that the sample would contain no more than one defective? To answer this question, we must find:*

$$P(X \leq 1|n = 10, p = .10) = F_b(1|10, .10)$$

From Appendix Table E, we find $F_b(1) = .7361$.

3. If a sample of fifteen items were taken from the output of the process, what is the probability that it would contain four or more defectives? The probability we wish to find is given by the expression:

$$P(X \geq 4|n = 15, p = .10) = 1 - F_b(3|15, .10)$$

Appendix Table E indicates that $F_b(3) = .9444$, *so our answer is* $1 - .9444 = .0556$.

Mean and Variance of the Binomial Distribution. From the individual probabilities in Table 9.4 we may calculate the mean and variance of the binomial distribution with $n = 4$, $p = .20$. Applying Formula (8.3) to compute the mean of the distribution, we obtain:

$$\mu = E(X) = 0(.4096) + 1(.4096) + 2(.1536) + 3(.0256) + 4(.0016) = .80$$

To obtain the variance of the distribution by using Formula (8.15), we first compute:

$$E(X^2) = [.4096(0^2) + .4096(1^2) + .1536(2^2) + .0256(3^2) + .0016(4^2)] = 1.28$$

Then the variance is:

$$\sigma^2 = V(X) = E(X^2) - \mu^2 = 1.28 - (.80)^2 = .64$$

By substituting the binomial mass function for $f(x)$ into Formula (8.3), it can be shown that the mean of the binomial distribution is:

$$\mu = E(X) = np \tag{9.7}$$

For example, for the binomial distribution in Table 9.4, in which $n = 4$ and $p = .20$, we obtain $\mu = E(X) = 4(.20) = .80$, which is the same result we obtained previously from applying Formula (8.3).

Similarly, by using Formula (8.15), the variance of a binomially distributed random variable X can be shown to be:

$$\sigma^2 = V(X) = np(1 - p) \tag{9.8}$$

For the binomial distribution in Table 9.4, Formula (9.8) gives $\sigma^2 = 4(.20)(.80) = .64$, which agrees with the result we obtained from Formula (8.15). In certain applications, we find it useful to recognize that, as the square root of the variance, the standard deviation of a binomially distributed random variable is given by:

$$\sigma = \sqrt{np(1 - p)} \tag{9.9}$$

Application of this formula yields: $\sigma = \sqrt{4(.20)(.80)} = \sqrt{.64} = .80$

9.2.2 The Negative Binomial Model

In our discussion of the binomial model, we mentioned that the parameters—n (the number of Bernoulli trials) and p (the probability of success on any single trial)—are fixed values, and that the random variable X is the number of successes in the n trials. Thus, in applying this model to a sampling experiment, the sampling plan specifies the number of trials to be observed. Then the experiment stops when the specified number of trials have been observed, regardless of how many successes have been obtained in those trials.

In contrast, let us now consider the sampling situation in which the number of trials to be observed is not fixed, so that the sampling continues until some specified number of successes is observed. In this situation, the number of successes is a fixed value and the number of trials is a random variable. Thus, following our convention of using capital letters to denote random variables and lower-case letters to indicate fixed values, the lower-case x will be used to denote the specified number of successes, and the capital letter N will be used to denote the number of trials required to obtain x successes.[3] In this case, the probability distribution of the random variable N is the *negative binomial model* rather than the *binomial model*.

Negative Binomial Probability Functions. To develop the negative binomial model, we begin by noting that the probability of needing exactly n independent trials to obtain x successes *in any particular sequence* of x successes and $n - x$ failures is given by:

$$p^x(1 - p)^{n-x}$$

Then if exactly n trials are required to obtain exactly x successes, the xth success must occur on the nth trial. This implies that there must have been exactly $(x - 1)$ successes and $[(n - 1) - (x - 1)] = (n - x)$ failures, occurring in some sequence, in the first $(n - 1)$ trials. The total number of such sequences is the number of distinct permutations of two kinds of objects when there are $(x - 1)$ objects of one kind and $(n - x)$ objects of the other kind. From Formula (3.4), we obtain:

$$\frac{(n - 1)!}{(x - 1)!(n - x)!}$$

From our previous discussion, we may recognize that this expression is also equal to the number of combinations specified by:

$$\binom{n - 1}{x - 1}$$

[3] This use of N should not be confused with its use in the hypergeometric model in which N is employed to denote the size of a finite population.

Thus, the total probability that exactly n trials will be required to obtain x successes is derived by multiplying the probability of a single sequence by the total number of equiprobable sequences. This gives us the negative binomial mass function, which is:

$$f_{nb}(n|x,p) = \begin{cases} \binom{n-1}{x-1} p^x(1-p)^{n-x} & \text{if } n = x, x+1, \ldots \\ 0 & \text{otherwise} \end{cases} \tag{9.10}$$

where $0 < p < 1$ and $x = 1, 2, \ldots$.

As a probability mass function, Formula (9.10) gives $P(N = n)$. That is, in sampling from a Bernoulli process, if the probability of a success on any single trial is equal to p, then the negative binomial mass function gives the probability that it will take exactly n trials to obtain x successes.

To illustrate the use of Formula (9.10), consider a market research agency that conducts interviews by telephone. From past experience, the agency has determined that there is a probability of .25 that any call made between 4:00 P.M. and 6:00 P.M. on a Saturday afternoon will be answered. An interviewer wishes to complete a quota of ten calls during this time period. What is the probability that he will have to make exactly eighteen calls in order to obtain ten answers? From Formula (9.10), the solution is:

$$\begin{aligned} P(N = 18|x = 10, p = .25) &= f_{nb}(18|10, .25) \\ &= \binom{18-1}{10-1}(.25)^{10}(.75)^8 \\ &= .0023 \end{aligned}$$

The negative binomial cumulative mass function is given by the expression:

$$F_{nb}(n|x,p) = \sum_{t=x}^{n} \binom{t-1}{x-1} p^x(1-p)^{t-x} \tag{9.11}$$

Formula (9.11) specifies a summation of individual negative binomial probabilities. In this formula, the dummy variable t represents possible values of N, taken in ascending order, and n represents the upper limit of the summation.

Obtaining Negative Binomial Probabilities from Binomial Tables. Direct computation of negative binomial probabilities from Formulas (9.9) and (9.10) can be quite cumbersome. Fortunately, negative binomial probabilities may be obtained from tables of the binomial distribution because of the existence of certain relationships between the binomial and negative binomial distributions. Specifically, it is possible to demonstrate the following relationships:

$$f_{nb}(n|x,p) = p\, f_b(x - 1|n - 1,p) \tag{9.12}$$

$$F_{nb}(n|x,p) = 1 - F_b(x - 1|n,p) \tag{9.13}$$

With a little ingenuity, we may grasp these relationships without going through any mathematical derivations. Formula (9.12) says that the probability of n trials being required to generate x successes is equal to the probability, p, of a single success times the probability of $(x - 1)$ successes in $(n - 1)$ trials. This makes sense when we recall that if exactly n trials are required to obtain exactly x successes, the xth success must occur on the nth trial and there must be $(x - 1)$ successes in the first $(n - 1)$ trials. To understand the relationship expressed by Formula (9.13), we first recognize that the probability of requiring at most n trials to generate x successes is the same as the probability of observing at least x successes in n trials. That is, $P(N \leq n|x,p) = P(X \geq x|n,p)$. We then note that $P(X \geq x|n,p) = 1 - F_b(x - 1|n,p)$. Finally, we observe that $P(N \leq n|x,p) = 1 - F_b(x - 1|n,p)$, which explains Formula (9.13).

To illustrate the application of Formulas (9.12) and (9.13), consider again the example of the market research agency. In that example, we used Formula (9.10) to calculate the probability that a telephone interviewer would have to make exactly eighteen calls to obtain ten answers. Applying Formula (9.12) to the same problem, we have:

$$f_{nb}(18|10, .25) = .25\, f_b(9|17, .25)$$

From Appendix Table D, we find that $f_b(9|17, .25) = .0093$. Thus,

$$f_{nb}(18|10, .25) = .25(.0093) = .0023.$$

Now suppose that the interviewer wishes to determine the probability that it will take at most eighteen calls to obtain ten answers. To compute this probability, $P(N \leq 18)$, we may apply Formula (9.13), which yields:

$$F_{nb}(18|10, .25) = 1 - F_b(10 - 1|18, .25)$$

From Appendix Table E, we find that $F_b(9|18, .25) = .9946$. Thus,

$$F_{nb}(18|10, .25) = 1 - .9946 = .0054.$$

Mean and Variance of the Negative Binomial Distribution. It can be shown mathematically that the mean of the negative binomial distribution is:

$$\mu = E(N) = \frac{x}{p} \tag{9.14}$$

The variance of the negative binomial distribution can be shown to be:

$$\sigma^2 = V(N) = \frac{x(1-p)}{p^2} \tag{9.15}$$

The Geometric Distribution. The negative binomial distribution with $x = 1$ is known as the *geometric distribution.* The geometric mass function is given by:

$$f_g(n|p) = p(1-p)^{n-1} \tag{9.16}$$

This mass function gives the probability of the number of trials required to obtain the *first* success. Since the geometric distribution is a special case of the negative binomial distribution, formulas for the negative binomial distribution are also applicable to the geometric distribution.

9.3 THE POISSON MODEL

We have conceptualized a Bernoulli process as one involving a series of discrete trials, each of which would result in one of the two possible outcomes—success or failure. We saw that the Bernoulli process has a parameter, p, which is the probability of obtaining a success on any single trial. When a particular number of Bernoulli trials is specified, in addition to a value of p, then the random variable of interest is the number of successes observed in the specified number of trials. We have seen that the behavior of this random variable is described by the binomial model.

We will now consider a different kind of random process, which differs from a Bernoulli process in two important respects:

1. Rather than consisting of discrete trials, the process operates continuously over some given amount of time, distance, area, or volume.
2. Rather than producing a sequence of successes and failures, the process simply produces successes, which occur at random points in the specified time, distance, area, or volume. These successes are commonly referred to as "occurrences."

Visualize, for example, a manufacturing process that produces a continuous flow of yard-wide textile. As the yardage flows from the process, burls will occur at random points. Here we can count the number of burls that occur in a given length of the textile, but we cannot count the number of burls that do not occur. As another example, consider the occurrence of machine breakdowns in a manufacturing plant. Over some continuous period of time, such as a week, machine breakdowns may occur at random points in that time interval. For a given week it is possible to count the number of breakdowns that occurred, but it is meaningless to ask how many breakdowns did not occur. To further clarify this type of random process, additional examples are listed below:

1. *Occurrences of an event in a unit of time*

   Telephone calls received in an hour at an office switchboard

   Articles received in a day at an airline's lost and found office

   Auto accidents in a month at a busy intersection

   Deaths due to a rare disease in a month

   Service calls required in a month for an installation of fifty office machines

   Arrivals of depositors at a bank in an hour on Monday morning

2. *Occurrences of an event in a unit of distance*

   Defects occurring in 50 yards of insulated wire

   Deaths occurring in 10,000 passenger miles

   Tire repairs for a truck in 1,000 miles traveled on the road

3. *Occurrences of an event in a given area*

   Surface blemishes in a square foot of a synthetic covering for shower walls

   Burls per square yard of woolen textile

   Bacteria on a square centimeter of a culture plate

4. *Occurrences of an event in a given volume*

   White cells in a cubic centimeter of blood

   Hydrogen atoms in a cubic light-year of intergalactic space

   Bacteria in a quart of Pasteurized milk

9.3.1 Characteristics of the Poisson Process

Over the years, it has been found that many phenomena, such as those listed above, occur as a *Poisson process*. The Poisson process derives its name from Simeon Denis Poisson (1781–1840), a French mathematician. As a random process that produces successes at random points in some continuous unit of time or space, the Poisson process has the following general characteristics:

1. The number of occurrences of a phenomenom in a unit of any specified size is independent of the number of occurrences in any other unit. Suppose in a large city, for example, that the monthly number of newly detected cases of gout occurs as a Poisson process. Then the probability of the occurrence of any given number of newly detected cases during a particular month would not be affected by the number of newly detected cases in any other month.

2. The mean number of occurrences of the phenomenon in a unit of specified size is proportional to the size of the unit. Suppose that sales of a particular item occur at a mean rate of two sales per day. Then there would be a mean rate of six sales for an interval of three days, or a mean rate of ten sales for an interval of five days.

3. The probability of two or more occurrences of the phenomenon in an infinitesimal (extremely small) unit is negligible. For example, if incoming calls arrive at a telephone switchboard according to a Poisson process at an average rate of two calls per minute, any number of calls may arrive during the course of a one-minute interval, but it is assumed that there is an essentially zero probability that more than one call would arrive during any given split second.

4. The probability of a single occurrence in an infinitesimal unit remains constant from one such unit to another. For example, if there is a .001 probability that a call will arrive at a switchboard during any given split second, that probability remains constant for every split second.

9.3.2 The Poisson Mass Function

It should be emphasized that the occurrences of a phenomenon in a unit of time or space need not necessarily behave as a Poisson process. However, if such occurrences possess the characteristics listed in the preceding section, then the probabilistic behavior of the number of occurrences in a specified unit of time or space is described by the Poisson probability model. If we let X denote the number of occurrences of a phenomenon that is generated by a Poisson process in a specified number of units of time or space, the probability mass function of X is given by:

$$f_P(x|\lambda,t) = \begin{cases} \dfrac{(\lambda t)^x \, e^{-\lambda t}}{x!} & \text{if } x = 0, 1, 2, \ldots \\ 0 & \text{otherwise} \end{cases} \tag{9.17}$$

where e = a constant, the natural logarithm base, approximately equal to 2.7183

λ = average rate of occurrence of the phenomenon in a single unit of time or space

t = number of contiguous units of time or space under consideration

This formula can be derived mathematically from the statement of the characteristics of the Poisson process in the preceding section. However, this same function was originally formulated by Poisson as a limiting form of the binomial function as n approaches infinity and p approaches zero. As such, the Poisson model has been associated with phenomena that can be described as "rare events." That is, for a given segment of time or space there usually will be few, if any, occurrences of such an event, although there is the possibility, with small probability, that there may be a very large number of occurrences of the event.

Table 9.5

Poisson Mass Function for Number of Flaws in Ten Yards of Wire (λt = 2)

Number of Flaws (x)	*Probability* $f(x)$
0	$\frac{2^0}{0!}e^{-2} = \frac{1}{1}(.135335) = .1353$
1	$\frac{2^1}{1!}e^{-2} = \frac{2}{1}(.135335) = .2707$
2	$\frac{2^2}{2!}e^{-2} = 2(.135335) = .2707$
3	$\frac{2^3}{3!}e^{-2} = \frac{8}{6}(.135335) = .1804$
4	$\frac{2^4}{4!}e^{-2} = \frac{16}{24}(.135335) = .0902$
5	$\frac{2^5}{5!}e^{-2} = \frac{32}{120}(.135335) = .0361$
6	$\frac{2^6}{6!}e^{-2} = \frac{64}{720}(.135335) = .0120$
7	$\frac{2^7}{7!}e^{-2} = \frac{128}{5{,}040}(.135335) = .0034$
8	$\frac{2^8}{8!}e^{-2} = \frac{256}{40{,}320}(.135335) = .0009$
9	$\frac{2^9}{9!}e^{-2} = \frac{512}{362{,}880}(.135335) = .0002$
10	$\frac{2^{10}}{10!}e^{-2} = \frac{1{,}024}{3{,}628{,}800}(.135335) = .0000+$

As a probability mass function, Formula (9.17) gives the probability of x occurrences of a phenomenon in t units of time or space when the mean number of occurrences in a single unit is λ. To illustrate the use of this formula, consider a manufacturing process that produces steel wire. Flaws occur in this wire as a Poisson process with a mean rate of $\lambda = 0.2$ flaws per yard. If a ten-yard length of this wire is taken from the output of the process, and the total number of flaws is counted, what is the probability that there will be a total of no flaws, one flaw, two flaws, and so on?

To answer this question, we need simply to apply Formula (9.17), computing $f(x)$ for $x = 0$, $x = 1$, $x = 2$, and so on, until we reach a value of X for which $f(x)$ is so negligibly small that it is essentially zero for all practical purposes. We have noted that, for this production process, the mean rate of occurrence is $\lambda = 0.2$ flaws per yard. However, our sample consists of $t =$ ten yards. Therefore, the mean rate of occurrence in a ten-yard length of wire is $\lambda t = (0.2)(10) = 2$ flaws. The calculations to obtain the desired probabilities are shown in Table 9.5.

The probability distribution presented in Table 9.5 is shown graphically in Figure 9.1. This figure indicates that the distribution is *asymmetrical*. Specifically, the distribution comes to a peak near the low end of the scale, with a long thin tail extending to the right. Such a distribution is said to be *positively skewed*. This positive skewness is typical of the Poisson distribution, indicating that, with extremely small probability, there is the possibility that a Poisson process will produce an indefinitely large number of occurrences in a segment of time or space, even though the mean rate of occurrence may be quite small.

Figure 9.1

Poisson Mass Function ($\lambda t = 2$)

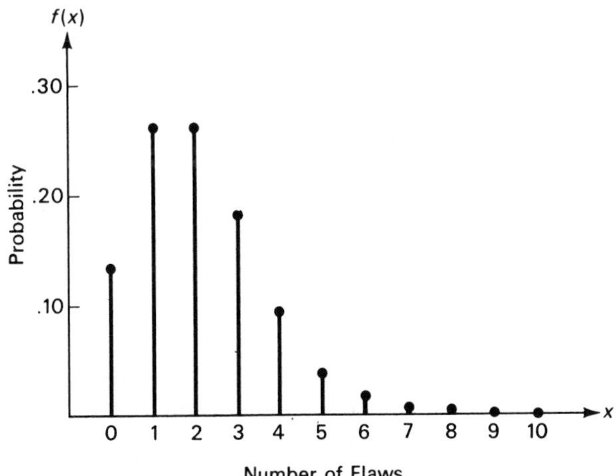

9.3.3 The Poisson Cumulative Mass Function

The Poisson cumulative mass function is given by:

$$F_P(x|\lambda,t) = \sum_{y=0}^{x} \frac{(\lambda t)^y e^{-\lambda t}}{y!} \tag{9.18}$$

In other words, the Poisson cumulative mass function is simply a summation of individual terms of the Poisson mass function. This is demonstrated by Table 9.6, which displays the individual terms determined in Table 9.5 along with the corresponding cumulative terms. For example, the probability that *exactly* three flaws will be observed in a sample of ten yards of wire is given by the individual term $f_P(3) = .1804$. Correspondingly, the probability of observing *three flaws or less* is given by the cumulative term:

$$\begin{aligned} F_P(3|.2,10) &= \sum_{y=0}^{3} \frac{2^y e^{-2}}{y!} \\ &= f_P(0|.2,10) + f_P(1|.2,10) + f_P(2|.2,10) + f_P(3|.2,10) \\ &= .1353 + .2707 + .2707 + .1804 = .8571 \end{aligned}$$

Table 9.6

Poisson Probability Distribution and Cumulative Distribution for Number of Flaws in Ten Yards of Wire ($\lambda t = 2.0$)

Number of Defects in the Sample (x)	*Individual Probability* $f(x)$	*Cumulative Probability* $F(x)$
0	.1353	.1353
1	.2707	.4060
2	.2707	.6767
3	.1804	.8571
4	.0902	.9473
5	.0361	.9834
6	.0120	.9955
7	.0034	.9989
8	.0009	.9998
9	.0002	1.0000
10	.0000	1.0000

9.3.4 The Use of Poisson Tables

Formulas (9.17) and (9.18) are useful for explaining the Poisson distribution and are sometimes necessary for actually calculating Poisson probabilities. For most practical purposes, however, Poisson probabilities are obtained from tables such as Appendix Table F (individual terms) and Table G (cumulative terms).[4] The following examples illustrate the use of these tables.

Example 1:

A manufacturer of vinyl wall covering has found that, with respect to surface blemishes, his manufacturing process operates as a Poisson process with an average of $\lambda = 0.2$ blemishes per square foot of manufactured material. Periodically, a one-square-yard sample of the material is inspected.

1. *What is the probability that a sample will contain exactly two blemishes? Since there are nine square feet in one square yard, $\lambda t = 0.2 \times 9 = 1.8$. Then, from Appendix Table F, $P(X = 2|\lambda t = 1.8) = f_P(2) = .2678$.*
2. *What is the probability that a sample will contain two or less blemishes? From Appendix Table G, $P(X \leq 2|\lambda t = 1.8) = F_P(2) = .7306$.*
3. *What is the probability that a sample will contain at least two blemishes? From Appendix Table G, $P(X \leq 1|\lambda t = 1.8) = F_P(1) = .4628$. Thus,*

$$\begin{aligned} P(X \geq 2|\lambda t = 1.8) &= 1 - P(X \leq 1|\lambda t = 1.8) \\ &= 1 - F_P(1) = 1 - .4628 = .5372 \end{aligned}$$

Example 2:

During peak hours, calls arrive at a switchboard according to a Poisson process with an average of $\lambda = 60$ calls per hour.

1. *During a given ten-minute period during the peak hours, what is the probability of receiving at least twelve calls? Since a ten-minute period is 1/6 hour, $\lambda t = 60 \times \frac{1}{6} = 10$. Then, using Appendix G, $P(X \geq 12)|\lambda t = 10) = 1 - P(X \leq 11|\lambda t = 10) = 1 - .6968 = .3032$.*
2. *During a given five-minute period, what is the probability of receiving at least three but no more than seven calls? Since a five-minute period is 1/12 hour, $\lambda t = 60 \times \frac{1}{12}, = 5$. Then, using Appendix Table G, we obtain:*

$$\begin{aligned} P(3 \leq X \leq 7|\lambda t = 5) &= P(X \leq 7|\lambda t = 5) - P(X \leq 2|\lambda t = 5) \\ &= F_P(7) - F_P(2) \\ &= .8666 - .1247 \\ &= .7419 \end{aligned}$$

[4] For more extensive tables see, E. C. Molina, *Poisson's Exponential Binomial Limit*, D. Van Nostrand, 1949.

9.3.5 Mean and Variance of the Poisson Distribution

If a random variable X has a Poisson distribution given by Formula (9.17), then the mean can be shown to be:

$$\mu = E(X) = \lambda t \tag{9.19}$$

It can be shown that the variance is:

$$\sigma^2 = V(X) = \lambda t \tag{9.20}$$

It is significant that the mean and the variance are equal. This is a unique feature of the Poisson distribution.

PROBLEMS

9.1 Before purchasing a batch of 60 light bulbs, a customer tests 8 of them. If the batch contains 12 defectives, what is the probability that the customer will find none of the defectives in the sample?

9.2 A box contains 50 bolts: 20 bolts with left-hand threads and 30 with right-hand threads. A sample of 10 bolts is drawn at random from the box. What is the probability that the sample will contain at least 2 bolts with left-hand threads?

9.3 Certain missile components are shipped in lots of 12. Three components are selected from each lot, and a particular lot is accepted if none of the three components selected is defective.
(a) What is the probability that a lot will be accepted if it contains 5 defectives?
(b) What is the probability that a lot will be rejected if it contains 9 defectives?
(c) Let X be a random variable denoting the number of defectives in a sample of 3 components selected randomly from one of the above lots. If the lot contains 4 defectives, specify the probability function $f(x)$. Present the probability distribution (i) as a mathematical expression, and (ii) in the form of a table.
(d) Under the conditions stated in (c) above, what is the expected number of defectives in a sample of 3 components?

9.4 An electronics firm purchases a particular component in lots of 50. As a routine acceptance sampling procedure, a random sample of 5 components from each lot are inspected, and the lot is rejected if the sample contains at least 2 defectives.
(a) If a particular lot contains 4 defectives, what is the probability that exactly 2 defectives will be found in the sample?
(b) If a lot contains 6 defectives, what is the probability that the sampling and inspection procedure will result in rejection of the lot?
(c) If X is a random variable denoting the number of defectives in a random sample of 5 items, and if a lot contains 4 defectives, specify the probability function $f(x)$. Present the probability distribution (i) as a mathematical expression, and (ii) in the form of a table.
(d) Determine the mean and variance of the random variable X in (c) above.

9.5 Five employees of the Vippo Corporation go to the company garage to check out cars. At the moment, there are 20 available cars, one-fourth of which are foreign-made. Cars are assigned to the employees at random from those that are available.

(a) What is the probability that a foreign-made car will be assigned to none of the employees? To 1 employee? To 2 employees? To 3 employees? To 4 employees? To all 5 employees?

(b) What is the expected number, out of the 5 employees, who will receive foreign-made cars?

9.6 With respect to quality of output ("defective" vs. "good" items), a particular production process operates as a Bernoulli process, with a long-run defective rate of .10. Suppose that a sample of 5 items is selected from the output of the process, and each of the 5 items is inspected. Let X be a random variable denoting the number of defectives in the sample.

(a) What is the probability that $X = 3$?

(b) What is the probability that $X \leq 3$?

(c) What is the probability that $X < 3$?

(d) What is the probability that $X \geq 3$?

(e) What is the probability that $X > 3$?

(f) What is the probability that $X \neq 3$?

9.7 A dichotomous population is composed of Munchkins and Hobbits. Of these, exactly 50 per cent are Munchkins, and the remainder are Hobbits. As everyone knows, Munchkins and Hobbits are very prolific and long-lived, so that the population may be considered effectively infinite for all practical sampling purposes. If 10 of these irascibly lovable creatures are selected at random from the population, what is the probability that:

(a) Exactly 4 are Hobbits?

(b) Exactly 4 are Munchkins?

(c) At least 2 are Hobbits?

(d) No more than 2 are Hobbits?

(e) Less than 2 are Munchkins?

9.8 The Zoomite Corporation has just accepted shipment on a new automatic machine that functions as a Bernoulli process. Anxious to try out the machine, the production manager decides to produce a sample of 10 items. Because the machine has not been adjusted, he does not know the defective rate, p, of the process. What is the probability that there will be 3 or more defectives in the sample if:

(a) $p = .05$?

(b) $p = .10$?

(c) $p = .20$?

(d) $p = .25$?

9.9 A personnel manager is in the process of designing a multiple-choice examination as an aid in screening applicants. He plans to use 20 questions, each to contain 1 correct and 3 incorrect answers. He has raised the following questions about the reliability of the examination:

(a) If an applicant guesses on every question, what is the probability of getting 5 or more answers correct?

(b) How many right answers should be required to ensure that the chance of an applicant obtaining that score or higher from pure guessing is approximately .10?

(c) What is the probability of obtaining a score of 8 by pure guessing?

(d) What is the probability of an applicant obtaining a score of 6 or less on the basis of random guessing?

9.10 If 40 per cent of the voters in a large state favor Candidate A, what is the probability that, in a random sample of 20 voters interviewed by a pollster, the majority will favor him?

9.11 In a large metropolitan county, 50 per cent of the registered voters are Democrats, and 40 per cent of the registered voters are men. Assume that political affiliation is independent of sex. If 5 registered voters are selected at random, what is the probability that the group will contain at least 3 Democrats and at least 4 men?

9.12 The Betty Drucker Food Corp. is recruiting housewives for its cake-mix testing panel. For this purpose it has devised the following test of the candidates' ability to discriminate. Each applicant is presented with 4 cakes, 3 of them baked from Mix A and one from Mix B. The applicant is to identify the cake from Mix B. This procedure is repeated 10 times, and the passing score is 7 or more correct choices.

(a) If a housewife cannot discriminate but guesses each time, what is the probability that she will be admitted to the panel?

(b) Let R be a random variable denoting the number of correct choices in the 10 trials. If a housewife cannot discriminate between the two mixes, what is the probability distribution of R? Specify the distribution (i) in terms of a mathematical function, and (ii) in the form of a table.

(c) Determine the mean and standard deviation of the random variable R in in (b) above.

9.13 A study by the Blacksky Oil Company reveals that there is an independent probability of .10 that any given offshore well will develop a leak when it is drilled. It is also determined that, within the drilling regulations, this probability is constant from well to well. If Blacksky is allowed to drill 18 wells:

(a) What is the probability that none of the wells will leak?

(b) What is the probability that between 4 and 6 wells, inclusive, will develop leaks?

(c) If Blacksky makes a profit of \$2 million on each well that does not leak, and loses \$5 million on each well that does leak, what is the expected profit from drilling the 18 wells?

9.14 The salesmen for a particular real estate firm receive their "leads" from the names of persons who have responded to advertisements that offer "further details on request." From past experience, it has been determined that, as a long-run figure, a salesman can make a sale to 1 out of 4 such leads.

(a) Regarding each follow-up sales call on one of these leads as a trial of a Bernoulli process:

(i) What is the probability that a salesman will make his first sale on his fourth call?

(ii) What is the expected number of calls a salesman will need to make in order to produce 6 sales?

(iii) What is the probability that a salesman will need at least 15 leads to produce 5 sales?

(b) In this particular situation, what specific assumptions are made if each sales call is regarded as a trial of a Bernoulli process?

9.15 A market research agency that conducts interviews by telephone has found that, from past experience, there is a .40 probability that a call made between 2:30 and 5:30 p.m. will be answered. Assume a Bernoulli process.

(a) What is the probability that an interviewer's tenth answer comes on his twentieth call?

(b) What is the expected number of calls necessary to obtain 7 answers?
(c) What is the probability that an interviewer will receive his first answer on his third call?

9.16 From past experience, a stock broker has found that 60 per cent of the calls he receives during the morning are orders, and the remaining calls are other business. Assume that these calls occur according to a Bernoulli process.
(a) What is the probability that exactly 6 of the first 10 calls that the broker receives on a particular morning are order calls?
(b) What is the probability that at least 6 of the first 10 calls that he receives on a particular morning are order calls?
(c) On a particular morning, what is the probability that his sixth order call will be the tenth call he receives?
(d) What is the expected number of calls he will accept in a morning, if he receives calls until he has obtained 6 order calls?
(e) In this situation, what specific assumptions are made if the separate calls are regarded as individual trials of a Bernoulli process. Are these assumptions reasonable?

9.17 The Lakeside Candle Company manufactures custom-designed candles that are produced to order in lot sizes specified by the customer. Ten per cent of the candles produced by the manufacturer's process are rejected because of flaws. Experience indicates that it is reasonable to regard this production process as a Bernoulli process.
(a) If an order is received for a lot of 65 candles, what is the probability that a production run of no more than 75 candles will be sufficient to fill the order?
(b) What is the probability that an order for 85 candles can be satisfied by producing no more than 100 candles?
(c) What is the expected number of candles that must be produced in order to obtain 20 acceptable candles?

9.18 A particular textile is produced in continuous rolls. Burls in the material occur according to a Poisson process at an average rate of 20 burls per 100 linear feet. If a 10-foot length of textile is cut from the roll, what is the probability that it will contain:
(a) Exactly 2 burls?
(b) At least 2 burls?
(c) Less than 2 burls?
(d) More than 2 burls?
(e) No more than 2 burls?

9.19 Industrial accidents commonly occur according to a Poisson process, so that the Poisson distribution is of particular interest to insurance actuaries. Suppose, for instance, that fatal accidents associated with a particular hazardous occupation occur at a rate of .0072 per man-year on the average. If a company insures 1,000 such employees for a given year, what is the probability that it will have to meet at least 10 fatality claims?

9.20 Suppose that the number of calls arriving at a company's switchboard during a 5-minute span has a Poisson distribution with $\lambda = 5.2$.
(a) What is the probability that there will be exactly 4 incoming calls during such a 5-minute span?
(b) What is the probability that there will be at most 4 calls?
(c) What is the probability that there will be at least 4 calls?
(d) Should the sum of the probability in (b) and (c) be equal to 1? Explain.

9.21 A telephone switchboard handles 600 calls, on an average, during a rush hour. The board can make a maximum of 20 connections per minute. Use the Poisson distribution to evaluate the probability that the board will be overtaxed during any given minute.

9.22 In a large manufacturing plant, machinists arrive at a tool crib for service at a mean rate of 1.5 arrivals per 5-minute period. It is reasonable to assume that these arrivals occur according to a Poisson process.
(a) What is the probability that exactly 3 machinists will arrive during a specified 5-minute period?
(b) What is the probability that no more than 2 machinists will arrive during a specified 1-minute period?
(c) What is the probability that at least 4 machinists will arrive during a specified 15-minute period?
(d) What is the expected number of arrivals in a 10-minute period?
(e) In this particular situation, what specific assumptions are made if the occurrence of the arrivals is regarded as a Poisson process?

9.23 Knot-holes in a particular kind of pine siding occur according to a Poisson process at a mean rate of 1.4 knot-holes per board-foot.
(a) What is the probability that 2 board-feet of this siding will contain exactly 3 knot-holes?
(b) What is the probability that 10 board-feet of this siding will contain less than 15 knot-holes?
(c) What is the probability that 5 board-feet of this siding will contain at least 8 knot-holes?
(d) What is the probability that a 6-inch length of this siding will contain fewer than 2 knot-holes?
(e) What is the expected number of knot-holes in 10 board-feet of this siding?
(f) What is the variance of the number of knot-holes in 10 board-feet of this siding? What is the standard deviation?

9.24 The Kitcheneze Manufacturing Company produces enamel kitchen utensils. One of the best-selling items produced by the firm is a two-quart enamel-coated sauce pan. With respect to surface blemishes, the manufacturing process may be described by the Poisson model with a mean rate of 3.9 blemishes per pan.
(a) What is the probability that 1 of these sauce pans will have no more than 5 surface blemishes?
(b) A customer purchases 6 of these sauce pans. What is the probability that exactly 3 of these pans will each have no more than 5 surface blemishes?
(c) For the customer in (b) above, what is the probability that no more than 3 of the 6 sauce pans will each have no more than 5 surface blemishes?

Chapter 10
Commonly Used Continuous Probability Models

In the preceding chapter, we examined several discrete probability models that are employed extensively in the analysis of decision problems. In this chapter, we will turn our attention to some of the most widely used continuous probability models. Continuous probability models demand our attention for two primary reasons. First, many real-life decision problems often involve random variables that are in fact continuous—such as time, distance, and weight. Second, in applying probability models to the analysis of decision problems, continuous models often are mathematically more tractable than discrete models.

10.1 THE NORMAL PROBABILITY MODEL

By far the best known and most commonly used of all probability models is the normal model. Not only is this model important in the analysis of decision problems, but its use pervades the social, physical, and biological sciences. Of course, we should recognize that the term "normal" model is not meant to imply that any other model is "abnormal."

10.1.1 The General Normal Probability Model

The general normal probability function was first published in 1733 by Abraham DeMoivre. However, this work was overlooked for nearly 200 years. Later, in the nineteenth century, the Marquis Pierre Simon de Laplace and Karl Friedrich Gauss each independently derived the same function as that presented in DeMoivre's earlier publication. Thus, the normal distribution has been variously referred to as the Gaussian, the Laplacean, and the Gauss-Laplace distribution.

For any continuous random variable X, the general normal probability density function, with parameters α and β, is given by:

$$f_n(x|\alpha,\beta) = \frac{1}{\beta\sqrt{2\pi}} e^{-\frac{1}{2}\left(\frac{x-\alpha}{\beta}\right)^2}, \qquad -\infty < x < \infty$$

where π = the familiar constant, approximately equal to 3.1416
e = another constant, the natural logarithm base, approximately equal to 2.7183

It can be shown mathematically that the mean of the normal distribution is equal to the parameter α and the standard deviation is equal to the parameter β. Therefore, the general normal density function is customarily expressed as:

$$f_n(x|\mu,\sigma) = \frac{1}{\sigma\sqrt{2\pi}} e^{-\frac{1}{2}\left(\frac{x-\mu}{\sigma}\right)^2}, \qquad -\infty < x < \infty \tag{10.1}$$

Because of the considerable importance of the normal distribution in statistical theory and application, it is worthwhile to examine some of the characteristics of the normal density function:

1. The normal density function describes a continuous, bell-shaped curve, as illustrated in Figure 10.1.

Figure 10.1

The Normal Density Curve

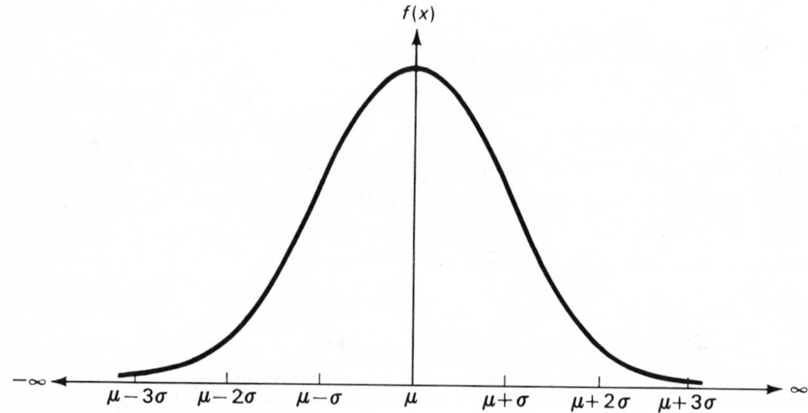

2. The normal curve is symmetrical about μ, at which point $f(x)$ reaches its maximum value. Since $f(x)$ reaches its maximum value for $x = \mu$, the mode is equal to the mean. Due to the symmetry, $P(X \leq \mu) = P(X \geq \mu) = .50$ for any normal distribution. Thus the median of any normal distribution also is equal to the mean.

3. In both directions from the mean, the curve is asymptotic to the horizontal axis. That is, as the tails of the distribution extend toward infinity in either direction, they approach the axis more and more closely without ever reaching it.

4. Within the central range of one standard deviation on either side of the mean, the curve is concave downward. Beyond that range—that is, for x-values less than $(\mu - \sigma)$ or greater than $(\mu + \sigma)$—the curve is concave upward. In other words, the inflection points of the curve lie at a distance of one standard deviation above and below the mean.

Figure 10.2

Comparison of Normal Curves with Selected Means and Standard Deviations

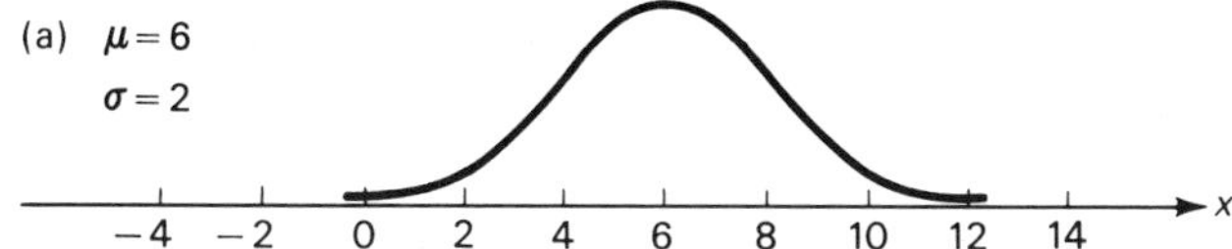

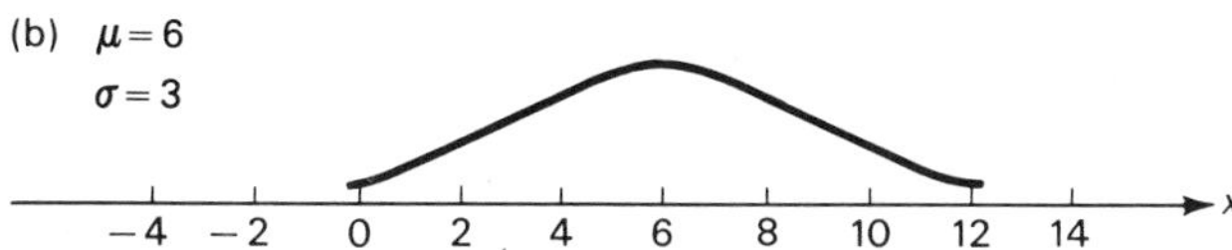

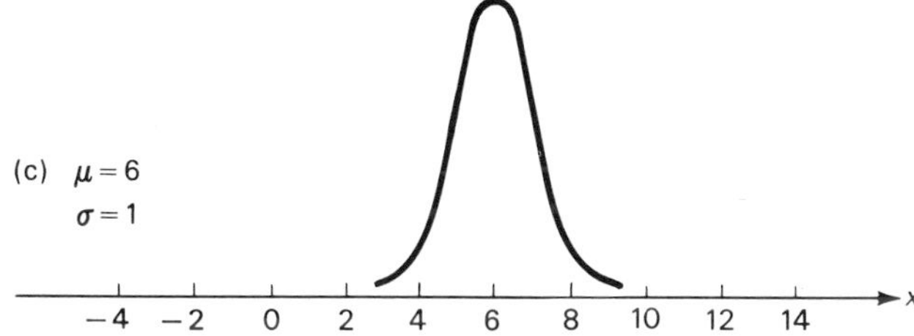

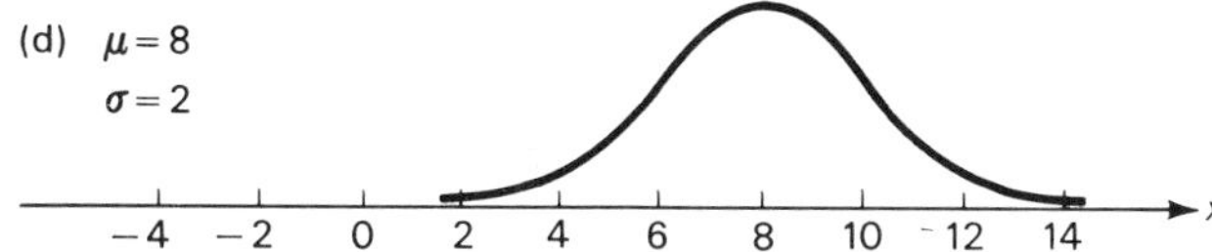

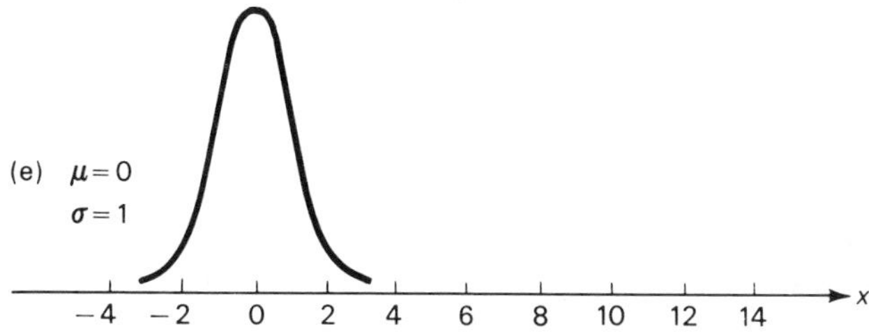

5. Any normal distribution is determined completely by specifying values for the parameters μ and σ. The *shape* of the curve—whether it is low and spread out, or tall and narrow—is determined by the value of σ. The *location* of the curve with respect to its position on the x-axis is determined by the value of μ. In Figure 10.2, the distributions (a), (b), and (c) have identical means ($\mu = 6$), so they all are centered at $x = 6$. However, these three distributions differ in shape since they have different standard deviations. Curves (a) and (d) have identical shapes because they have identical standard deviations, but they are located in different positions owing to their different means. Likewise, curves (c) and (e) have identical standard deviations (and hence identical shapes) but different means (and hence different locations). The fact that the exact shape of a normal distribution is determined if we know the value of σ, even though the value of μ may be unknown, is extremely important in applying the normal model to decision problems.

The normal cumulative density function may be expressed as:

$$F_n(x|\mu,\sigma) = \int_{-\infty}^{x} \frac{1}{\sigma\sqrt{2\pi}} e^{-\frac{1}{2}\left(\frac{t-\mu}{\sigma}\right)^2} dt \tag{10.2}$$

The cumulative function corresponding to the density function in Figure 10.1 is shown graphically in Figure 10.3. As this figure indicates, the normal cumulative density function describes a smooth sigmoid (S-shaped) curve that is sometimes called the *normal ogive*.

Figure 10.3

The Normal Cumulative Density Function

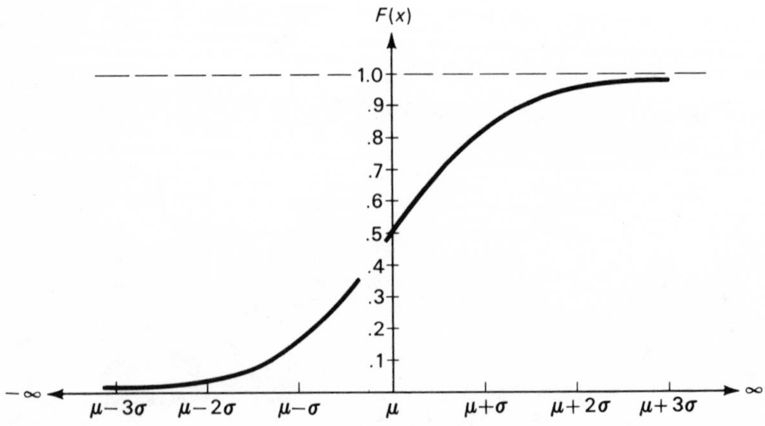

10.1.2. The Standard Normal Probability Model

In order to use the general normal probability model to calculate probabilities, it would be necessary to perform the integration indicated in Formula (10.2). Unfortunately, this integral cannot be evaluated to obtain an *exact* solution. Nevertheless, various complex methods are available for obtaining approximate evaluations of normal integrals. To avoid this laborious procedure in practice, we must rely on tabulated values of normal cumulative probabilities. However, since there are an infinite number of possible combinations of values of μ and σ, Formula (10.2) describes an infinite number of integrals. Thus, if we were to work with the general normal model, it would be necessary to use elaborate tables listing cumulative normal probabilities for an extremely large number of combinations of values of μ and σ. Fortunately, this difficulty has been resolved by the discovery that there is one specific normal integral from which evaluations of any other normal integrals may be easily obtained. This specific integral is the integral of that normal density function for which $\mu = 0$ and $\sigma = 1$.

When a random variable has a normal density function with $\mu = 0$ and $\sigma = 1$, Formula (10.1) is reduced to:

$$f_n(x|\mu = 0,\sigma = 1) = \frac{1}{\sqrt{2\pi}} e^{-\frac{1}{2}x^2}$$

This particular density function is called the *standard normal* density function. To distinguish the standard normal density function from the general normal density function, we will use the notation $f_N(z)$ in place of $f_n(x|\mu,\sigma)$. Using this notation, the standard normal density function is given by:

$$f_N(z) = \frac{1}{\sqrt{2\pi}} e^{-\frac{1}{2}z^2} \tag{10.3}$$

Then, the standard normal integral (the standard normal cumulative density function) is:

$$F_N(z) = \int_{-\infty}^{z} \frac{1}{\sqrt{2\pi}} e^{-\frac{1}{2}t^2}\, dt \tag{10.4}$$

It is important to observe the following distinctions in notation between the standard normal and the general normal probability models.

1. The random variable in the standard normal model is symbolized by Z, whereas the random variable in the general normal model is denoted by X.

2. The subscript N is used to designate a standard normal function, whereas the subscript n is used to denote a general normal function.

3. The parameters μ and σ are omitted from the standard normal notation since, by definition, μ must be equal to 0 and σ must be equal to 1.

The standard normal integral in Formula (10.4) has been evaluated for selected values of z. The results are tabulated in Appendix Table H. This table gives $F_N(z)$ for selected values of z. In other words, this table permits us to obtain standard normal probabilities without evaluating the integral in Formula (10.4). The reader should familiarize himself with Table H by verifying the following probabilities:

$$
\begin{aligned}
P(Z < 1.00) &= F_N(1.00) = .8413 \\
P(Z < 2.17) &= F_N(2.17) = .9850 \\
P(Z < 0.04) &= F_N(0.04) = .5160 \\
P(Z < -1.55) &= F_N(-1.55) = .0606 \\
P(Z < -0.28) &= F_N(-0.28) = .3897
\end{aligned}
$$

10.1.3 The Use of the Standard Normal Integral Table

The key to using the standard normal integral table to obtain probabilities for any normal distribution lies in the fact that any normal distribution can be converted into the standard normal distribution. Specifically, if X is a normally distributed random variable with mean μ and standard deviation σ, then the quantity $\dfrac{X - \mu}{\sigma}$ is a random variable that has the standard normal distribution. This relationship between the general normal variable X and the standard normal variable Z may be expressed as follows:

$$Z = \frac{X - \mu}{\sigma} \tag{10.5}$$

By means of this formula, values of any normally distributed random variable X may be converted into corresponding values of Z, such that there is a direct linear relationship between the values on the two scales. For instance, suppose that a random variable X is normally distributed with $\mu = 100$ and $\sigma = 10$. Then, applying Formula (10.4) to selected x-values:

$$\text{if } x = 100, \qquad z = \frac{100 - 100}{10} = 0$$

$$\text{if } x = 110, \qquad z = \frac{110 - 100}{10} = +1.0$$

$$\text{if } x = 125, \qquad z = \frac{125 - 100}{10} = +2.5$$

$$\text{if } x = 90, \qquad z = \frac{90 - 100}{10} = -1.0$$

$$\text{if } x = 85, \qquad z = \frac{85 - 100}{10} = -1.5$$

This relationship between corresponding values of X and Z is illustrated in Figure 10.4.

Figure 10.4

Relationship Between Corresponding Values of X and Z for a Normal Distribution ($\mu = 100$, $\sigma = 10$)

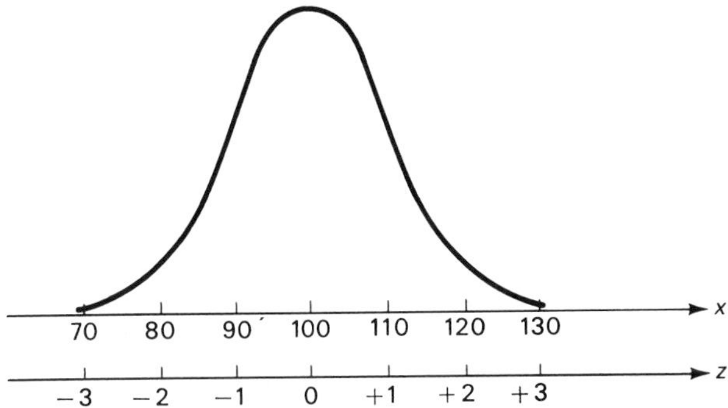

Because of the direct linear relationship between X and Z, the general normal integral may be converted to the standard normal integral in the following manner:

$$F_n(x|\mu,\sigma) = F_N\left(\frac{x - \mu}{\sigma}\right) = F_N(z) \tag{10.6}$$

For instance, let X be a random variable that is normally distributed with $\mu = 100$ and $\sigma = 10$. Then the probability that a value of X will be less than 125 may be computed as follows:

$$P(X < 125) = F_n(125|100,10) = F_N\left(\frac{125 - 100}{10}\right) = F_N(2.5).$$

From Appendix Table H, we find that $F_N(2.5) = .9938$. Thus, $P(X < 125)$ is equal to .9938.

In a similar manner, the probability that a random variable X will lie in the interval between a and b may be obtained from the standard normal integral table by using the following relationship:

$$\begin{aligned} P(a \leq X \leq b) &= P(X \leq b) - P(X \leq a) \\ &= F_n(b|\mu,\sigma) - F_n(a|\mu,\sigma) \\ &= F_N\left(\frac{b-\mu}{\sigma}\right) - F_N\left(\frac{a-\mu}{\sigma}\right) \end{aligned} \tag{10.7}$$

For example, if X is normally distributed with $\mu = 100$ and $\sigma = 10$, the probability that a value of X will lie between 85 and 110 is obtained as follows:

$$\begin{aligned} P(85 \leq X \leq 110) &= F_N\left(\frac{110-100}{10}\right) - F_N\left(\frac{85-100}{10}\right) \\ &= F_N(1.0) - F_N(-1.5) \end{aligned}$$

From Appendix Table H, we find $F_N(1) = .8413$ and $F_N(-1.5) = .0668$. Therefore,

$$P(85 \leq X \leq 110) = .8413 - .0668 = .7745$$

Example 1:

Let X be a normally distributed random variable with $\mu = 60$ and $\sigma = 4$.

1. *What is the probability that a randomly observed value of X will be less than 58? Using Formula (10.6) together with Appendix Table H, we may obtain this probability as indicated below:*

$$\begin{aligned} P(X < 58) &= F_n(58|60,4) \\ &= F_N\left(\frac{58-60}{4}\right) = F_N(-.50) = .3085 \end{aligned}$$

The above probability is shown graphically by the shaded area under the normal curve in Figure 10.5a.

2. *What is the probability that a randomly observed value of X will be greater than 65? This probability is computed as follows:*

$$\begin{aligned} P(X > 65) &= 1 - F_n(65|60,4) \\ &= 1 - F_N\left(\frac{65-60}{4}\right) = 1 - F_N(1.25) \\ &= 1 - .8943 = .1057 \end{aligned}$$

The probability obtained is represented by the shaded area under the normal curve in Figure 10.5b.

3. *What is the probability that a randomly observed value of X will lie between 59 and 66? Using Formula (10.7) together with Appendix Table H, we obtain:*

$$\begin{aligned} P(59 \leq X \leq 66) &= F_N\left(\frac{66 - 60}{4}\right) - F_N\left(\frac{59 - 60}{4}\right) \\ &= F_N(1.50) - F_N(-.25) \\ &= .9332 - .4013 = .5319 \end{aligned}$$

This probability is shown by the shaded area under the normal curve in Figure 10.5c.

Figure 10.5

Areas Under Normal Curve ($\mu = 60$, $\sigma = 4$)

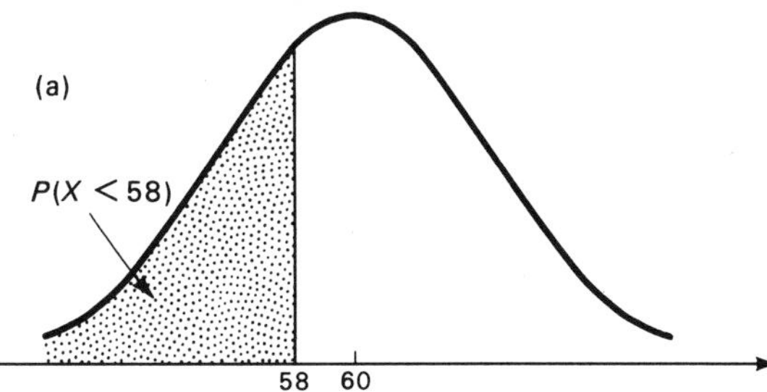

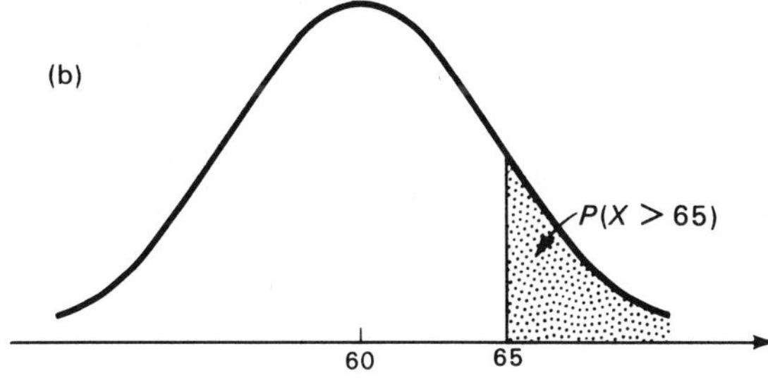

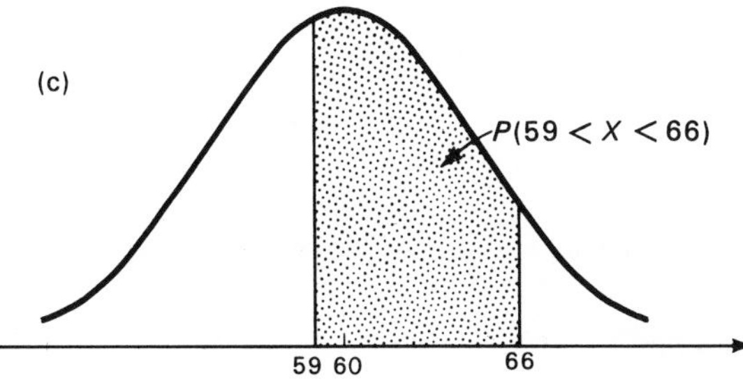

Example 2:

A photographic studio purchases a particular type of floodlamp from a distributor who claims that the lives of these particular lamps are normally distributed with a mean of eighty hours and a standard deviation of two hours. If the distributor's claim is correct, what is the probability that a given lamp will last as long as eighty-one hours?

Let H be a random variable denoting the number of hours that a floodlamp will last. A hasty reading of the problem might incorrectly suggest that the question is: What is $P(H = 81)$? However, this would be a meaningless question, since H is a continuous variable and therefore $P(H = 81)$ must equal zero. The expression "as long as eighty-one hours" is meant to denote "at least eighty-one hours." Thus the question really is: What is $P(H \geq 81)$? But since $P(H = 81)$ is zero, we might just as well say: What is $P(H > 81)$? Then we have:

$$\begin{aligned} P(H > 81) &= 1 - F_n(81|80, 2) \\ &= 1 - F_N\left(\frac{81 - 80}{2}\right) \\ &= 1 - F_N(.50) = 1 - .6915 = .3085 \end{aligned}$$

Example 3:

A machine produces ball bearings that are considered to meet specification if their diameters are within the limits $12.00 \pm .04$ centimeters. Suppose that the diameters of the bearings produced by the machine are normally distributed with a mean of 11.98 centimeters and a standard deviation of .02 centimeter. What is the probability that a bearing selected randomly from the machine's output will meet the specification?

Let D be a random variable denoting the diameter of a bearing. Since a bearing is within specification if its diameter is between 11.96 and 12.04 centimeters, the desired probability may be expressed as $P(11.96 \leq D \leq 12.04)$. This probability is calculated as follows:

$$\begin{aligned} P(11.96 \leq D \leq 12.04) &= F_n(12.04|11.98, .02) - F_n(11.96|11.98, .02) \\ &= F_N\left(\frac{12.04 - 11.98}{.02}\right) - F_N\left(\frac{11.96 - 11.98}{.02}\right) \\ &= F_N(3) - F_N(-1) \\ &= .9986 - .1587 = .8399 \end{aligned}$$

Example 4:

Assume that the net weights of packages of breakfast cereal produced by an automatic process are normally distributed such that the probability is .33 that the net weight of a package will exceed 16.06 ounces. If the standard deviation of the distribution of weights is .10 oz., what is the average net weight per package?

Let W be a random variable denoting the net weight of a package. We are told that $P(W > 16.06) = .33$, *which implies that* $P(W < 16.06) = .67$. *Thus,*

$$F_N\left(\frac{16.06 - \mu}{.10}\right) = .67$$

From Appendix Table H, we can see that if $F_N(z) = .67$, *then* $z = .44$. *Consequently,*

$$\frac{16.06 - \mu}{.10} = .44$$

By simple algebra,

$$\mu = 16.06 - .44(.10) = 16.016$$

Thus the mean net weight per package produced by the process is 16.016 ounces.

Example 5:

Suppose that the breaking strengths of steel wires of a particular type are normally distributed such that there is a .281 probability that a wire will have a breaking strength less than twenty pounds and a .879 probability that a wire will have a breaking strength less than fifty pounds. Determine the mean and standard deviation of the distribution of breaking strengths.

In this example, we have two unknowns, μ *and* σ. *Thus, we must generate a pair of simultaneous equations in order to solve the problem. Let B denote the random variable representing the breaking strength, in pounds, of a wire. We are told that there is a .281 probability that a wire will have a breaking strength less than twenty pounds; that is,* $P(B < 20) = .281$. *From Appendix Table H, if* $F(z) = .281$, *then* $z = -0.58$. *Thus,*

$$-0.58 = \frac{20 - \mu}{\sigma}$$

Similarly, we are told that $P(B < 50) = .879$. *From Appendix Table H, if* $F(z) = .879$, *then* $z = 1.17$. *This gives us the equation:*

$$1.17 = \frac{50 - \mu}{\sigma}$$

Thus we have the following pair of simultaneous equations:

$$\begin{cases} \mu - \ .58\sigma = 20 \\ \mu + 1.17\sigma = 50 \end{cases}$$

It is left to the student to verify that, if these two equations are solved simultaneously, we obtain $\mu = 29.94$ *and* $\sigma = 17.14$.

10.1.4 Remarks on the Normal Model

The normal density function first came into popular use during the nineteenth century as a model for the distribution of errors of measurement in making scientific observations. In fact, at that time it was commonly referred to as the "normal law of error." As scientists discovered an increasing number of natural phenomena that appeared to fit this model, they began to regard this law with almost a mystical reverence, as if it represented some universal truth. This awe is reflected in the following remarks by Sir Francis Galton (1822–1911), one of the intellectual superstars of the nineteenth century:

> I know of scarcely anything so apt to impress the imagination as the wonderful form of cosmic order expressed by the "Law of Frequency of Error." The law would have been personified by the Greeks and deified, if they had known of it. It reigns with serenity and in complete self-effacement amidst the wildest confusion. The huger the mob and the greater the apparent anarchy, the more perfect is its sway. It is the supreme law of Unreason.[1]

For a period of time, the growth of probability theory was actually retarded by this phenomenon, which has been called the "normal mystique," since there was little motivation to search for new laws of probability as long as there was a prevalent feeling that the normal model was universally true.[2]

Even today, because the normal distribution is highly emphasized in statistics and in the sciences in general, there is a tendency to regard this "law" with undue reverence. It is true that many natural phenomena—such as the IQ's of ten-year-olds, the tensile strengths of synthetic fibers produced by an automatic process, the heights of adult males in a particular population, or the velocities of molecules in a cubic centimeter of gas—tend to be approximately normally distributed. But it must be realized that the distributions of such phenomena are only *approximately* normal. For instance, there is no record of any human's being as tall as the Empire State Building, but the normal distribution, if it applied to human statures, would imply that we might expect to encounter occasional individuals even taller. Within a practical

[1] Sir Francis Galton, as quoted in John I. Griffin, *Statistics: Methods and Applications*, Holt, Rinehart and Winston, 1962, pp. 4–5.

[2] See James V. Bradley, *Distribution-Free Statistical Tests*, McGraw-Hill, 1968.

range, however, it is convenient for us to regard the distribution of the heights of human beings as being approximately normal. Actually, it is only through such simplifications that we can reach any understanding of nature, for a complete comprehension of "reality" in all its intricacies and subtleties is far beyond the capacity of the human intellect.

10.2 THE GAMMA PROBABILITY MODEL

We have mentioned that the normal model is the most extensively used of all probability models. However, many real-life random variables cannot be described accurately by the normal model even though these random variables may in fact be continuous. Suppose, for example, that a queue of fifteen customers has formed at a particular check-out stand in a supermarket. Suppose further that, for the checker at this station, the time required to service fifteen customers is a random variable with a mean of one hour and a standard deviation of one hour. If we were to assume a normal distribution for the service time, we would find that there would be a .1587 probability that the service time would be *less than zero*. Clearly, it is impossible for the checker to serve fifteen customers in *less than no time*. A negative amount of time is not only impossible but also meaningless. Therefore, other types of probability models have been developed to handle such situations. One such model is the *gamma* probability model.

In our discussion of models based on the Bernoulli process, we observed that the negative binomial distribution is used to determine the probability that a specified number of trials will be required in order to obtain x successes in a sequence of trials produced by a Bernoulli process. Similarly, it has been shown that the gamma model can be used to determine the probability that a specified interval of time will be required in order to observe x occurrences of a phenomenon produced by a Poisson process. In other words, just as the negative binomial distribution is the converse of the binomial distribution, so the gamma distribution is the converse of the Poisson distribution.

The reader may recall from the discussion of the Poisson model in Section 9.3 that λ (the rate of occurrence in a single unit of time or space) and t (the number of units of time or space) are fixed values. The random variable of interest, then, is X, which is the number of occurrences in t units of time or space. We now consider the converse situation in which the number of occurrences is a fixed value and the number of units of time or distance is a random variable. Thus, following our notational convention, the lower-case x will be used to denote the specified number of occurrences, and the capital letter T will be used to denote the number of units of time or distance required to generate x Poisson occurrences.

The gamma density function for the random variable T, with parameters x and λ, is

$$f_\gamma(t|x,\lambda) = \begin{cases} \dfrac{\lambda e^{-\lambda t}(\lambda t)^{x-1}}{(x-1)!} & \text{if } t \geq 0 \\ 0 & \text{if } t < 0 \end{cases} \tag{10.8}$$

In this expression, λ may be any positive value. Furthermore, although our preceding discussion of T as the time or distance required to generate x Poisson occurrences implies that x is a positive integer, the gamma density function actually holds for all positive x-values, integer or noninteger.

Regardless of the values of the parameters λ and x, the gamma density function approaches zero as t approaches infinity. However, the general shape of the function does depend on the value of x. If $x < 1$, the function approaches infinity as t approaches zero, producing a J-shaped curve. This is illustrated by the curve for $x = \frac{1}{2}$ in Figure 10.6. If $x = 1$, the curve is also

Figure 10.6

Gamma Density Functions ($\lambda = 1$)

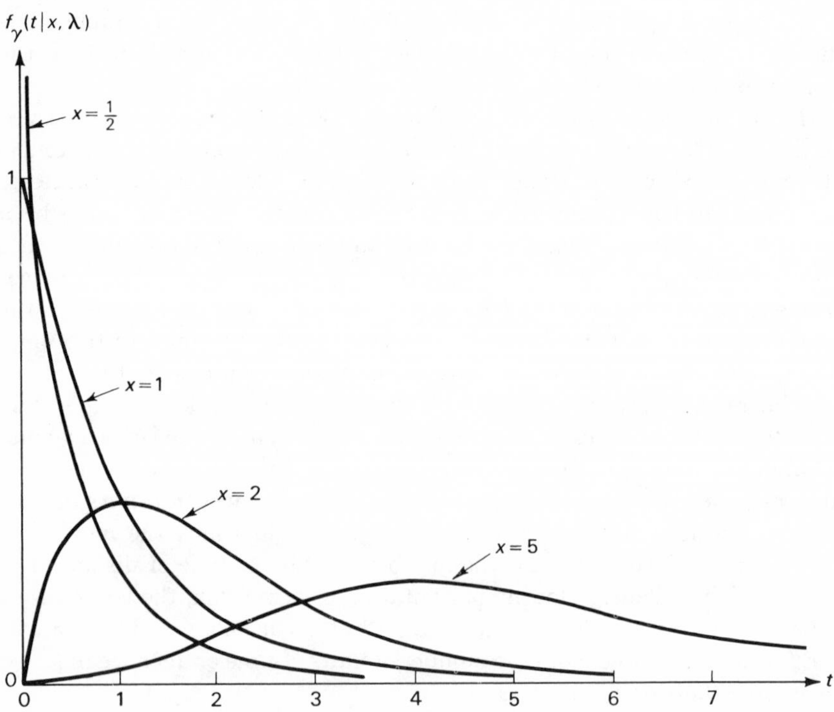

J-shaped, but the function is equal to λ when $t = 0$. This is illustrated by the curve for $x = 1$ in Figure 10.6. If $x > 1$, the gamma distribution approaches a bell-shaped curve. Unlike the bell-shaped curve of the normal distribution, which is symmetrical, the gamma distribution is skewed to the right. That is, the curve starts at zero on the left and extends indefinitely to the right. This is shown by the curves for $x = 2$ and $x = 5$ in Figure 10.6.

The gamma cumulative density function for $t \geq 0$ is given by the integral:

$$F_\gamma(t|x,\lambda) = \int_0^t \frac{\lambda e^{-\lambda y}(\lambda y)^{x-1}}{(x-1)!}\,dy \tag{10.9}$$

The mean and variance of the gamma distribution can be shown to be:

$$\mu = E(T) = \frac{x}{\lambda} \tag{10.10}$$

$$\sigma^2 = V(T) = \frac{x}{\lambda^2} \tag{10.11}$$

10.2.1 The Erlang Distribution

The gamma density function is difficult to manipulate mathematically, particularly if x is not an integer. However, in many practical applications of the gamma distribution, x is an integer. This special case of the gamma distribution in which x is limited to *positive integers* is called the *Erlang* distribution. This distribution was formulated by A. K. Erlang in 1917 as a result of his study of congestion in telephone traffic. The Erlang distribution is particularly useful in analyzing the waiting times that customers spend in lines at service facilities, such as tellers' windows in banks and checkout counters in supermarkets.

As a special case of the gamma distribution, the Erlang density function $f_E(t|x,\lambda)$ is identical to the general gamma density function $f_\gamma(t|x,\lambda)$ given in Formula (10.8). However, since x is an integer in the Erlang distribution, it is possible to obtain cumulative Erlang probabilities without integrating the density function. Specifically, cumulative Erlang probabilities may be obtained by using tables of cumulative Poisson probabilities. This is possible because the Erlang cumulative density function $F_E(t|x,\lambda)$ is related to the Poisson cumulative mass function $F_P(x|\lambda,t)$ as follows:

$$F_E(t|x,\lambda) = 1 - F_P(x-1|\lambda,t) \tag{10.12}$$

To grasp this relationship, we first observe that if the time or distance required to generate x occurrences is at most t, then there must be at least x

occurrences in t. Thus, $P(T \leq t|x,\lambda) = P(X \geq x|\lambda,t)$. Then to understand Formula (10.12), we simply recognize that $P(T \leq t|x,\lambda) = F_E(t|x,\lambda)$ and that $P(X \geq x|\lambda,t) = 1 - F_P(x - 1|\lambda,t)$.

Example:

A building contractor has signed a contract to erect sixteen pre-fabricated cabins at a mountain resort. He has agreed to complete the project within forty working days. On the average, the contractor can erect one cabin in two days. If the cabins are completed according to a Poisson process, what is the probability that the contract will be fulfilled within forty days?

To apply the gamma probability model to this problem, we first note that $x = 16$, which is the number of cabins to be built. Furthermore, $\lambda = \frac{1}{2} = .5$, which is the average number of cabins built per day. Then using Formula (10.12), we have:

$$P(T \leq 40) = F_E(40|16, .5) = 1 - F_P(16 - 1|.5, 40)$$

From Appendix Table G, we find $F_P(15|.5, 40) = .1565$. Thus,

$$P(T \leq 40) = 1 - .1565 = .8435$$

which is the probability that the contract will be fulfilled within forty days.

10.2.2 The Exponential Distribution

The *exponential* distribution is a special case of the Erlang distribution in which $x = 1$. Since the Erlang distribution is a special case of the gamma distribution, the exponential distribution is also a special case of the gamma distribution. When $x = 1$, all terms containing x vanish from the gamma density function in Formula (10.8), which then becomes:

$$f_e(t|\lambda) = \begin{cases} \lambda e^{-\lambda t} & \text{if } t \geq 0 \\ 0 & \text{if } t < 0 \end{cases} \tag{10.13}$$

This particular form of the gamma density function is the exponential density function. Because the exponential density function is a special case of the gamma distribution when $x = 1$, the exponential distribution has the general shape illustrated by the curve for $x = 1$ in Figure 10.6. Also since $x = 1$, the random variable T in the exponential model represents the time or distance required to generate a *single* Poisson occurrence. Alternatively, T may be regarded as the time or distance between two successive Poisson occurrences.

The exponential cumulative density function is given by the integral:

$$F_e(t|\lambda) = \int_0^t \lambda e^{-\lambda y}\, dy$$

This integral is easily solved, and it is left to the reader to demonstrate that, when the integration is performed, the exponential cumulative density function becomes:

$$F_e(t|\lambda) = 1 - e^{-\lambda t} \tag{10.14}$$

Thus, using Formula (10.14), exponential probabilities can be easily computed with the aid of a table of exponential functions, such as Appendix Table C.

Example:

An amusement arcade has a booth in which, for the small price of a quarter, a person may be photographed in four flattering poses. The machine operates automatically and takes exactly two minutes to complete its cycle from the time the customer inserts his quarter in the slot. Suppose that customers arrive at the booth according to a Poisson process at a mean rate of .4 customers per minute. If a customer has just dropped his quarter in the slot, what is the probability that the next customer will arrive at the booth before the machine completes its cycle? To answer this question, we need to determine:

$$P(T < 2) = F_e(2|.4) = 1 - e^{-(.4)(2)} = 1 - e^{-.8}$$

From Appendix Table C, $e^{-.8} = .4493$. *Therefore,*

$$P(T < 2) = 1 - .4493 = .5507$$

It is left as an exercise for the reader to show that the mean of the exponential distribution is:

$$\mu = E(T) = \frac{1}{\lambda} \tag{10.15}$$

and that the variance is:

$$\sigma^2 = V(T) = \frac{1}{\lambda^2} \tag{10.16}$$

10.3 THE RECTANGULAR PROBABILITY MODEL

In our discussion of the normal model we saw that the domain of the normal density function is $\{x| -\infty < x < \infty\}$, so that the curve extends to infinity in both directions. In the case of the gamma model, we observed that the domain of the function is $\{t|0 \leq t < \infty\}$ so that the curve extends indefinitely to the right. In other words, the domains of both the normal and gamma probability functions are *infinite.*

We will now examine another probability model, the *rectangular* distribution, which has a *finite* domain. If a random variable X has a rectangular density function, the values of X can lie only within an interval bounded by two finite values. Denoting these two values by a and b (where $a < b$), we may express the domain of X as $\{x|a \leq x \leq b\}$. Then the rectangular density function of X is

$$f_r(x|a,b) = \begin{cases} \dfrac{1}{b-a} & \text{if } a \leq x \leq b \\ 0 & \text{otherwise} \end{cases} \tag{10.17}$$

As Formula (10.17) indicates, the rectangular density function has a constant value of $\dfrac{1}{b-a}$ over the interval from a to b. Thus the graph of this function is a rectangle with altitude $\dfrac{1}{b-a}$, as shown in Figure 10.7.

Figure 10.7

The Rectangular Density Function

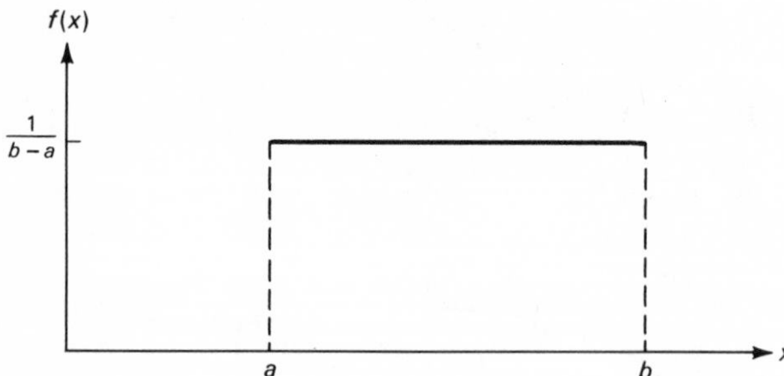

Unlike the normal and gamma distributions, the rectangular distribution has the property that events which are defined by subintervals of equal length always have equal probabilities. For example, let X be a random variable which has a rectangular distribution over the interval from 2 to 12. Since $(12 - 2) = 10$, the p.d.f. of X is:

$$f(x) = \begin{cases} \dfrac{1}{10} & \text{if } 2 \leq x \leq 12 \\ 0 & \text{otherwise} \end{cases}$$

Now consider the following three events which are defined as subintervals of the domain of X:

$$A = \{x|3 \leq x \leq 5\}$$
$$B = \{x|6 \leq x \leq 8\}$$
$$C = \{x|10 \leq x \leq 12\}$$

Notice that all three events are defined by intervals of equal length—two units. Then, as the following integrals verify, these three events have equal probabilities:

$$P(A) = \int_3^5 \frac{1}{10}\,dx = .20$$
$$P(B) = \int_6^8 \frac{1}{10}\,dx = .20$$
$$P(C) = \int_{10}^{12} \frac{1}{10}\,dx = .20$$

These three probabilities are represented by the shaded areas of Figure 10.8. Because the rectangular density function assigns uniform probabilities to equal intervals, it is often referred to as the *continuous uniform* distribution.

Figure 10.8

Three Events Defined on a Rectangularly Distributed Random Variable

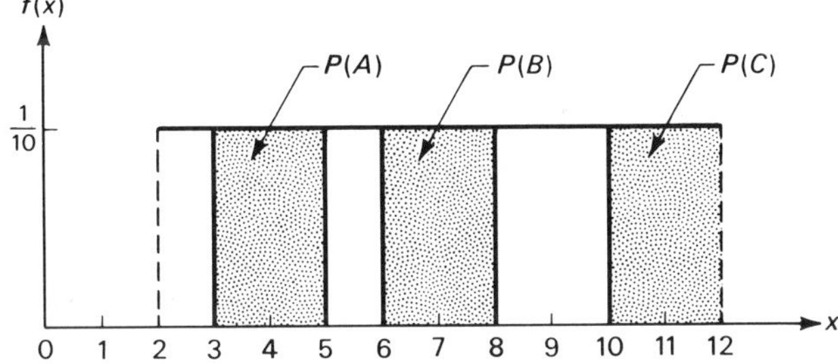

As the reader can verify by integrating the density function in Formula (10.17), the rectangular cumulative density function is given by

$$F_r(x|a,b) = \begin{cases} 0 & \text{if } x < a \\ \dfrac{x - a}{b - a} & \text{if } a \leq x \leq b \\ 1 & \text{if } x > b \end{cases} \tag{10.18}$$

As the reader can further verify, the mean and variance of the rectangular distribution are

$$\mu = E(X) = \frac{a + b}{2} \tag{10.19}$$

$$\sigma^2 = V(X) = \frac{(b - a)^2}{12} \tag{10.20}$$

Example:

A business executive commutes to his office in Metropolis from his home in the outskirts of the suburban community of Dullsville. Each morning he drives from his home to the Dullsville dock, where he boards a ferry that takes him across the river to Metropolis. Suppose his driving time from home to the dock is a random variable T which has a rectangular distribution over the interval between 10 minutes and 30 minutes.

In this example, the domain of T is $\{t|10 \leq t \leq 30\}$. *Applying Formula (10.17), the probability density function of the executive's driving time to the dock is*

$$f_r(t|10{,}30) = \begin{cases} \dfrac{1}{20} & \text{if } 10 \leq t \leq 30 \\ 0 & \text{otherwise} \end{cases}$$

From Formula (10.18), the corresponding cumulative density function is

$$F_r(t|10{,}30) = \begin{cases} 0 & \text{if } t < 0 \\ \dfrac{t - 10}{20} & \text{if } 10 \leq t \leq 30 \\ 1 & \text{if } t > 30 \end{cases}$$

Suppose that the ferry leaves the dock precisely at 7:00 a.m. If the executive leaves his home at 6:45 on a particular morning, what is the probability that he will miss the boat? To answer this question, we first observe that, if he leaves home at 6:45, the executive must take no more than 15 minutes to reach the dock if he is going to catch the 7:00 ferry. Thus, if $T \leq 15$ *minutes, he will catch the ferry. However, he will miss the boat if* $T > 15$, *which has the probability*

$$\begin{aligned} P(T > 15) &= 1 - P(T \leq 15) \\ &= 1 - F_r(15|10, 30) \\ &= 1 - \frac{15 - 10}{20} = 1 - .25 \\ &= .75 \end{aligned}$$

This solution is illustrated diagramatically in Figure 10.9.

Figure 10.9

Diagram Illustrating Probability that Commuter Will Miss the Boat

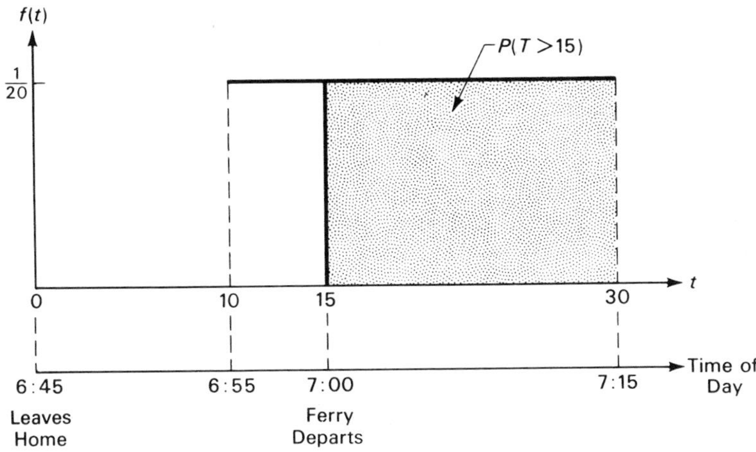

Applying Formula (10.19) to this problem, we find that the mean time required by the executive to drive to the dock is

$$E(T) = \frac{10 + 30}{2} = 20 \text{ minutes}$$

Applying Formula (10.20), the variance of driving time is

$$V(T) = \frac{(30 - 10)^2}{12} = \frac{400}{12} = 33.33 \text{ square minutes}$$

Thus the standard deviation of the driving time is $\sqrt{33.33} = 5.77$ *minutes.*

PROBLEMS

10.1 Let $f(z)$ represent the standard normal density function. Using a standard normal integral table, evaluate each of the following expressions:

(a) $\int_{.60}^{\infty} f(z)\,dz$ (b) $\int_{-\infty}^{-2.06} f(z)\,dz$ (c) $\int_{.20}^{.97} f(z)\,dz$

(d) $\int_{-.20}^{.97} f(z)\,dz$ (e) $\int_{-1.20}^{-1.02} f(z)\,dz$ (f) $\int_{-1.96}^{1.96} f(z)\,dz$

10.2 Let M be a normally distributed random variable with $\mu = 30$ and $\sigma = 2$. Find the following probabilities:
(a) $P(M \leq 34)$
(b) $P(M \leq 26)$
(c) $P(M \geq 31)$
(d) $P(M \geq 33.5)$
(e) $P(29.5 \leq M \leq 33.0)$

10.3 Y is a normally distributed random variable with $\mu = 14$ and $\sigma = 25$. For a single observation, find:
(a) $P(Y < 0)$
(b) $P(Y > 50)$

10.4 Let H be a normally distributed random variable with $\mu = .025$ and $\sigma = .0015$. Find the following probabilities:
(a) $P(H \leq .03)$
(b) $P(H \leq -.022)$
(c) $P(H \geq .027)$
(d) $P(.024 \leq H \leq .026)$

10.5 Assume that the time required for a production crew to produce a custom-made Wombat Special limousine is a normally distributed random variable with a mean of 1,200 man-hours with a standard deviation of 200 man-hours. If a customer orders one of these limousines, what is the probability that its production will require between 1,140 man-hours and 1,520 man-hours?

10.6 An electric utility company has 200,000 domestic subscribers. For the month of October, the mean electric bill was $\mu = \$8.50$ and the standard deviation of these bills was $\sigma = \$1.25$.
(a) If it were reasonable to assume that these bills were normally distributed, how many of these bills would lie between \$6.00 and \$11.00?
(b) How reasonable would it be to assume that these bills are normally distributed? Explain.

10.7 Assume that the net weight of a crate of apples is a normally distributed random variable with a mean of 60 pounds and a standard deviation of three-quarters of a pound. If a grocer buys a crate of these apples, what is the probability that its net weight will be:
(a) above 62 pounds?
(b) less than 61 pounds?
(c) above 59.5 pounds?

10.8 Two different processes, A and B, are available for producing a particular type of chain links. The links produced by Process A have a mean breaking strength of 50 pounds with a standard deviation of 5 pounds. The links produced by Process B have a mean breaking strength of 60 pounds with a standard deviation of 12 pounds. Suppose the distributions of breaking strength are normal for both processes.
(a) Which process will yield the greater proportion of links with breaking strengths above 45 pounds?
(b) Which process will yield the greater proportion of links with breaking strengths less than 40 pounds?

10.9 Let H be a random variable denoting the life (in hours) of a particular type of volume-produced battery designed for transistor radios. Assume that H is normally distributed with $\mu = 26$ and $\sigma = 5$.
(a) Of this type of battery, 2.5 per cent will have lives less than what value of H?

(b) Of this type of battery, 67 per cent will have lives greater than what value of H?

10.10 Let D be a normally distributed random variable with $\sigma = 1$. Suppose that $P(D > 0) = .877$. Find μ.

10.11 Let G be a normally distributed random variable with $\mu = 10$ and $P(G > 9) = .975$. Find σ.

10.12 An investigation of the life cycle of a particular type of telephonic equipment disclosed that there was a .10 probability that a piece of this equipment would last 25 years or longer, and a .67 probability that it would last 17 years or longer. Assume that the life of a piece of this equipment is a normally distributed random variable.
(a) What is the mean life of this type of equipment?
(b) What is the standard deviation of the life of this type of equipment?
(c) What percentage of the installations of this equipment last between 12 and 22 years?

10.13 Let X be a normally distributed random variable with $\mu_X = 60$ and $\sigma_X = 10$. Furthermore, let Y be a normally distributed random variable, independent of X, with $\mu_Y = 10$ and $\sigma_Y = 8$. Determine $P[(X > 73) \cap (Y > 0)]$.

10.14 The lengths of steel pins produced by an automatic process are normally distributed with a mean of 1.96 millimeters and a standard deviation of .10 millimeter.
(a) If 1 of these pins is selected at random, what is the probability that it will have a length between 1.94 millimeters and 1.99 millimeters?
(b) If 3 of these pins are selected at random, what is the probability that exactly 1 of them has a length less than 1.95 millimeters?

10.15 An automatic process produces steel shafts that must meet specifications with respect to both length and diameter. The lengths of the shafts produced by the process are normally distributed with a mean of 8 centimeters and a standard deviation of .20 centimeter. The diameters are normally distributed with a mean of 28 millimeters and a standard deviation of .05 millimeter. If a shaft is randomly selected from the output of the process, and if the lengths and diameters of the shafts are independent, what is the probability that it will be a shaft with a length less than 7.70 centimeters and a diameter greater than 28.15 millimeters?

10.16 In a particular industrial plant, fatal accidents occur according to a Poisson process at a mean rate of $\lambda = .05$ accidents per month. A fatal accident has just occurred.
(a) What is the probability that at least 24 months will elapse before a total of 2 more of these accidents occur?
(b) What is the expected amount of time that will elapse before a total of 2 more of these accidents occur?

10.17 Suppose that the number of calls arriving at a company's switchboard during a 5-minute span has a Poisson distribution with $\lambda = 4.8$.
(a) If a call arrives at exactly 10:00 A.M., what is the probability that it will be no later than 10:10 A.M. by the time a total of 8 additional calls arrive?
(b) What is the expected amount of time that will elapse between the moment when one call arrives and the moment that the eighth succeeding call arrives?

10.18 The Beehive Wireworks produces a particular type of wire by a continuous automatic process. Suppose that "pin-holes" occur in this wire according to a Poisson process with a mean of 12 pin-holes per 1,000 yards. At a particular point in the output of this process, a pin-hole occurs. What is the probability that the process will produce at most 100 additional yards of this wire when 3 additional pin-holes occur?

10.19 The Hi Speed Plastic Products Company has a very efficient machine for producing a special toy. However, due to the failure of a critical part, machine breakdowns occur as a Poisson process at a mean rate of 1 breakdown in 4 months. If there are 3 spare parts on hand after a breakdown is repaired, what is the probability that it will take at least 8 months to use up this supply of spare parts?

10.20 The owner of a charter boat operating from a Florida fishing pier makes the following deal with a client: The client will pay the boat operator a fee of $50 to take him marlin fishing. From the moment when the boat passes a particular deep-water buoy, the boat will stay out 8 hours before turning back toward shore. However, if as many as 3 marlin "strikes" should occur, the boat will turn and begin heading toward shore when the third strike occurs. Furthermore, if less than 3 strikes occur within the 8 hours after the boat passes the buoy, the operator agrees to refund $10 of the client's fee. Assume that marlin strikes in these waters occur according to a Poisson process with a mean rate of .5 strike per hour.

(a) What is the probability that the boat will turn for shore before the 8 hours elapse?

(b) If the boat stayed out until 3 marlin strikes occur, regardless of how long that might take, what would be the expected amount of time that would elapse between the moment that it passes the buoy and the moment that it turns for shore?

(c) What is the expected value of the fee that the boat owner will receive?

10.21 The number of flaws in reels of ribbon produced by an automatic machine occur as a Poisson process with an expected number of .2 flaws per foot. A seamstress purchases 1 yard of this ribbon. Suppose that, as she unwinds the ribbon, the seamstress finds a flaw at the end of the first 6 inches. What is the probability that she will unwind at least another 6 inches before she finds another flaw?

10.22 Between the hours of 2:00 P.M. and 4:00 P.M., calls arrive at the switchboard of a large city welfare department as a Poisson process with an expectation of 3 calls per minute. If a call arrives at exactly 3:10 P.M., what is the probability that the next call will not arrive until at least 3:12 P.M.?

10.23 The reliability of a component of an electronic system is commonly defined as the probability that the component will function continuously for at least t units of time. Suppose that the reliability of a particular component is defined as the probability that it will last at least 6 days. Suppose further that the lifetime of this type of component is exponentially distributed with an expected life of 10 days. What is the reliability of the component?

10.24 The life in hours, H, of a particular kind of electronic component is an exponentially distributed random variable with an expected value of 4 hours.

(a) Graph the probability density function of H.

(b) State the corresponding cumulative density function.

(c) Graph the c.d.f. obtained in (b).

(d) What is the standard deviation of the lives of this type of component?

(e) What is the probability that the life of one of these components will exceed expectation?

(f) An assembly contains 3 of these components. Assuming that the life spans of these components are mutually independent, what is the probability that exactly 2 of the components in an assembly will live beyond expectation?

10.25 For the exponential model, prove:

(a) $F_e(t|\lambda) = 1 - e^{-\lambda t}$

(b) $\mu = E(T) = \dfrac{1}{\lambda}$

(c) $\sigma^2 = V(T) = \dfrac{1}{\lambda^2}$

10.26 Let X be a random variable distributed uniformly over the interval between 2 and 5.

(a) Find the following probabilities:

(i) $P(2 \leq X \leq 3)$

(ii) $P(X < 3)$

(iii) $P(X \geq 4)$

(b) Determine the mean of the distribution.

(c) Determine the variance of the distribution.

10.27 During the evening rush hour, an express bus to Pine Rapids leaves the Central City terminal promptly every 10 minutes. Suppose that passengers arrive at the terminal randomly during the rush hour. What is the probability that a person arriving at the terminal will have to wait at least 6 minutes before the next express bus departs for Pine Rapids?

10.28 Suppose the time required by a commuter to travel from his home to the Toonerville monorail terminal is a random variable which has a rectangular distribution over the interval between 15 minutes and 25 minutes. If he leaves home precisely at 7:20 A.M., what is the probability that he will arrive at the terminal in time to catch the monorail coach which leaves the terminal promptly at 7:42 A.M.?

10.29 The signal lights controlling the flow of east-west traffic on 7th Street at 4th Avenue operates on a 100-second cycle. Each cycle consists of 56 seconds green, 10 seconds yellow, and 34 seconds red. Suppose motorists travelling west on 7th Street arrive at this particular intersection at random.

(a) What is the probability that the traffic light is green when a motorist arrives at this intersection?

(b) What is the probability that the traffic light is yellow when a motorist arrives at this intersection?

(c) What is the probability that the traffic light is red when a motorist arrives at this intersection?

10.30 For the rectangular model, prove:

(a) $F_r(x|a,b) = \dfrac{x - a}{b - a}$ if $a \leq x \leq b$

(b) $\mu = E(X) = \dfrac{a + b}{2}$

(c) $\sigma^2 = V(X) = \dfrac{(b - a)^2}{12}$

Chapter 11
Some Useful Probability Approximations

Chapters Nine and Ten considered several commonly used probability models, each of which is applicable under certain carefully defined conditions. If the specific conditions of a probability model are satisfied, then that model may be used to compute the *exact* probabilities of particular events. Moreover, due to certain mathematical relationships that exist among various models, it frequently is possible to use a related model to compute *approximate* probabilities of the same events. Indeed, when practical considerations are taken into account, the computation of approximate probabilities may well be preferable to the computation of exact probabilities. For example, suppose that we wish to determine the probability of obtaining at least 5,600 "successes" in 10,000 trials of a Bernoulli process for which $p = .55$. In this case, the exact probability may be obtained from the binomial model. Unfortunately, no available binomial tables contain probabilities for such a large value of n. We could, of course, calculate the necessary binomial cumulative probability using Formula (9.6), but this would be a burdensome, time-consuming, and expensive task. We then should consider the use of some alternate probability model that, although it may provide probabilities that are only approximate, may be much more practical to employ under the circumstances.

In this chapter we will be particularly concerned with certain relationships among the hypergeometric, binomial, Poisson, and normal probability models. Rather than focusing on the detailed mathematical demonstration of these relationships, we will concentrate on the use of these relationships in deriving approximate solutions for problems in which exact solutions are overly complex or cumbersome. Specifically, this chapter will consider the following three approximation procedures:

1. Poisson approximation to the binomial
2. Normal approximation to the binomial
3. Binomial approximation to the hypergeometric

In many cases, the above approximation procedures are considerably easier to apply than the exact models. Obviously, if this is true, it would be reasonable to use an approximation procedure, *if* the approximate probabilities are sufficiently accurate for our needs. This proviso concerning the magnitude of error incurred in using an approximation is central to the use of approximation methods for calculating probabilities. Unfortunately, there are no hard-and-fast rules for deciding whether a particular approximation is adequate in a specific case, but this fact does not mitigate our need to be aware of this problem. Lack of such awareness can, and frequently does, lead to flagrant misuse of approximation procedures. In the absence of strict rules for specifying the conditions under which an approximation might be adequate, this chapter will present some general rules of thumb. In the final analysis, however, the decision to employ an approximation procedure must rest on substantive judgment concerning whether it is economically sound to tolerate the inaccuracy introduced by the approximation method.

11.1 EFFECT OF THE BINOMIAL PARAMETERS ON THE SHAPE OF THE DISTRIBUTION

To understand how alternate probability models may be used to approximate binomial probabilities, it is helpful to consider how the values of the parameters n and p affect the shape of the binomial distribution. Figure 11.1 presents a comparison of various binomial distributions for selected values of n and p. From careful study of these graphs, we can make the following generalizations:

1. As shown by distributions (c), (f), and (i) in Figure 11.1, when $p = .50$ the graph of the binomial is symmetrical regardless of the size of n.
2. If the value of p is not .50, the binomial distribution is skewed. The further p is from .50 in either direction, the greater the skewness. For example, distributions (a), (b), and (c) all represent binomials with $n = 10$ but with different values of p. Distribution (c), with $p = .50$, is symmetrical, whereas distribution (b), with $p = .25$, is skewed, and distribution (a), with $p = .05$, is skewed even more. A similar observation may be made by comparing distributions (d), (e), and (f), or by comparing distributions (g), (h), and (i).
3. If the value of p is not .50, the degree of skewness of the binomial distribution is more pronounced when n is small than when n is large. For example, distributions (a), (d), and (g) all represent binomials with $p = .05$ but with different values of n. Distribution (a), with $n = 10$, is markedly skewed, whereas distribution (d), with $n = 30$, is less so, and distribution (g), with $n = 100$, approaches symmetry.
4. As n increases, so that there is an increased number of terms in the binomial, the differences in probability values between successive terms become increasingly small so that, when n is large, the graph approaches the appearance of a continuous curve. This may be seen from Figure 11.1 by comparing the distributions in the bottom row with the corresponding distributions in the top row.

Figure 11.1

Different Shapes of the Binomial Distribution with Variations in *n* and *p*

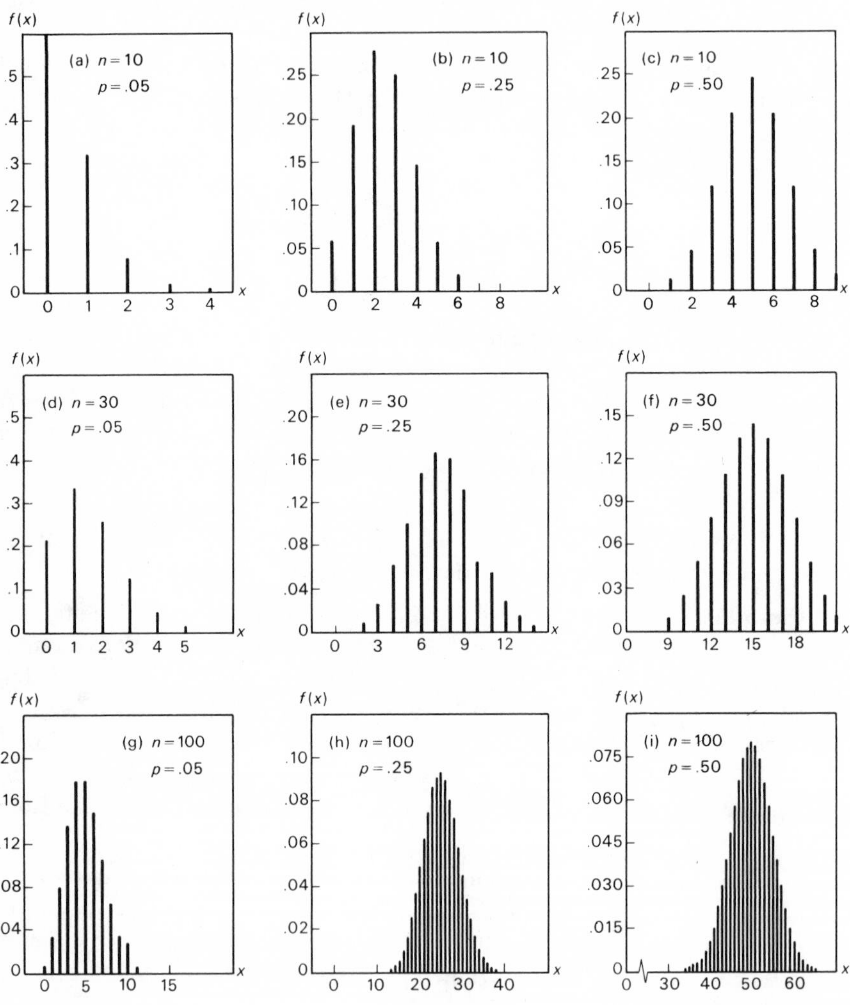

Distributions (a) and (i) provide a comparison of extreme conditions. In distribution (a), with small n and p far from .50, we have a markedly skewed, obviously discrete distribution. In distribution (i), with large n and $p = .50$, we have a symmetrical distribution that approaches the appearance of a continuous curve.

11.2 POISSON APPROXIMATION TO THE BINOMIAL

In our discussion of the Poisson distribution, we pointed out that this model has been associated with "rare events." In its general shape, for relatively small values of λt, the Poisson distribution appears as a discrete function that is markedly skewed. If we are dealing with a Bernoulli process in which p, the probability of a success, is extremely small, we may regard the occurrence of such a success as a rare event. We then might expect that, for a series of Bernoulli trials with very small p, the binomial distribution for the number of successes would resemble a Poisson distribution. It is indeed true that, when p is very small, the binomial distribution has the same general

Figure 11.2

Comparison of Binomial Distribution ($n = 10$, $p = .05$) with Corresponding Poisson Distribution ($\lambda = .50$, $t = 1$)

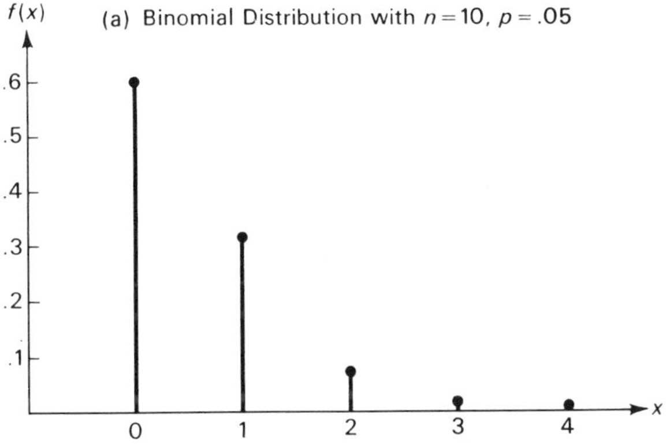

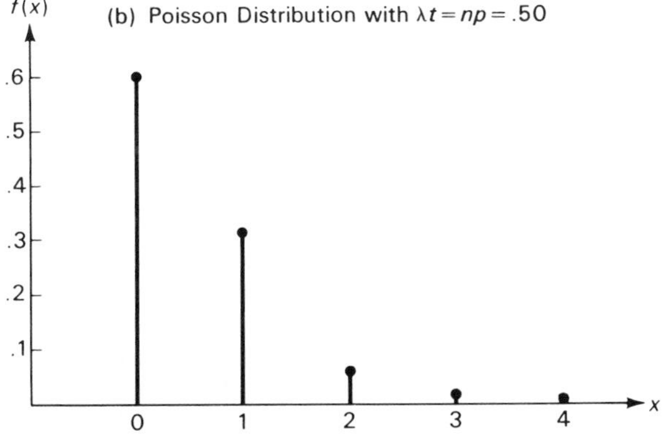

appearance as the Poisson distribution. This resemblance is illustrated in Figure 11.2, which compares a binomial distribution for $n = 10$ and $p = .05$ with a Poisson distribution for $\lambda = .50$ and $t = 1$, or $\lambda t = .50$. This resemblance is more than superficial since it may be demonstrated mathematically that the Poisson distribution is the limit of the binomial distribution as p approaches zero and n approaches infinity.

11.2.1 Mechanics of the Poisson Approximation

In our discussion of the binomial distribution, we mentioned that the parameters of the distribution are n and p, and that the mean of the distribution is given by np. Similarly, we noted that the Poisson distribution has the parameters λ and t, and the mean of that distribution is λt. Thus, when p is close to zero[1], binomial probabilities are approximately the same as the corresponding probabilities given by the Poisson distribution with $\lambda t = np$.

To illustrate the use of the Poisson approximation to the binomial, suppose that the H. L. Glad Manufacturing Company is considering entering into a contract with Feeder Enterprises to purchase a standard part as a component for one of its large-selling products. Feeder maintains that the defective rate of its production process does not exceed .05. To support this claim, Feeder agrees to produce a sample of 100 parts, which H. L. Glad will be permitted to inspect. Glad wishes to determine the probability that he will find no more than three defectives in the sample, assuming that the production process operates as a Bernoulli process with a defective rate of $p = .05$. Working under the assumption that the production process meets the conditions of a Bernoulli process, Glad realizes that the binomial model should be used to compute the exact probability. After searching the company library, however, he is unable to find a binomial table that includes $n = 100$. Suppose, however, that he is able to locate a fairly extensive volume of tabulated values of the Poisson distribution. In order to save time and the expense of making binomial calculations, or obtaining more complete binomial tables, and since p is fairly close to zero, he decides to derive the required probability by using the Poisson approximation to the binomial. To do this, he enters the Poisson table with $\lambda t = np = 100(.05) = 5$, and finds:

$$P(X \leq 3|\lambda t = 5) = F_P(3|\lambda t = 5) = .2650$$

Thus, the desired probability is approximately equal to .2650.

[1] The Poisson approximation to the binomial may also be employed when p is close to 1, such that (1-p) is close to zero. That is, if the probability of a "success" is close to 1, then the probability of a "failure" will be close to zero, and the Poisson approximation may be used by working in terms of "failures" rather than "successes."

11.2.2 Accuracy of the Poisson Approximation

In the above discussion, we observed that $P(X \leq 3)$ is *approximately* equal to .2650. Actually, how accurate is this particular approximation? This question may be answered readily, since the exact cumulative binomial probabilities for $n = 100$ and $p = .05$ are available in Appendix Table E, where we find that $P(X \leq 3|n = 100, \text{p} = .05) = F_b(3|100, .05) = .2578$. Thus the error in the approximate answer is $(.2650 - .2578) = .0072$. That is, the approximate probability is .0072 greater than the exact probability. In relative terms, the error may be expressed as:

$$\frac{.2650 - .2578}{.2578} = .0279 = 2.79\%$$

In other words, the approximate probability is 2.79 per cent greater than the exact probability.

So far, we have been concerned with the Poisson approximation to the binomial probability for the one particular event that $X \leq 3$. To obtain a somewhat broader perspective on the accuracy of the Poisson approximation to the binomial, let us examine Table 11.1, which compares the individual terms of the binomial mass function for $n = 100$ and $p = .05$ with the corresponding terms of the Poisson mass function for $\lambda t = 5$. For this binomial distribution, the mean is $E(X) = np = 5$. In general, as suggested by the

Table 11.1

Comparison of Binomial Mass Function ($n = 100$, $p = .05$) with Poisson Mass Function ($\lambda t = 5$)

(1) x	*(2)* *Poisson* $f_p(x\|\lambda t = 5)$	*(3)* *Binomial* $f_b(x\|n = 100, p = .05)$	*(4)* *Error* *(2) − (3)*
0	.0067	.0059	+.0008
1	.0337	.0312	+.0025
2	.0842	.0812	+.0030
3	.1404	.1396	+.0008
4	.1755	.1781	−.0026
5	.1755	.1800	−.0045
6	.1462	.1500	−.0038
7	.1044	.1060	−.0016
8	.0653	.0649	+.0004
9	.0363	.0349	+.0014
10	.0181	.0167	+.0014
11	.0082	.0072	+.0010
12	.0034	.0028	+.0006
13	.0013	.0010	+.0003
14	.0005	.0003	+.0002
15	.0002	.0001	+.0001

signs of the errors in Column (4) of the table, the Poisson tends to underestimate the binomial probabilities for the x-values in the vicinity of the mean, and to overestimate the binomial probabilities for x-values in the tails of the distribution. Moreover, in terms of the *magnitude* of error, the Poisson model is least accurate in approximating the individual probabilities in the neighborhood of the mean, and most accurate in approximating the individual probabilities in the tails of the distribution.

Table 11.2 compares the cumulative terms of the binomial distribution for $n = 100$ and $p = .05$ with the corresponding cumulative terms of the Poisson distribution for $\lambda t = 5$. As suggested by the signs of the errors in Column (4), there is a general tendency for the Poisson approximation to overestimate cumulative probabilities in the lower tail of the binomial, and to underestimate cumulative probabilities in the upper tail.[2]

Overall comparison of the distributions shown in Tables 11.1 and 11.2 indicates that the values of the corresponding terms of the two distributions are "reasonably" close. As we have noted, however, there is no steadfast rule to tell us how close is "close enough." That will depend on the accuracy required for the problem at hand.

Table 11.2

Comparison of Binomial Cumulative Function ($n = 100$, $p = .05$) with Poisson Cumulative Function ($\lambda t = 5$)

(1) x	*(2)* *Poisson* $F_P(x\|\lambda t = 5)$	*(3)* *Binomial* $F_b(x\|n = 100, p = .05)$	*(4)* *Error* (2) − (3)
0	.0067	.0059	+.0008
1	.0404	.0371	+.0033
2	.1247	.1183	+.0064
3	.2650	.2578	+.0072
4	.4405	.4360	+.0045
5	.6160	.6160	.0000
6	.7622	.7660	−.0038
7	.8666	.8720	−.0054
8	.9319	.9369	−.0050
9	.9682	.9718	−.0036
10	.9863	.9885	−.0022
11	.9945	.9957	−.0012
12	.9980	.9985	−.0005
13	.9993	.9995	−.0002
14	.9998	.9999	−.0001
15	.9999	1.0000	−.0001

[2] For a more detailed technical discussion of the Poisson approximation to the binomial see T. W. Anderson and S. M. Samuels, "Some Inequalities among Binomial and Poisson Probabilities," *Fifth Berkeley Symposium on Mathematical Statistics and Probability*, University of California Press, 1967.

11.3 NORMAL APPROXIMATION TO THE BINOMIAL

As we pointed out in our discussion of the effect of the values of n and p on the general shape of the binomial distribution, a graph of the distribution approaches the appearance of a symmetrical, smooth curve as n becomes quite large. In fact, the binomial distribution in Figure 11.1(i) looks strikingly similar to a normal distribution. This is more than mere coincidence. As early as 1733, Abraham DeMoivre demonstrated that the formula for the normal density function is the limit of the binomial distribution as n approaches infinity. Thus we find that, under particular circumstances, the normal distribution may be used to obtain approximations of binomial probabilities.

We have seen that the Poisson approximation to the binomial is primarily applicable when p deviates markedly from .50. In contrast, *when n is very large, particularly when p is close to .50,* the normal distribution may be used to provide adequate approximations of binomial probabilities.

11.3.1 Mechanics of the Normal Approximation

In the case of the Poisson approximation to the binomial, the mechanics of making the approximation were quite simple, since both the Poisson and the binomial are discrete distributions. In the case of the normal approximation to the binomial, however, the mechanics are somewhat more involved since we are using a *continuous* distribution to approximate a *discrete* distribution. To illustrate the procedure for the normal approximation to the binomial, consider a simplified case in which we wish to determine the probability of observing six successes in eight trials of a Bernoulli process with $p = .50$. For this particular problem, the normal approximation is not needed, since n is so small that the exact binomial probability may be readily obtained from binomial tables. However, this simple case will be particularly useful for illustrating the approximation procedure.

The *first* step in performing the approximation is to determine μ and σ of the binomial distribution with parameters n and p. This may be accomplished by using Formulas (9.7) and (9.9), which we have already examined in our discussion of the binomial distribution. For our example, we obtain:

$$\mu = np = 8(.50) = 4$$
$$\sigma = \sqrt{np(1 - p)} = \sqrt{8(.50)(1 - .50)} = 1.414$$

The *second* step is to adopt a normal distribution whose mean is set equal to the binomial mean, and whose standard deviation is set equal to the binomial standard deviation. That is, for our example, we will use a normal distribution with $\mu = 4$ and $\sigma = 1.414$.

The *third* step is to convert the definition of the event under consideration from discrete terms into continuous terms. This is necessary since we are approximating a discrete distribution by a continuous distribution. The rationale underlying this step is illustrated in Figure 11.3, which shows the normal distribution superimposed on the binomial distribution. The first important point to note in this figure is that the various terms of the binomial have been represented by contiguous bars rather than by separate vertical lines. Each bar extends one-half unit above and one-half unit below the integer value that it represents. That is, each integer value, x, is redefined as an interval extending between the *limits* of $(x - .5)$ and $(x + .5)$. By this redefinition, each of the bars has unit width, so that the area of each bar is equal to the height of the bar. Thus we are able to represent the various binomial probabilities as areas of rectangles, just as normal probabilities are represented by areas under the curve. As shown in Figure 11.3, the event $(X = 6)$ corresponds to the interval on the x-axis from 5.5 to 6.5. By inspection, it may be seen that the shaded area under the normal curve between 5.5 and 6.5 is approximately equal to the area under the bar representing $P(X = 6)$.

The *fourth* step in performing the approximation is to obtain the area under the normal curve over the interval on the x-axis corresponding to the limits of the redefined event. For our example, this step amounts to computing $P(5.5 \leq X \leq 6.5)$, which is represented by the shaded area in Figure 11.3.

Figure 11.3

Comparison of Binomial Distribution ($n = 8, p = .50$) with corresponding normal distribution ($\mu = 4, \sigma = 1.414$)

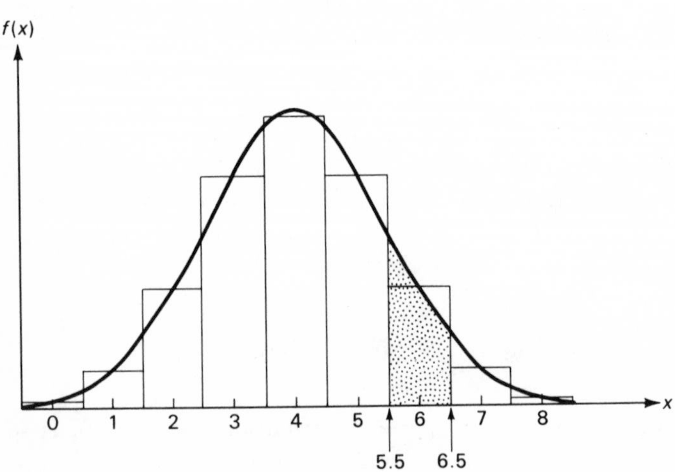

Bearing in mind that we are dealing with a normal distribution with $\mu = 4$ and $\sigma = 1.414$, we compute this probability as follows:

$$\begin{aligned} P(5.5 \leq X \leq 6.5) &= P\left(\frac{5.5 - 4}{1.414} \leq Z \leq \frac{6.5 - 4}{1.414}\right) \\ &= P(1.06 \leq Z \leq 1.77) \\ &= F_N(1.77) - F_N(1.06) \\ &= .9616 - .8554 \\ &= .1062 \end{aligned}$$

That is, the normal approximation to the binomial probability $P(X = 6)$ is .1062.

From Appendix Table D, we find that the *exact* binomial probability is:

$$P(\mathrm{X} = 6) = f_b(6|n = 8, p = .50) = .1094$$

Thus, the error of approximation in this case is $.1062 - .1094 = -.0032$. In cases in which n is much larger, the magnitude of error resulting in such approximations is considerably smaller, and it is only with large values of n that the approximation generally should be applied in actual practice.

Example:

A particular type of component is produced by a new automatic machine that behaves as a Bernoulli process with a defective rate of .20. Suppose that the machine produces a lot of 225 of these components.

(a) What is the probability that there are exactly forty defectives in the lot?

Step 1: Compute the mean and standard deviation of the binomial distribution:

$$\mu = np = 225(.20) = 45$$
$$\sigma = \sqrt{np(1 - p)} = \sqrt{225(.20)(.80)} = 6$$

Step 2: Adopt a normal distribution with mean equal to 45 and standard deviation equal to 6.

Step 3: The integer value 40 is redefined as an interval extending between the limits of $(40 - .5)$ and $(40 + .5)$. Thus, for purposes of approximation, $P_b(X = 40) \approx P_n(39.5 \leq X \leq 40.5)$

Step 4: Since we have adopted a normal distribution with $\mu = 45$ *and* $\sigma = 6$*, we compute:*

$$P_n(39.5 \leq X \leq 40.5) = P_N\left(\frac{39.5 - 45}{6} \leq Z \leq \frac{40.5 - 45}{6}\right)$$
$$= P_N(-.917 \leq Z \leq -.750)$$
$$= F_N(-.750) - F_N(-.917)$$
$$= .2266 - .1796 = .0470$$

(b) *What is the probability that there are no more than forty defectives in the lot?*

Steps 1 and 2: Same as above.

Step 3: The event under consideration is $\{X \leq 40\}$*, which includes the value 40. The upper limit of 40 is 40.5, as indicated in (a) above. Therefore,* $P_b(X \leq 40) \approx P_n(X \leq 40.5)$.

Step 4: For a normal distribution with $\mu = 45$ *and* $\sigma = 6$*, we obtain:*

$$P_n(X \leq 40.5) = F_n(40.5) = F_N\left(\frac{40.5 - 45}{6}\right)$$
$$= F_N(-.75) = .2266$$

(c) *What is the probability that there are less than forty defectives in the lot?*

Steps 1 and 2: Same as (a) above.

Step 3: The event under consideration is $\{X < 40\}$*, which does not include 40 but does include 39. The upper limit of 39 is 39.5. Therefore,* $P_b(X < 40) = P_b(X \leq 39) \approx P_n(X \leq 39.5)$

Step 4: For the same normal distribution, we compute:

$$P_n(X \leq 39.5) = F_n(39.5) = F_N\left(\frac{39.5 - 45}{6}\right)$$
$$= F_N(-.917) = .1796$$

(d) *What is the probability that the number of defectives will be between 41 and 49 inclusive?*

Steps 1 and 2: same as (a) above.

Step 3: The event under consideration is $\{41 \leq X \leq 49\}$.

$$\text{Thus, } P_b(41 \leq X \leq 49) = P_b(X \leq 49) - P_b(X \leq 40)$$
$$\approx P_n(X \leq 49.5) - P_n(X \leq 40.5)$$
$$\approx F_n(49.5) - F_n(40.5)$$

Step 4: Using the above normal distribution, we have:

$$F_n(49.5) - F_n(40.5) = F_N\left(\frac{49.5 - 45}{6}\right) - F_N\left(\frac{40.5 - 45}{6}\right)$$
$$= F_N(.75) - F_N(-.75)$$
$$= .7734 - .2266 = .5468$$

Generalizing the procedures employed in the above example, we may state the following formulas for obtaining normal approximations to binomial probabilities:

$$f_b(x|n,p) \approx F_N\left(\frac{x + .5 - np}{\sqrt{np(1 - p)}}\right) - F_N\left(\frac{x - .5 - np}{\sqrt{np(1 - p)}}\right) \qquad (11.1)$$

$$F_b(x|n,p) \approx F_N\left(\frac{x + .5 - np}{\sqrt{np(1 - p)}}\right) \qquad (11.2)$$

Frequently, in dealing with practical problems concerning a Bernoulli process, we may focus our attention on the *proportion* of successes in the sample, rather than on the *number* of successes. In such cases, we are concerned with the random variable $\frac{X}{n}$ rather than the random variable X. For each value of X there is a corresponding value of $\frac{X}{n}$. For example, suppose that a sample of ten observations is taken from a Bernoulli process with $p = .40$. If two successes are observed in the sample (i.e., $x = 2$), this is equivalent to a proportion of successes equal to $\frac{x}{n} = 2/10 = .20$. Obviously, whether we express the sample outcome in terms of *number* of successes or *proportion* of successes, the probability of corresponding outcomes will be the same. That is, $P\left(\frac{X}{n} = .20|n = 10, p = .40\right) = P(X = 2|n = 10, p = .40) = .1209$. The point to be made is that, if a Bernoulli experiment is conducted, the probability distribution for *proportion* of successes will be identical to the binomial distribution for *number* of successes except that the outcomes are expressed on a "proportion" scale rather than a "number" scale. This is illustrated by Figure 11.4, which shows the complete probability distributions for number of successes and proportion of successes for a Bernoulli experiment with $n = 10$ and $p = .40$. Notice that these two binomial distributions are identical except for the scales on the horizontal axis. Because of this identity, a problem concerning a proportion may be solved by converting the proportion of successes to the corresponding number of successes, and then applying the binomial model.

Figure 11.4

Comparison of Binomial Distributions for "Number of Successes" and "Proportion of Successes" with $n = 10$ and $p = .40$

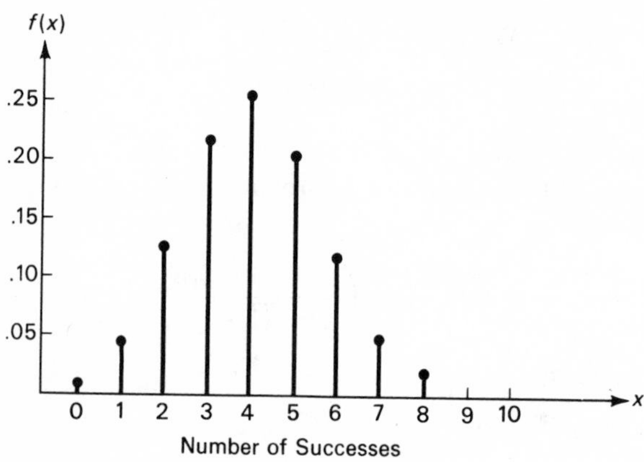

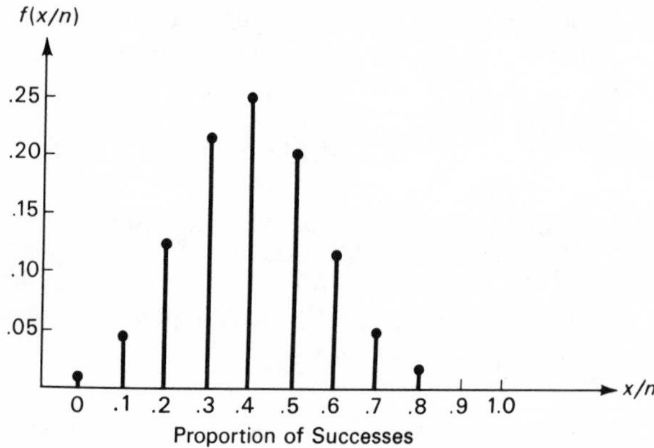

Example:

Suppose that we conduct a Bernoulli sampling experiment by taking a sample of 150 independent observations from a dichotomous population in which the probability of a success on a single observation is .40. What is the approximate probability that the proportion of successes in the sample will be no greater than .38?

From the statement of the problem, we wish to determine $P\left(\frac{X}{n} \leq .38\right)$ *Since* $n = 150$*, this probability expression is identical to* $P[X \leq 150(.38)]$ *or* $P(X \leq 57)$*. To approximate this binomial probability by the normal distribution, we follow the four steps described earlier.*

Step 1: Compute the mean and standard deviation of the binomial distribution:

$$\mu = np = 150(.40) = 60$$

$$\sigma = \sqrt{np(1-p)} = \sqrt{150(.40)(.60)} = 6$$

Step 2: Adopt a normal distribution with mean equal to 60 and standard deviation equal to 6.

Step 3: The event under consideration is $\{X \leq 57\}$, which includes the value 57. The upper limit of 57 is 57.5. Thus, $P_b(X \leq 57) \approx P_n(X \leq 57.5)$.

Step 4: $P_n(X \leq 57.5) = F_n(57.5) = F_N\left(\frac{57.5 - 60}{6}\right)$

$$= F_N(-.4167) = .3384$$

11.3.2 Accuracy of the Normal Approximation to the Binomial

As we have noted, the use of the normal approximation to the binomial is most appropriate if n is large and particularly if p is close to .50. Under these conditions, the discrepancies are smallest between exact binomial probabilities and their corresponding normal approximations. The degree of closeness of the approximation involves an interaction between the values of n and p, as we might suspect from careful observation of Figure 11.1. Raff[3] has demonstrated that, regardless of the values of n and p, the error introduced in using the normal model to approximate a *cumulative* binomial term will never exceed $.140/\sqrt{np(1-p)}$. Thus, for a fixed value of n, this upper limit on the possible error is smallest when $p = .50$; and for a fixed value of p, the upper limit on the possible error decreases as n is increased. This interactive effect of the values of n and p on the accuracy of the approximation may be seen in Table 11.2, which gives the actual maximum discrepancies between exact cumulative binomial probabilities and their corresponding normal approximations for selected values of n and p.

Notice that the actual maximum errors in Table 11.2 for specific values of n and p are less than the theoretical upper limit given by Raff's formula. For most specified combinations of n and p, this will be the case. That is, Raff's formula is a general statement of the *theoretical upper limit* of error that is true for any values of n and p, but in any specific case, the *actual* maximum error generally will be less.

[3] Morton S. Raff, "On Approximating the Point Binomial," *Journal of the American Statistical Association,* June 1956.

Table 11.2

Maximum Error of the Normal Approximation to the Cumulative Binomial Distribution

Values of p	*Value of n*				
	5	*10*	*25*	*50*	*100*
.10	.158	.106	.060	.040	.027
.20	.086	.054	.032	.022	.015
.30	.054	.032	.019	.013	.009
.40	.024	.016	.009	.006	.004
.50	.011	.005	.002	.001	.001

Adapted and abridged from Morton S. Raff, "On Approximating the Point Binomial," *Journal of the American Statistical Association*, June 1956.

11.4 POISSON OR NORMAL APPROXIMATION: WHICH?

In situations in which the use of an approximation procedure for the binomial model is being considered, the problem may arise whether to select the Poisson approximation or the normal approximation, if either. If our criterion of choice is to adopt the approximation procedure that has the smaller maximum error, we may say that, regardless of sample size, the Poisson approximation is generally preferable to the normal approximation if p is less than about .075 or greater than .925. Also, regardless of sample size, the normal approximation is preferable to the Poisson approximation if p is between .25 and .75. For other values of p, the choice between these two approximations depends on the value of n. A rough guide for the choice between the two approximations is provided by the chart in Figure 11.5. As examples, we can see from this chart that the normal approximation is preferred to the Poisson approximation if:

n is at least 200 when p is as small as .08 (or as great as .92)
n is at least 100 when p is as small as .10 (or as great as .90)
n is at least 50 when p if as small as .12 (or as great as .88)
n is at least 25 when p is as small as .16 (or as great as .84)

It is important to realize that these rules simply indicate the conditions under which either the Poisson or normal approximation is preferred to the other. It should not be inferred, however, that either approximation is necessarily justified under a given set of conditions simply because it is preferable. That is, although one approximation might be preferable to the other under particular conditions, it might be that neither of the two approximations is necessarily satisfactory under those conditions. For example, if $p = .20$ and $n = 100$, Figure 11.5 indicates that the normal approximation is preferred to

Figure 11.5

A Rough Guide for Choosing Between Normal and Poisson Approximations to the Binomial

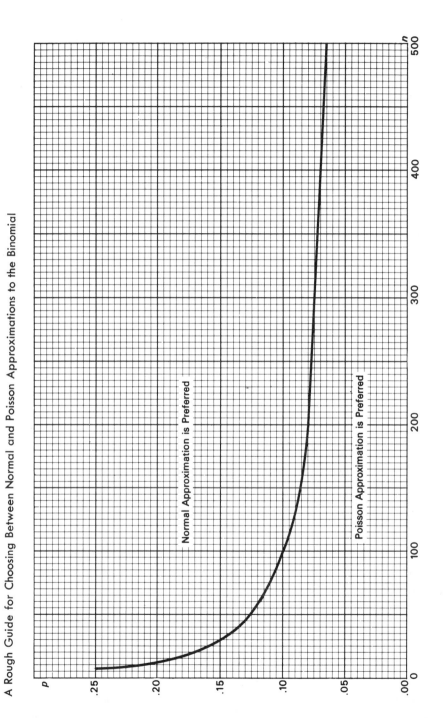

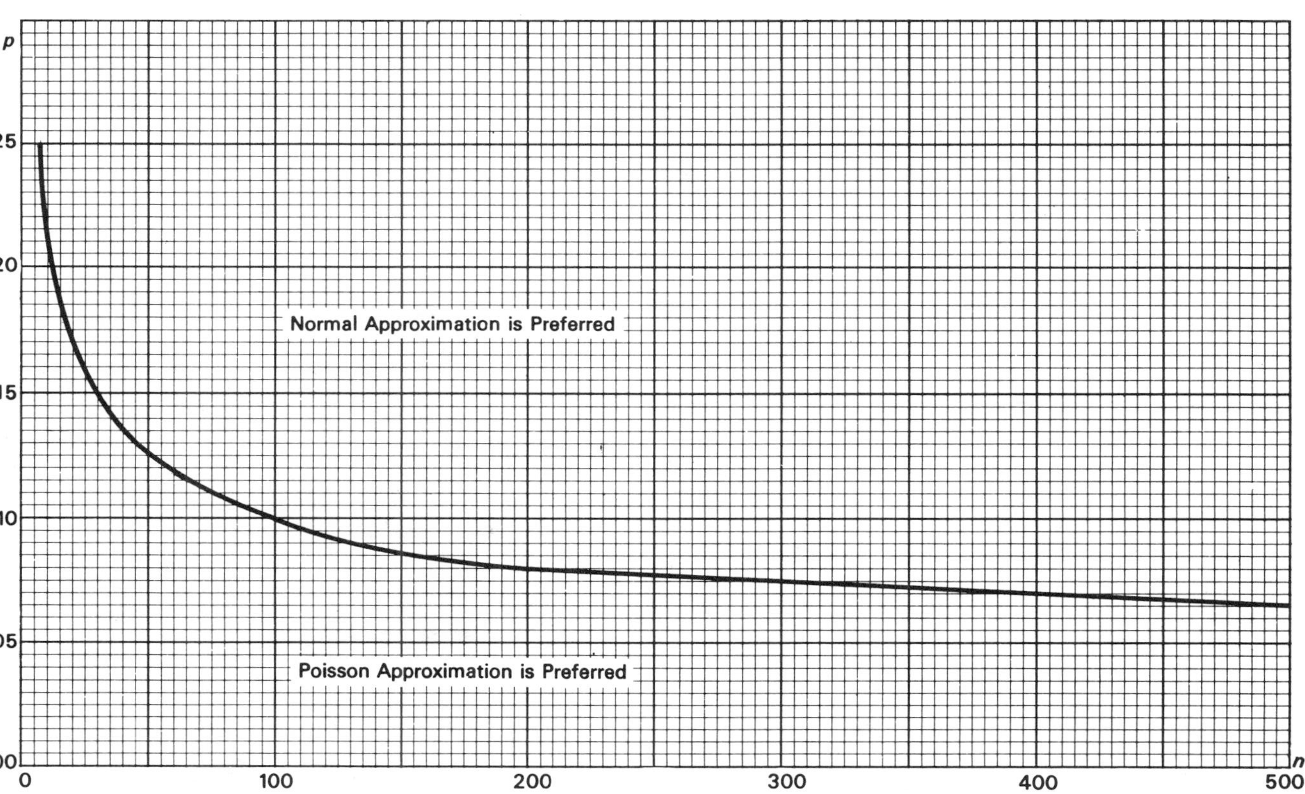

the Poisson approximation. However, reference to Table 11.2 indicates that, with $n = 100$ and $p = .20$, the maximum error resulting from the normal approximation is .015. If this level of error is greater than our accuracy requirements permit, we would not use either of the two approximations.

As long as p is quite small, particularly .02 or less, the Poisson provides a fairly close approximation to binomial probabilities by most commonly accepted standards. The accuracy of the Poisson approximation is essentially unaffected by the sample size, whether $n = 10$ or $n = 10{,}000$.

In dealing with the normal approximation to the binomial, statisticians have become increasingly insistent in recent years that the sample size should be substantial in order to obtain acceptable accuracy. If p = .50, most practitioners prescribe a minimum sample size between 25 and 50. As p departs from .50, the necessary sample size for an adequate approximation may be considerably larger. For example, if p is as small as .10 or as large as .90, a sample size of several hundred cases usually is recommended in order for the normal approximation to provide adequate accuracy.

Because the Poisson approximation to the binomial is suitable only for extreme values of p, and because the normal approximation for moderate values of p requires a substantial value of n, there is a rather wide range of combinations of n and p values for which neither approximation is suitable for many practical applications. Other, more accurate, approximation procedures are available, particularly the arcsine approximation, the Poisson Gram-Charlier approximation, and the Camp-Paulson approximation. Because of their mathematical complexity, these more accurate approximation methods seldom have been used in actual practice.

11.5 BINOMIAL APPROXIMATION TO THE HYPERGEOMETRIC DISTRIBUTION

In our discussion of the use of the hypergeometric and binomial models in Chapter Nine, we observed that (1) both models are applicable to sampling from dichotomous populations, but that (2) the hypergeometric model applies to situations involving sampling without replacement from finite populations, whereas the binomial model applies to situations involving sampling from infinite populations or sampling with replacement from finite populations. The essential point in this distinction between the two models was that the probability of observing a success remains constant from observation to observation when the sample is drawn from an infinite population, whereas the probability of obtaining a success on any particular observation taken from a finite population without replacement varies conditionally on the outcome of previous observations. We also noted that, if a finite population were large, and if the sample size were very small relative to the population size, then the population could be considered "effectively infinite" and the binomial model would be adequate. Actually, what this means is that, if a random sample of size n is selected from a finite, dichotomous population of size N (where N is large), and if the sampling fraction n/N is quite small (particularly below .10, as a rough rule of thumb), then the binomial distribution will provide reasonably accurate approximations of hypergeometric probabilities.

Due to some intricate mathematical relationships between the hypergeometric and binomial distributions, the binomial model may be employed in several different ways to obtain hypergeometric approximations. In this book we will limit our consideration to situations in which (1) the sampling fraction n/N is less than .50, and (2) the population "success ratio" D/N is also less than .50. This limitation is assumed in the following discussion.[4]

If D (the number of successes in the population) is *greater* than n (the number of cases in the sample), the generally most accurate binomial approximation to the hypergeometric is given by:

$$f_h(x|N,n,D) \approx f_b\left(x|n,p = \frac{D}{N}\right) \tag{11.3}$$

For example, suppose that a box contains 100 transistors, of which 15 are defective. If a random sample of 5 transistors are taken without replacement from the box, what is the probability that the sample will contain exactly 2 defectives? In this situation we have a hypergeometric problem with $N = 100$, $n = 5$, $D = 15$, and $x = 2$. The exact hypergeometric solution to the problem is:

$$f_h(2|N = 100, n = 5, D = 15) = \frac{\binom{15}{2}\binom{85}{3}}{\binom{100}{5}} = .1378$$

Notice in this case that the number of defectives in the box ($D = 15$) is *greater* than the size of the sample ($n = 5$). Thus if we wish to approximate this probability by the binomial mass function, we may apply Formula (11.3), with $n = 5$ and $p = 15/100 = .15$. This approximate solution is:

$$f_b(2|n = 5, p = .15) = \binom{5}{2}(.15)^2(.85)^3 = .1382$$

[4] For a discussion of binomial approximations to the hypergeometric with high sampling fractions and success ratios, see Lieberman and Owen, *Tables of the Hypergeometric Probability Distribution*, Stanford University Press, 1961; also see H. D. Brunk, J. E. Holstein and F. Williams, "A Comparison of Binomial Approximations to the Hypergeometric Distribution," *The American Statistician*, February 1968.

Thus, in this particular case, the magnitude of error is .1382 − .1378 = .0004.

If the number of successes in the population is *less* than the number of cases in the sample ($D < n$), the preferred binomial approximation to the hypergeometric is given by;

$$f_h(x|N,n,D) \approx f_b\left(x|n = D, p = \frac{n}{N}\right) \tag{11.4}$$

For example, suppose that a box contains 100 transistors, of which 5 are defective. If a random sample of 10 transistors are taken without replacement from the box, what is the probability that the sample will contain exactly 2 defectives? Now we have a hypergeometric problem with N = 100, $n = 10$, $D = 5$, and $x = 2$. The exact hypergeometric solution to the problem is:

$$f_h(2|N = 100, n = 10, D = 5) = \frac{\binom{5}{2}\binom{95}{8}}{\binom{100}{10}} = .0702$$

In this case, the number of defectives in the box ($D = 5$) is *less* than the size of the sample ($n = 10$). Thus, if we wish to solve this problem by the binomial approximation to the hypergeometric, we may use Formula (11.4). To apply this formula, we let $n = D = 5$, and $p = n/N = 10/100 = .10$. The approximate binomial solution is:

$$f_b(2|n = 5, p = .10) = \binom{5}{2}(.10)^2(.90)^3 = .0729$$

In this instance, the error of approximation is .0729 − .0702 = .0027. If we had solved this problem by the "standard" approximation of Formula (11.3), we would have obtained:

$$f_b(2|n = 10, p = .05) = \binom{10}{2}(.05)^2(.95)^8 = .0746$$

The error of approximation then would have been .0746 − .0702 = .0044, which is noticeably greater than the error obtained with the use of Formula (11.4). This observation stresses the fact that, when $D < n$, Formula (11.4) tends to yield more accurate approximations of hypergeometric probabilities than the commonly used Formula (11.3). Of course, if $D > n$, Formula (11.3) tends to be more accurate. For the case in which $D = n$, the two approximation formulas will give the same result.

PROBLEMS

11.1 Using the Poisson approximation, evaluate each of the following binomial expressions:
(a) $f_b(4|200, .003)$
(b) $f_b(2|280, .01)$
(c) $f_b(8|400, .05)$
(d) $f_b(5|10{,}000, .0002)$
(e) $F_b(7|500, .006)$
(f) $F_b(6|1200, .004)$.

11.2 Consider the following binomial expressions:

$$f_b(1|20, .01)$$
$$f_b(1|50, .01)$$
$$f_b(1|100, .01)$$

(a) Find the exact probability corresponding to each of the above expressions.
(b) Determine the Poisson approximation corresponding to each of the probabilities found in (a) above.
(c) For each of the approximate probabilities obtained in (b) above, determine the algebraic error of approximation by subtracting the exact probability from the approximate probability.
(d) For each of the approximate probabilities obtained in (b) above, determine the *relative* error of approximation by dividing the algebraic error by the exact probability.

11.3 Let X represent the number of failures observed in 50 trials of a Bernoulli process for which the failure rate is p.
(a) Using the Poisson approximation to the binomial, determine the approximate probability that $X = 2$ if:
(i) $p = .001$
(ii) $p = .01$
(iii) $p = .02$
(iv) $p = .05$
(v) $p = .10$
(vi) $p = .20$
(b) Determine the exact probability corresponding to each of the approximations obtained in (a) above.
(c) Prepare a table that shows, for each value of p, the Poisson approximation of $P(X = 2)$, the corresponding exact probability, and the error of approximation computed by subtracting the exact probability from the approximate probability.

11.4 The Heart and Soul Corporation, makers of musical instruments and supplies, manufactures guitar picks by an automatic machine that produces defective picks according to a Bernoulli process at a rate of $p = .03$. An inspector takes a sample of 200 picks from the output of this process. Using the Poisson approximation to the binomial, determine the probability that the sample will contain:
(a) no defectives
(b) exactly 4 defectives
(c) no more than 4 defectives
(d) at least 4 defectives
(e) more than 4 defectives
(f) more than 5 but less than 10 defectives.

11.5 Let X represent the number of successes observed in 50 trials of a Bernoulli process for which the success rate is $p = .01$.
(a) Using the Poisson approximation to the binomial, determine $P(X \leq 3)$.
(b) Determine the exact probability that was estimated in (a) above.
(c) Determine the error of the approximation by subtracting the exact probability from the approximate probability.

11.6 The Dexter Pharmaceutical Company has determined that 2/10 of 1 per cent of the people who are injected with Z-13 type serum suffer an unfavorable reaction. In a community immunization drive, a clinic has innoculated 3,000 volunteers with this serum. Using the Poisson approximation to the binomial, determine the probability that the number of these volunteers who suffer unfavorable reactions will be:
(a) exactly 5
(b) more than 7
(c) less than 8
(d) at least 4 but no more than 10.

11.7 From records compiled over a period of years, the credit manager of Wood's Department Store has determined that the long-run percentage of accounts receivable that are uncollectible is 1.4 per cent. Currently, the store has 350 accounts receivable. Determine the approximate probability that the number of these accounts that are uncollectible will be:
(a) 5 or more
(b) exactly 5
(c) 3 or less
(d) between 3 and 9, inclusive.

11.8 The management of Black Opal Cosmetics, Ltd., estimates that 5 per cent of the women who are regular users of hairspray prefer their Lovelox brand to all other brands. In a marketing survey, 400 women indicate that they are regular hairspray users. If Black Opal's estimate is correct, what is the approximate probability that the number of these hairspray users who prefer Lovelox will be:
(a) more than 29
(b) exactly 20
(c) less than 11
(d) more than 14 but less than 26.

11.9 Let X represent the number of successes observed in 300 trials of a Bernoulli process for which the success rate is $p = .95$. Using the Poisson approximation to the binomial, determine the following probabilities:
(a) $P(X = 285)$
(b) $P(X \geq 295)$
(c) $P(X \leq 280)$
(d) $P(278 < X < 292)$.

11.10 One-tenth of 1 per cent of the oysters that grow in Akutiki lagoon contain black pearls. A seafood packing firm has just received a shipment of 5,000 oysters that were harvested from the lagoon.
(a) In this shipment what is the expected number of oysters containing black pearls?
(b) What is the approximate probability that, in this shipment, the number of oysters containing black pearls will be less than the expected number?

11.11 Using the normal approximation, evaluate each of the following binomial expressions:
(a) $f_b(100|200, .5)$
(b) $f_b(210|500, .4)$

(c) $f_b(245|300, .8)$
(d) $F_b(125|600, .2)$
(e) $F_b(550|800, .7)$

11.12 Let X represent the number of successes observed in 100 trials of a Bernoulli process with a success rate of $p = .50$.
(a) Using the normal approximation, find the following probabilities:
(i) $P(X \leq 35)$
(ii) $P(X \leq 38)$
(iii) $P(X \leq 42)$
(iv) $P(X \leq 48)$
(v) $P(X \leq 50)$
(vi) $P(X \leq 55)$
(vii) $P(X \leq 61)$
(viii) $P(X \leq 64)$
(b) Find the exact probability corresponding to each of the approximate probabilities determined in (a) above.
(c) For each of the approximate probabilities found in (a) above, determine the *algebraic* error of approximation by subtracting the exact probability from the approximate probability.
(d) For each of the approximate probabilities, determine *relative* error by dividing the algebraic error by the exact probability.

11.13 Tom Eager, the marketing director of Worldwide Motors Corporation, has requested his dealers to ask each new car purchaser to answer a short questionnaire within a few days after purchase. One of the questions is: "Have you ever owned another Worldwide Motors cars before the one you have just purchased?" Mr. Eager estimates that 60 per cent of Worldwide's new car sales are to former owners of Worldwide cars. If this estimate is correct, and questionnaires are received from 1,000 respondents, what is the approximate probability that the number of "yes" answers to the previous ownership question will be:
(a) 620 or more
(b) 619 or less
(c) exactly 600
(d) exactly 400
(e) at least 500 but no more than 590
(f) at least 550 but no more than 650.

11.14 Calvin Kindly, credit manager of the Longreen Federal Savings and Loan Association, has determined from past records that 36 per cent of applicants for home loans are making their first home purchase. If a sample of 400 home loan applications is selected for analysis, what is the approximate probability that the sample will contain fewer than 235 applicants who are purchasing their first homes?

11.15 In an upcoming state election, over 10 million voters are expected to cast their votes on a proposition to open publicly owned timberlands to limited logging operations. The research director of the State Taxpayers Association estimates that 50 per cent of the voters in the state are in favor of this proposition, but he decides to conduct a sample survey of 500 potential voters to confirm his judgment. If the research director's estimate is correct, what is the approximate probability that the number of voters in the sample who favor the proposition will be:
(a) exactly 250
(b) less than 275

(c) greater than 240 but less than 260
(d) greater than 245.

11.16 Over the years, the best-selling product of Liberated Ladieswear, Inc., has been their Holdtight girdle. Each girdle is inspected, and the defective girdles are marked as irregulars, which wholesale at a lower price. On the average, the irregular girdles account for 12 per cent of the total girdles produced. On a particular production run, 341 of these girdles are produced. What is the probability that there will be 30 irregular girdles at most in this run?

11.17 Suppose that 55 per cent of the registered voters in a large state are planning to vote for Burton House as U.S. senator in an upcoming election.
(a) What is the approximate probability that, if a random sample of 582 registered voters is polled, a majority of the sample will indicate that they do *not* intend to vote for Burton House?
(b) Why did you choose the particular approximation procedure that you used?

11.18 A widely accepted rule of thumb is that the normal model provides satisfactory approximations to binomial probabilities if both np and $n(1 - p)$ are 5 or greater. However, unless n is very large, this rule of thumb is not dependable. To demonstrate the unreliability of the rule of thumb, perform the following exercise:
(a) From Appendix Table E, find the following exact binomial probabilities:
(i) $F_b(5|50, .10)$
(ii) $F_b(10|50, .20)$
(iii) $F_b(5|100, .05)$
(iv) $F_b(10|100, .10)$
(v) $F_b(20|100, .20)$
(b) Determine the normal approximation to each of the probabilities found in (a) above.
(c) Determine the Poisson approximation to each of the probabilities found in (a) above.
(d) Prepare a table as indicated below:
(i) In Column (1), enter the exact probabilities obtained in (a) above.
(ii) In Column (2), enter the corresponding normal approximations obtained in (b) above.
(iii) In Column (3), subtract the entry in Column (1) from the entry in Column (2).
(iv) In Column (4), divide the entry in Column (3) by the entry in Column (1).
(v) In Column (5), enter the Poisson approximations corresponding to the binomial probabilities in Column (1).
(vi) In Column (6), subtract the entry in Column (1) from the entry in Column (5).
(vii) In Column (7), divide the entry in Column (6) by the entry in Column (1).
(e) Explain the meaning of the entries in Columns (3), (4), (6), and (7).
(f) Using the evidence contained in the table you have constructed, what conclusions do you draw concerning the rule of thumb stated at the beginning of this exercise?

11.19 The Betatron Electronics Company purchases transistors from the Feedem Corporation in lots of 10,000 units. Upon receipt of a lot, a sample of 100 transistors is selected at random and carefully tested. If a lot actually contains 500 defective transistors, what is the approximate probability that the sample from that lot will contain exactly 4 defective transistors?

11.20 The Bimonthly Book Club has a membership of 1,000,000 readers. The marketing manager estimates that a book it is considering for an upcoming selection would be accepted by 250,000 of these readers. Nevertheless, he decides to send questionnaires to a random sample of 100 members of the club, asking them to indicate whether they would buy the book if it were offered as a selection. If all 100 of these questionnaires are returned, and if management's estimate is correct, what is the approximate probability that the number of respondents who indicate that they would buy the book will be:

(a) exactly 25?
(b) 25 or less?
(c) at least 20 but no more than 30?

11.21 The Softline Corporation produces ballpoint pens in production runs of 10,000 pens. As a check on the quality of the production process, a sample of 200 pens is selected from each run and carefully inspected.

(a) If a particular production run produces only 10 defectives, what is the approximate probability that the sample will contain no defectives?
(b) If a production run produces 50 defectives, what is the approximate probability that the sample will contain no defectives?
(c) If a production run produces 100 defectives, what is the probability that the sample will contain:
  (i) no defectives?
  (ii) 2 defectives or less?
  (iii) 3 or more defectives

Chapter 12
Decision Conditions and Criteria

The primary purpose of the present chapter is to introduce the reader to the fundamentals of the modern quantitative approach to decision making. In particular, we first will examine various kinds of situations under which decisions are made. These situations, or *conditions,* are certainty, risk, uncertainty, and conflict. We then will discuss *criteria* for decision making under these different conditions. These two factors, decision conditions and decision criteria, are crucial in distinguishing one quantitative method of decision making from the other.

12.1 CONDITIONS UNDER WHICH DECISIONS ARE MADE

The awareness of a decision problem begins with the realization that there are two or more possible alternative courses of action that are available to achieve some goal or objective. In such a situation, the decision maker attempts to choose the best course of action according to some predetermined *criterion* or standard of judgment. This choice is further complicated by the fact that in many decision situations, the decision maker does not know exactly what consequence will result from each of the available courses of action. This is because the consequence of an action may be the result not only of the course of action that is taken, but also of other factors, or *events,* that are not under the control of the decision maker. In other words, if there are two or more possible events, one of which will occur, the particular consequence resulting from any specific course of action depends on which of the two or more possible events actually occurs. As the reader may recall from Chapter One, these possible events or uncontrollable factors usually are called *states of nature.*

To illustrate the above points, consider the case of Joe Adams, a student at a university that is located near the center of a metropolitan city. He commutes to the university each morning from his home in the suburbs. He has found that two routes, Freeway and Boulevard, are superior to all others. The travel time on the Freeway route is highly variable. When traffic is moving openly, the travel time is fifteen minutes. When it is jammed, however, the travel time is as much as thirty-five minutes. The Boulevard route requires a fairly consistent travel time of twenty-five minutes. In this example, there are two alternative courses of action:

(i) to travel via the Freeway (denoted as a_1), or
(ii) to travel via the Boulevard (denoted as a_2).

There are two possible states of nature:

(i) Freeway traffic is open (denoted as θ_1), or
(ii) Freeway traffic is jammed (denoted as θ_2).

The consequence of this decision problem depends not only on which of the alternative courses of action would be chosen, but also on which of the two states of nature might occur. Thus, there can be four different consequences associated with the four possible action-state pairs. In our particular example, two of the four consequences are identical, and hence there are three different consequences, as shown in Table 12.1. This table shows that the consequence for each available act is conditional on each possible state of nature that might prevail. Such a table thus may be called a *conditional consequence table.* The consequences shown in Table 12.1 are the required travel times in minutes under various situations. Obviously, the decision maker (Joe) prefers a smaller amount of travel time to a larger one.

Table 12.1

Conditional Consequence Table for Joe Adams (travel time in minutes)

State of Nature	a_1 (*To travel via Freeway*)	a_2 (*To travel via Boulevard*)
θ_1 (open)	15	25
θ_2 (jammed)	35	25

The above example demonstrates some basic elements of a decision problem. Notice that the decision maker in this example has two alternative actions available. In a decision problem, whatever the number of available alternatives, such a set of all possible actions that can be chosen is called the *action space.* Notice also that there are two possible states of nature. Such a set of all possible states of nature that will determine the consequence of any course of action is called the *state space.* Notice finally that various consequences will result depending on which action Joe takes and what the Freeway situation is. Such a set of all possible consequences that might result from the various combinations of actions and states of nature is called the *consequence space.*

In our example, although Joe knows exactly what each of the possible consequences is, he still cannot uniquely associate each course of action with a particular consequence unless he knows which state of nature is the actual one at the time. Thus, his uncertainty stems from his imperfect knowledge of reality—of the relevant factors that he does not control, but which nevertheless affect the consequences. Depending on the extent and kind of information available about the states of nature, decision conditions can be classified into the following categories:

(i) Certainty
(ii) Risk
(iii) Uncertainty
(iv) Conflict

12.1.1 Decision Making Under Certainty

The type of decision situation in which the decision maker has *complete information* concerning all relevant factors affecting the consequence is called decision making under *certainty.* When a decision is made under certainty, several possible alternative courses of action may be available to the decision maker. However, since the decision maker has complete information, he is certain about which particular state of nature is prevailing. Therefore, he is able to determine with certainty the consequence of each of the alternative actions available to him. Then the solution to the decision problem is simply one of computing the consequence of each course of action for the known state of nature and selecting that course of action that would result in the *optimal* (most desirable) consequence.

Consider, for example, the case of Bob Baker, who has an amount of $10,000 to invest for one year. Initially, Bob considers two alternatives: either (1) to purchase a particular state bond issue that matures in a year's time, or (2) to open a savings account with a reliable bank. If he buys the bonds at an annual interest rate of 4.5 per cent, he will receive a profit of $450. If he deposits the amount in the savings account at 5 per cent annual interest rate, he will make $500. Clearly, if Bob's objective is to maximize profit, the solution of this decision problem is for him to deposit the money in the savings account, which will result in the optimal consequence.

12.1.2 Decision Making Under Risk

In decision making under *risk*, the decision maker has only *partial information* concerning the possible state of nature that will affect the consequence of the decision. Specifically, there are two or more possible states of nature associated with each alternative course of action. The decision maker does not know the true state of nature, but he does know the probability distribution of the occurrences of the states of nature. The solution to the decision problem requires, first, the selection of a decision criterion; second, the evaluation of alternative courses of action; and, finally, the identification of the optimal action according to the criterion selected.

As an illustration, let us continue the example of Bob Baker, who has $10,000 to invest for one year. Bob has a good friend, John Carter, who is a successful stock broker. John has advised Bob to purchase the stock of Elden, Inc. This company is developing a new product. According to John, the price of this stock will go up 10 per cent in one year if the new product has been successfully developed, and will drop 1 per cent otherwise. After some discussion with Bob, John estimates the odds are 3 to 1 that the new product will be successfully developed. In other words, the probability is .75 that the product will be successfully developed, and .25 that it will not. Now, although Bob has decided not to invest in bonds, the intervention of the stock broker confronts him with a new alternative to depositing his money in a savings account. Thus, he again has two possible courses of action available:

(i) deposit the money in the bank (denoted as a_1), or
(ii) purchase the stock (denoted as a_2).

There are two possible states of nature:

(i) the new product is successfully developed (denoted as θ_1), or
(ii) the new product is not successfully developed (denoted as θ_2).

The consequence of Bob's decision depends on which of the alternative courses of action he will choose and on which of the two possible states of nature will occur. The various consequences associated with the four possible action-state pairs are shown in Table 12.2. This table shows, for each available act, the profit or monetary consequence conditional on each possible state of nature prevailing, and thus is a *conditional profit table*. However, this table often is called a *conditional payoff table* or simply a *payoff table*.

Table 12.2

Conditional Payoff (Profit) Table for Bob Baker

States of Nature (Is New Product Developed?)	a_1 *(To deposit in the Bank)*	a_2 *(To purchase the Stock)*	*Probability of State of Nature*
θ_1 (yes)	$500	$1,000	.75
θ_2 (no)	500	−100	.25

In order for Bob to analyze his problem for the purpose of determining the optimal course of action, he first must specify his decision criterion. Suppose, for example, that he would like to be *assured* of a minimum profit of \$500. Then he should choose a_1, since there is a .25 probability that a_2 could result in a \$100 loss. Yet, if he would like to select the action that could possibly lead to the highest profit, then clearly he should choose a_2. We will discuss various decision criteria in Section 12.2.

12.1.3 Decision Making Under Uncertainty

The term *uncertainty,* as used to describe a decision condition, may be interpreted in both a narrow sense and a broad sense. In the *narrow sense,* uncertainty refers to the situation in which the true state of nature is completely unknown to the decision maker. That is, under this interpretation of uncertainty, the decision maker not only does not know which state of nature actually will occur, but he also does not even know the probabilities of occurrences of the various possible states of nature. The case of Bob Baker, described in Section 12.1.2, would fall into this category if the probabilities of the two possible states of nature were unknown.

In the *broad sense,* decision making under uncertainty refers to the situation in which the true state of nature simply is not known with certainty. This can be either a case in which the true state of nature is partially unknown (i.e., decision making under risk), or a case in which the true state of nature is completely unknown (i.e., decision making under uncertainty in the narrow sense as described above).

For most administrative problems, decision makers do have partial information regarding the states of nature, which can be expressed in terms of subjective or objective probabilities. Consequently, the condition of strict uncertainty seldom exists. Therefore, we shall adopt the broad definition of decision making under uncertainty and use this term interchangeably with decision making under risk.

12.1.4 Decision Making Under Conflict

In decision making under *conflict,* the consequence of a specific course of action chosen by the decision maker depends on the reaction or counteraction taken by his opponent. This opponent is an *adverse intellect* rather than nature. Each of the opposing decision makers attempts to optimize his decision at the other's expense. Thus, neither of them knows exactly which consequence will eventually materialize. Situations such as this may arise in business and other competitive situations.

As an example, consider a simplified cola market in which there exist only two competing brands, *A* and *B*. The producers of the two brands have been contacted by a can manufacturer offering a new cola can design. Unfortunately, this new design is more expensive than the present one. The producer of Brand *A* figures as follows: "I am making about $10,000 on cola now. It looks to me as if the consumer will like this new design. If I go ahead with it and the competitor doesn't, I stand to make about $11,000. On the other hand, if he does and I don't, I'll probably fall off to $9,500. If both of us use it, we'll all edge our price up a little, which would likely cause some reduction in sales so that we would come out at about $9,800." From this statement we obtain the conditional payoff table presented in Table 12.3. This table clearly shows that the consequence of *A*'s decision not only depends on the specific course of action chosen by *A*, but also depends on the specific counter-action taken by *B*. Similarly, we could construct a conditional payoff table for *B* that would show that the consequence of *B*'s decision depends on both his choice of course of action and his opponent's reaction.

Table 12.3

Conditional Payoff Table for Brand A

B's Action	*A's Action*	
	Not to Adopt the New Can	*To Adopt the New Can*
Not to Adopt the New Can	$10,000	$11,000
To Adopt the New Can	9,500	9,800

To summarize, we have described various conditions under which decisions are made. The analysis of decision problems under certainty may require mathematical methods such as calculus and linear programming to identify optimal solutions. Consequently, decision problems of this type are treated in books on applied calculus and mathematical programming. The body of the techniques used for analysis of decision problems under conflict is known as game theory. The analysis of decision problems under risk or uncertainty is the subject of applied probability and statistics.

12.2 CRITERIA FOR DECISION MAKING UNDER CERTAINTY

In discussing decision making under certainty in Section 12.1.1, we used maximization of profit as the criterion for selecting the optimal action. There are, however, other criteria that may be used for decision making under certain circumstances.

To illustrate this point, let us return to the case of Bob Baker. He considered two alternatives: either (1) purchasing state bonds to earn an interest of $450, or (2) opening a savings account with a reliable bank to earn an interest of $500. The decision was to choose the second alternative since his objective was to maximize his profit.

Now, suppose that Bob were an extremely loyal supporter of the governor and his decision criterion were to maximize his contribution to the state. Then he should choose to purchase the bonds. On the other hand, Bob might be opposed to the state's deficit financing. Thus his decision criterion would be to minimize his direct contribution to the state. Then he should deposit his money in the bank.

In this text, since we are concerned with business and industrial problems in which the objective is primarily to maximize profits,[1] we will not use such personal criteria, but instead will adopt maximization of profit as the criterion for decision making under certainty.

12.3 CRITERIA FOR DECISION MAKING UNDER RISK, UNCERTAINTY, OR CONFLICT

When a decision is made under certainty, the actual consequence should be exactly as determined by the analysis on which the decision is based. But when a decision is made under the condition of risk, uncertainty, or conflict (i.e., *when the true state of nature is not known with certainty,* regardless of whether it is partially unknown or completely unknown, and regardless of whether this uncertainty is due to the decision maker's ignorance of reality or the counter-action of his opponent), it is impossible for the decision maker to predict exactly the consequence of his decision. Under such conditions, the decision maker is forced to choose a specific course of action from a set of alternatives, hoping that the action chosen will yield the maximum profit but knowing that it may not result in maximum payoff after a particular state of nature has actually occurred. In effect, the decision maker is forced to gamble against nature or his opponent. He is in a position wherein he must place bets, hoping that he will win but knowing that he may lose. Different decision makers may use different criteria for placing the bets (i.e., for selecting a specific course of action), even though they all may have the same ultimate objective of maximizing profits.

To illustrate various criteria used in decision making under conditions of risk, uncertainty, or conflict, we will analyze an inventory-stocking problem. Consider a large newsstand that carries various newspapers and magazines. One of the items that the dealer stocks is the LAX Sunday paper. He pays 30 cents for each copy of the paper, which he prices at 50 cents a copy. Copies

[1] The reader should recognize, of course, that optimization of the firm's profit in the long run generally requires allowance for a variety of socio-economic factors.

remaining unsold at the end of the day are non-returnable and have no value. To help to decide the number of copies of the newspaper to stock every Sunday, the dealer has examined the sales record for a period of the last 100 weeks. This record, as shown in Table 12.4, indicates that during the period examined, the dealer has *never* sold more than twenty-four or less than sixteen copies. Since the dealer foresees no substantial changes in demand in the near future, he wishes to limit his future weekly stock to some number of copies between sixteen and twenty-four, inclusive.

Table 12.4

Weekly Demand for LAX Sunday Paper

Number of Copies Demanded per Week	*Number of Weeks*
16	5
17	10
18	12
19	16
20	10
21	20
22	16
23	6
24	5
	100

The adoption of some symbolic notation will aid our analysis of the news dealer's problem. We shall use the symbols defined below:

a: a particular course of action (a particular number of copies that the dealer may choose to stock). All possible courses of action ($a =$ 16, 17, . . . , 24) constitute the action space.

Θ: a general expression for the state of nature. The reader may recognize that Θ is a random variable in the extended sense discussed in Chapter Six. This is because, for any given decision problem under uncertainty, the decision maker does not know which one of the two or more possible states of nature will actually occur. Following our notational convention, we shall use the corresponding lower-case θ to denote any particular state of nature or specific value of the random variable Θ. Thus, in our example, the random variable Θ denotes demand for the Sunday paper and θ represents a particular number of copies demanded. Thus, Θ is a random variable whose possible values are $\theta = 16, 17, \ldots, 24$.

Using the above notation, the monetary value (conseqence) of a given course of action a is a function of the random variable Θ; this *function* will be denoted as $v(a,\Theta)$. We shall use $v(a,\theta)$ to denote the *particular* monetary value resulting from a *particular* course of action a (number of copies stocked) when a *specific* state of nature θ occurs (number of copies demanded). A positive value of $v(a,\theta)$ represents a gain or profit; a negative value represents a loss or negative profit. For our present example, we may express $v(a,\theta)$ as a function of a and θ in the following form:

$$v(a,\theta) = \begin{cases} .50\theta - .30a & \text{if } \theta \leq a \\ .50a - .30a & \text{if } \theta > a \end{cases}$$

where $a = 16, 17, \ldots, 24$, and $\theta = 16, 17, \ldots, 24$. This equation tells us the following:

1. If the dealer decides to stock a copies, and if the demand θ is no greater than supply a, his profit will be 50 cents for each copy *sold*, less 30 cents for each copy *stocked*.
2. If demand θ exceeds supply a, so that all copies in stock are sold, then $\theta = a$. Thus the profit will be 50 cents for each copy *stocked* (all of which are sold), less 30 cents for each copy *stocked*.

For any given course of action chosen, the monetary value is conditional on the occurrence of a specific state of nature. Since there are nine possible states of nature ($\theta = 16, 17, \ldots, 24$), by systematically varying values of θ in the above equation we obtain the conditional monetary values (conditional payoffs) for the particular action as shown in the corresponding column of

Table 12.5

Conditional Payoff Table for LAX Sunday Paper

State of Nature (Number of Copies Demanded): θ	*Action (No. of Copies Stocked)*								
	$a = 16$	$a = 17$	$a = 18$	$a = 19$	$a = 20$	$a = 21$	$a = 22$	$a = 23$	$a = 24$
$\theta = 16$	\$3.20	\$2.90	\$2.60	\$2.30	\$2.00	\$1.70	\$1.40	\$1.10	\$0.80
$\theta = 17$	3.20	3.40	3.10	2.80	2.50	2.20	1.90	1.60	1.30
$\theta = 18$	3.20	3.40	3.60	3.30	3.00	2.70	2.40	2.10	1.80
$\theta = 19$	3.20	3.40	3.60	3.80	3.50	3.20	2.90	2.60	2.30
$\theta = 20$	3.20	3.40	3.60	3.80	4.00	3.70	3.40	3.10	2.80
$\theta = 21$	3.20	3.40	3.60	3.80	4.00	4.20	3.90	3.60	3.30
$\theta = 22$	3.20	3.40	3.60	3.80	4.00	4.20	4.40	4.10	3.80
$\theta = 23$	3.20	3.40	3.60	3.80	4.00	4.20	4.40	4.60	4.30
$\theta = 24$	3.20	3.40	3.60	3.80	4.00	4.20	4.40	4.60	4.80

Table 12.5. This procedure is repeated for each of the available courses of action ($a = 16, 17, \ldots, 24$) to obtain entries for each column of the table. The reader is urged to verify each of the entries in the table.

The reader may recall that, when a decision is made under certainty, the true state of nature is known. Under this condition, the decision maker considers only a single consequence for each act and then chooses the act with the most desirable consequence. When a decision is made under uncertainty, however, the true state of nature is not known for sure. Thus, each act may have two or more possible consequences corresponding to the various possible states of nature. For example, Table 12.5 shows that each alternative act has nine possible monetary consequences corresponding to the nine possible states of nature ($\theta = 16, 17, \ldots, 24$). When a decision must be made under such conditions, the problem then arises about how to compare the alternative acts when each act has more than one possible consequence. This is a complex problem that has no single answer. Rather, a number of different criteria have been proposed for choosing among alternative acts when this problem of uncertainty exists. Some of these criteria entail the explicit use of all the possible consequences of each act, without considering how likely each of these consequences is to occur. That is, such criteria disregard the probabilities of the various possible states of nature occurring. In contrast, there are other criteria that make rather limited use of monetary consequences of the acts, but which do stress the probabilities of the various states of nature. There are still other criteria that combine full use of the monetary consequences with the probabilities of the states of nature. Hence, the various criteria for decision making under uncertainty may be classified into the following three categories:

1. criteria based on monetary consequences alone
2. criteria based primarily on the probabilities of states of nature
3. criteria based on monetary consequences combined with probabilities of states of nature

12.3.1 Decision Criteria Based on Monetary Consequences Alone

In this section, we will discuss four decision criteria that belong to the first category listed above. All of these criteria formally employ the monetary consequences associated with every possible combination of acts and states of nature. However, none of these criteria makes use of any probability information. A decision maker may determine to use one of these criteria either because the necessary probability information is not available or simply because he chooses to ignore such information. For purposes of illustration, let us continue with the LAX Sunday paper example.

1. *Maximin Criterion.* A *pessimistic* decision maker might seek the best payoff that he can be *assured* of, regardless of which state of nature should occur. He can accomplish this by determining the minimum possible

payoff for each act and then selecting that act for which this minimum possible payoff is best. For example, considering the conditional payoffs in Table 12.5, the news dealer might reason as follows: "If I stock twenty-four copies, the worst payoff that can happen (the minimum of the figures in the column for $a = 24$) is 80 cents. If I stock twenty-three copies, the worst payoff that can happen is $1.10. If I stock sixteen copies, the worst payoff that can happen is $3.20. Similarly, I can obtain the minimum payoff for each of the other possible acts as shown in the first row of Table 12.6. Comparing the minimum payoffs in that row, I will, therefore, choose the act of stocking 16 copies, because I am assured of a payoff of at least $3.20."

Table 12.6

Use of Monetary Consequences to Select a Course of Action

	Action (Number of Copies Stocked)								
	$a = 16$	$a = 17$	$a = 18$	$a = 19$	$a = 20$	$a = 21$	$a = 22$	$a = 23$	$a = 24$
(1) Minimum Payoff	$3.20	$2.90	$2.60	$2.30	$2.00	$1.70	$1.40	$1.10	$0.80
(2) Maximum Payoff	3.20	3.40	3.60	3.80	4.00	4.20	4.40	4.60	(4.80)
(3) Hurwicz Weighted Average with $c = .2$	3.20	3.30	3.40	3.50	3.60	3.70	3.80	3.90	(4.00)
(4) Maximum Regret	1.60	1.40	1.20	(1.00)	1.20	1.50	1.80	2.10	2.40

This *maximin* criterion identifies the act that *maxi*mizes the *min*imum payoffs. Under this criterion, the decision maker selects the act with the maximum payoff that can be *assured*. The above solution of stocking sixteen copies guarantees a profit of $3.20 regardless of what happens, and hence is considered to be the least risky or the most conservative action available. This action, however, precludes any profit higher than $3.20.

2. *Maximax Criterion.* An *optimistic* decision maker, on the other hand, might seek the maximum payoff that can possibly be obtained. He can accomplish this by determining the maximum possible payoff for each act and then selecting that act for which this maximum possible payoff is the greatest. For example, considering the conditional payoffs in Table 12.5, the news dealer might reason as follows: "If I stock twenty-four copies, the highest

payoff that can be obtained (the maximum of the figures in the column for $a = 24$) is \$4.80. If I stock twenty-three copies, the highest payoff that can be obtained is \$4.60. Similarly, I can find the maximum payoff for each of the other possible acts as shown in the second row of Table 12.6. Comparing the maximum payoffs in that row, I will, therefore, choose the act of stocking twenty-four copies, which can possibly yield a maximum payoff of \$4.80." This *maximax* criterion selects the act that *max*imizes the *max*imum possible payoffs. The above solution of stocking twenty-four copies makes it possible for the news dealer to enjoy the largest payoff, but it ignores the possible low payoffs. This action is considered to be the most risky alternative in the sense that the possible profits range from 80 cents (when only sixteen copies are sold) to \$4.80 (when all twenty-four copies are sold).

3. *Hurwicz Criterion (Pessimism-Optimism Coefficient).* The maximin criterion is completely pessimistic in that, relative to each act, it concentrates on the state having the worst consequence. The maximax criterion, on the other hand, is entirely optimistic in that, relative to each act, it concentrates on the state having the best consequence. Hurwicz[2] suggests a pessimism-optimism coefficient that emphasizes a weighted combination of the worst and the best. For the ith act, a_i, let m_i be the minimum and M_i be the maximum of the monetary consequences. The minimum and maximum payoffs for various possible acts for the news dealer's problem are shown, respectively, in the first and second rows of Table 12.6. Let a coefficient c between 0 and 1, called the pessimism-optimism coefficient, be given. For each possible act, we may compute the Hurwicz weighted average:

$$H_i = cm_i + (1 - c)M_i \tag{12.1}$$

Suppose that the news dealer adopts .2 for the value of c. Then we can calculate the weighted average for each of the possible acts. For example, the weighted average for the act of stocking twenty-four copies is: $H_{24} = .2(\$.80) + (1 - .2)(\$4.80) = \$4.00$. The weighted averages for all possible acts are similarly calculated and are shown in the third row of Table 12.6. Comparing the figures in that row, we see that the optimal act is to stock twenty-four copies, because it has the highest weighted average.

Notice that, if $c = 1$, the pessimism-optimism coefficient criterion becomes the maximin criterion, whereas if $c = 0$, it becomes the maximax criterion. Thus, the larger the magnitude of c that the decision maker selects, the more pessimistic he appears to be. In short, c may be considered the coefficient of pessimism, and $(1 - c)$ the coefficient of optimism. It should be stressed that, in applying the Hurwicz criterion, the value of c must be supplied by the decision maker himself as an expression of his personal degree

[2] Leonid Hurwicz, "Optimality Criteria for Decision Making Under Ignorance," Cowles Commission Discussion Paper, *Statistics,* No. 370, 1951 (mimeographed).

of pessimism. In the above example, the pessimism coefficient of $c = .2$, with the resulting optimism coefficient of $(1 - c) = .8$, implies that the news dealer feels quite optimistic. Since the c value is not zero, we would expect the solution of the pessimism-optimism coefficient criterion to be different from that of the maximax criterion. However, in this example, the optimal act obtained by this criterion turns out to be identical to that determined by the maximax criterion. This is simply a coincidence because of the particular payoff table involved.

4. *Minimax Regret Criterion.* For decision making under uncertainty, the decision maker is forced to choose a specific course of action from a set of alternatives before the fact (prior to knowing the true state of nature). However, that action that appears to be optimal before the fact may not necessarily turn out, after the fact, to be the optimal action that should have been taken. For instance, suppose that the news dealer in our example has decided to stock sixteen copies, which is the optimal act according to the maximin criterion. Suppose further, however, that the actual demand turns out be be twenty copies. As may be seen from Table 12.5, if he had known that the demand actually would be twenty copies, he could have maximized his profit at $4.00 by stocking twenty copies. However, by stocking sixteen instead of twenty copies, he has made a profit of only $3.20 rather than $4.00. Thus, he suffers an *opportunity loss* of 80 cents, which is the difference between the $4.00 that he could have made and the $3.20 that he actually made. More generally, if the action chosen does not yield the maximum profit that the decision maker could have made if he had known the true state of nature, then he suffers an opportunity loss that represents the difference between (1) the maximum profit he had the opportunity of obtaining, and (2) the profit he has actually obtained.

The concept of opportunity loss (regret) is basic to the *minimax regret* criterion, which was introduced by Leonard Savage. The application of this criterion requires the computation of the opportunity loss for each possible act conditional on each state of nature. The conditional opportunity losses for the news dealer problem are shown in Table 12.7. The opportunity losses shown in this table are derived, row by row, from the conditional payoffs given in Table 12.5. For any given row, the opportunity losses are obtained by subtracting each entry in the corresponding row of the payoff table from the largest entry in that row. For example, consider the first row of the payoff table showing the conditional payoffs for the various acts when $\theta = 16$. For this row, the act "$a = 16$" has the highest conditional payoff; namely, $3.20. For all other acts, the conditional payoffs in this row are smaller, ranging from $2.90 to 80 cents. The difference between the maximum conditional payoff of $3.20 and the conditional payoff associated with each of the other acts is the conditional opportunity loss (COL) associated with that act. For example, if $\theta = 16$, the COL for "$a = 17$" is ($3.20 − $2.90) = 30 cents, the COL for

"$a = 18$" is (\$3.20 − \$2.60) = 60 cents, and so on. The conditional opportunity losses for all other rows of Table 12.7 are obtained in a similar manner. The reader is urged to verify these figures.

Table 12.7

Use of Monetary Consequences to Select 9 Course of Action

State of Nature (Number of Copies Demanded): θ	*Action (Number of Copies Stocked)*								
	a = 16	*a = 17*	*a = 18*	*a = 19*	*a = 20*	*a = 21*	*a = 22*	*a = 23*	*a = 24*
16	\$ 0	\$ .30	\$ .60	\$ .90	\$1.20	\$1.50	\$1.80	\$2.10	\$2.40
17	.20	0	.30	.60	.90	1.20	1.50	1.80	2.10
18	.40	.20	0	.30	.60	.90	1.20	1.50	1.80
19	.60	.40	.20	0	.30	.60	.90	1.20	1.50
20	.80	.60	.40	.20	0	.30	.60	.90	1.20
21	1.00	.80	.60	.40	.20	0	.30	.60	.90
22	1.20	1.00	.80	.60	.40	.20	0	.30	.60
23	1.40	1.20	1.00	.80	.60	.40	.20	0	.30
24	1.60	1.40	1.20	1.00	.80	.60	.40	.20	0

Thus far we have shown how to find conditional opportunity losses associated with possible courses of action given a particular state of nature. In other words, in constructing the conditional opportunity loss table, as shown in Table 12.7, we obtained the entries row by row. Now let us switch our attention to the entries in each column. Since each column represents one of the alternative acts, the entries in a column indicate the opportunity losses associated with that act. For example, the entries in the column for $a = 16$ indicate that the opportunity losses associated with this act range from \$0 to \$1.60 depending on the particular state of nature (the specific value of θ). That is, whenever the news dealer stocks sixteen copies, his opportunity loss will be \$0 if demand is sixteen copies, \$.20 if demand is seventeen copies, and so on. To determine the optimal act under the minimax regret criterion, we first find the maximum opportunity loss (i.e., the maximum regret) for each act by locating the maximum figure in each column of the COL Table. The maximum opportunity loss is \$1.60 for $a = 16$, \$1.40 for $a = 17$, and so on. These maximum opportunity losses for the various alternative acts are shown in Row (4) of Table 12.6. Then, comparing the maximum opportunity losses in this row, we see that the minimum figure is \$1.00 for $a = 19$. Thus, if the news dealer adopts the minimax regret criterion, he will stock nineteen copies, since this is the act for which the maximum possible opportunity loss is the smallest. In short, the *minimax* regret criterion selects the act that *mini*mizes the *max*imum regret.

12.3.2 Decision Criteria Based Primarily on Probabilities

So far we have examined four criteria, all of which belong to one family, in the sense that they all are based on monetary consequences alone. We will now consider another family of criteria that devote primary consideration to the probabilities of the various states of nature, giving only limited attention to monetary consequences. We will continue with the LAX Sunday paper example to illustrate three decision criteria of this kind.

1. *Maximum Likelihood Criterion.* Since he is operating under uncertainty, the news dealer does not know how many copies of the Sunday paper actually will be sold. However, his historical demand record, which was summarized in Table 12.4, indicates that the demand has been between sixteen and twenty-four copies. From this record, the relative frequencies of various demand levels are calculated, with the results shown in the third column of Table 12.8. Since the news dealer anticipates no significant changes in demand in the near future, he is willing to accept the relative frequencies as probabilities for future demand levels.

Table 12.8

Demand Distribution and Expected Demand for LAX Sunday Paper

(1) Number of Copies Demanded Per Week: θ	*(2)* Number of Weeks	*(3)* Relative Frequency $f(\theta)$	*(4)* Calculation of Expected Demand $\theta f(\theta)$
16	5	.05	.80
17	10	.10	1.70
18	12	.12	2.16
19	16	.16	3.04
20	10	.10	2.00
21	20	.20	4.20
22	16	.16	3.52
23	6	.06	1.38
24	5	.05	1.20
Sum	100	1.00	20.00 = $E(\Theta)$

Column (3) of Table 12.8 shows clearly that some of the demand levels are more likely to occur than others. This column further indicates that the demand level $\theta = 21$ is most likely to occur, since it has the highest relative frequency. In other words, $\theta' = 21$ has the maximum likelihood (highest probability) of occurring. To apply the maximum likelihood criterion, the news dealer will consider the monetary consequences associated only with this particular demand level. Referring to Table 12.5, he will limit his consideration of monetary consequences to the conditional payoffs in the row corresponding to $\theta = 21$. Then, comparing the figures in this row, the highest payoff is \$4.20. This is the payoff for $a = 21$. Thus, using the maximum likelihood criterion, the news dealer should choose to stock twenty-one copies. Generally speaking, the *maximum likelihood* criterion advises the decision maker to identify the state of nature that has the *maximum likelihood* of occurring, and then to select the act that has the most desirable monetary consequence for that state of nature.

2. *Expected State of Nature Criterion.* Unlike the maximum likelihood criterion, which focuses on the *modal* state of nature (i.e., the state of nature most likely to occur), the *expected state of nature criterion* focuses on the *mean* state of nature. Using this criterion, the decision maker first computes the mean state of nature and then selects the act that has the most desirable consequence under the assumption that the actual state of nature will be approximately equal to the mean. The computation of the mean state of nature depends on whether the state space is discrete or continuous. If the state of nature Θ is a *discrete* random variable, the expected state of nature may be calculated by using Formula (8.3), which is restated in the following form:

$$E(\Theta) = \sum_{\text{all } \theta} \theta\, f(\theta) \tag{12.2}$$

If the state of nature θ is a *continuous* random variable, the expected state of nature may be obtained from Formula (8.4), which is restated in the following form:

$$E(\Theta) = \int_{-\infty}^{\infty} \theta\, f(\theta)\, d\theta \tag{12.3}$$

In our Sunday paper example, the set of possible states of nature consists of the discrete demand levels: $\theta = 16, 17, \ldots, 24$. Therefore, to compute the expected state of nature, the news dealer would use Formula (12.2). These calculations are shown in Column (4) of Table 12.8. As shown by the sum of this column, the expected demand is $E(\Theta) = 20$ copies. The news dealer will then consider the monetary consequences associated only with $\theta = 20$, under the assumption that this is the state of nature that actually will occur. Referring to the row corresponding to $\theta = 20$ in Table 12.5, the highest payoff is \$4.00. This is the payoff for $a = 20$. Hence, if the news dealer uses the expected state of nature criterion, he will choose to stock twenty copies.

In the above illustration, the mean demand is identical to one of the possible demand levels. This is merely a coincidence. However, when the state space is discrete, the mean state of nature more likely will not coincide exactly with any of the possible states of nature. In this case, the mean demand is rounded to the closest value of a state of nature. Of course, when the state space is continuous, this difficulty does not arise.

The mean or expected value of the state of nature can be calculated and used as a decision criterion only if the various states of nature can be described on a meaningful numerical scale such as in this inventory example. There are, however, decision problems in which the states of nature cannot be naturally expressed in numerical terms. Consider, for example, the case of Bob Baker as presented in Table 12.2, in which the state of nature denotes whether or not the new product will be successfully developed. For such a problem, the expected value of the state of nature is meaningless and hence cannot be used as a decision criterion.

12.3.3 Decision Criteria Based on Monetary Consequences and Probabilities Combined

We have already examined one family of decision criteria based on monetary consequences alone, and another family of criteria that is primarily concerned with the probabilities of the various states of nature. Finally, we will now consider a third family of decision criteria that *explicitly* take into account *both* monetary consequences *and* probabilities in making decisions. We shall, in particular, discuss two decision criteria of this nature: (1) the expected monetary value (EMV), and (2) the expected opportunity loss (EOL) criteria.

1. *EMV Criterion.* Using the EMV criterion, the decision maker computes the expected monetary value (EMV) for each alternative act and then chooses the act which has the largest expected monetary value. The computation required by this decision criterion explicitly considers both the monetary consequences and the probabilities of obtaining these consequences.

Bearing in mind that Θ is a random variable denoting the state of nature, the reader may recall that $v(a,\Theta)$ represents the monetary value of an act. Since $v(a,\Theta)$ is a function of Θ, the expected monetary value of an act may be denoted by $E[v(a,\Theta)]$. It may be recognized that $E[v(a,\Theta)]$ is the expected value of a function of a random variable, as discussed in Chapter 8. Thus, if the random variable Θ is discrete, the expected monetary value of an act may be computed by applying Formula (8.5). For our present purposes, this formula may be restated in the following form:

$$E[v(a,\Theta)] = \sum_{\text{all } \theta} v(a,\theta)\, f(\theta) \tag{12.4}$$

This formula says that the expected monetary value of an act is computed by multiplying (i) the monetary value for that act conditional on the occurrence of each state of nature by (ii) the probability of that state of nature occurring, and then summing the products.

In Chapter 8, the expected value of a function of a continuous random variable was given by Formula (8.6). A restatement of that formula provides the following formula for computing the EMV of an act when the state of nature is a *continuous* random variable:

$$E[v(a,\Theta)] = \int_{-\infty}^{\infty} v(a,\theta)\, f(\theta)\, d\theta \tag{12.5}$$

To illustrate the EMV criterion, let us now return to the LAX Sunday paper problem. Consider, for example, the act of stocking twenty-four copies weekly. As shown in Table 12.5, the monetary values (conditional payoffs) associated with this action range from 80 cents to $4.80, depending on how many copies actually are sold. The probabilities of possible numbers of copies demanded have been shown in Table 12.8. For convenience, the conditional payoffs for the act of stocking twenty-four copies, together with the probabilities of the various demand levels, are now exhibited in Table 12.9. The computation of the expected monetary value (the expected payoff) for this act, is shown in Column (4) of this table. Using Formula (12.4), this computation is performed by multiplying the payoff conditional on the occurrence of each demand level by the probability of that demand level occurring, and then summing these products. The resulting EMV for $a = 24$ is $2.80.

Table 12.9

Conditional and Expected Payoffs for the Act of Stocking 24 Copies of LAX Sunday Paper

(1) *State of Nature (Number of Copies Demanded):* θ	*(2)* *Conditional Payoff:* $v(24,\theta)$	*(3)* *Probability of State of Nature:* $f(\theta)$	*(4)* *Calculation of Expected Payoff* $v(24,\theta)\cdot f(\theta)$
16	$0.80	.05	$.040
17	1.30	.10	.130
18	1.80	.12	.216
19	2.30	.16	.368
20	2.80	.10	.280
21	3.30	.20	.660
22	3.80	.16	.608
23	4.30	.06	.258
24	4.80	.05	.240
Sum		1.00	$2.800 = EMV

By repeating the same general procedure shown in Table 12.9, the EMV for each of the other courses of action is obtained. The EMV's for all available course of action are summarized in Table 12.10. The reader may wish to verify the entries in the table. From this table, we see that the act of stocking nineteen copies yields the maximum EMV of $3.565. Therefore, using the EMV criterion, the dealer should stock nineteen copies.

Table 12.10

Expected Monetary Values of Alternative Actions*

	Action (Number of Copies Stocked)								
	$a = 16$	$a = 17$	$a = 18$	$a = 19$	$a = 20$	$a = 21$	$a = 22$	$a = 23$	$a = 24$
EMV:	$3.200	$3.375	$3.500	$3.565	$3.550	$3.485	$3.320	$3.075	$2.800
				Maximum EMV					

*As an example, the calculation of the EMV for $a = 24$ is shown in Table 12.9.

Notice that, although the expected monetary value for stocking nineteen copies is equal to $3.565, this amount of profit can never occur on any one Sunday. As pointed out in Section 8.1.3, the word "expected" does *not* mean that the decision maker actually expects that amount of profit. For any given Sunday, the actual profit resulting from stocking nineteen copies will be one of the monetary values (ranging from $2.30 to $3.80) shown in the column labeled $a = 19$ of Table 12.5. The EMV is simply a weighted average, the weights being the probabilities of the various states of nature. It is the average profit that is to be "expected" in the long run. In other words, it is the average profit that will result if the decision is repeated many times, and each time the decision maker chooses the same alternative. In the news dealer example, if the decision is to be repeated week after week, the act of stocking nineteen copies should produce the highest average weekly profit in the long run.

The EMV criterion explicitly considers both the monetary values and the probabilities, and in that sense may be regarded as superior to the previously discussed criteria that emphasize only one or the other of these two factors. The EMV criterion chooses the act that will yield the maximum average payoff in the long-run, and hence, has immediate intuitive appeal for businessmen. We will make use of this decision criterion in the next chapter. However, we must realize that this criterion is accepted largely because it *seems* to be reasonable, since cogent scientific evidence to support its acceptance is lacking. As Russell Ackoff has pointed out:

> The principal argument which has been used to defend the criterion of maximization of expected value is based on its apparent "reasonableness." It seems obvious to many that this *ought* to be the criterion employed by rational beings. In fact, use of this criterion is often cited as a necessary (if not sufficient) condition for rational choice. Despite the widespread acceptance by decision theorists of the superiority of this criterion, science is ultimately obliged to find less

intuitive grounds on which to support its selection of a criterion of choice, whatever that criterion may be. Purely intuitive or rational grounds for selecting such a criterion are ultimately doomed in science. Such grounds must eventually have empirical as well as rational content.[3]

2. *EOL Criterion.* Another criterion that belongs to the same family as EMV is the *Expected Opportunity Loss* (EOL) criterion. Both of these criteria combine the probabilities of the various states of nature with the monetary consequences of the alternative acts to arrive at an optimal decision. In treating monetary consequences, however, the EMV criterion works with the monetary *values* of the acts, whereas the EOL criterion works with *opportunity losses* of the acts. A decision maker who adopts the EOL criterion will compute the expected opportunity loss for each alternative act and then choose the act that has the *smallest* expected opportunity loss.

Just as $v(a,\theta)$ is used to denote the monetary value resulting from a particular course of action a when a specific state of nature θ occurs, so the expression $\ell(a,\theta)$ is used to designate the corresponding opportunity loss. Then $\ell(a,\Theta)$ represents the opportunity loss of an act as a function of the random variable Θ. Thus the expected opportunity loss of an act, $E[\ell(a,\Theta)]$, is the expectation of a function of the random variable Θ. Hence this expectation may be computed by the same general procedure employed in calculating the EMV of an act. Specifically, if the state of nature is a *discrete* random variable, the EOL of an act may be obtained from the following restatement of Formula (8.5):

$$E[\ell(a,\Theta)] = \sum_{\text{all } \theta} \ell(a,\theta)\, f(\theta) \tag{12.6}$$

In words, the EOL of an act is computed by multiplying (i) the opportunity loss for that act conditional on the occurrence of each state of nature, by (ii) the probability of that state of nature occurring, and then summing the products. In an analogous manner, if the state of nature is a *continuous* random variable, the EOL of an act may be computed from the following restatement of Formula (8.6):

$$E[\ell(a,\Theta)] = \int_{-\infty}^{\infty} \ell(a,\theta)\, f(\theta)\, d\theta \tag{12.7}$$

To apply the EOL criterion to the LAX Sunday paper problem, let us begin by referring back to the opportunity losses shown in Table 12.7. Now, using Formula (12.6), we may compute the *expected* opportunity loss of each act by multiplying each of the conditional opportunity losses in the column representing that act by the probability of the corresponding state of nature occurring, and then summing the products. As an example, the calculation of the EOL for the act of stocking sixteen copies (a = 16) is shown in Table

[3] Russell L. Ackoff, *Scientific Method*, New York, John Wiley & Sons, 1962, p. 38.

12.11. Column (1) contains the possible states of nature. Column (2) simply reproduces the figures in the column labeled $a = 16$ of Table 12.7. Each of the figures in this column is the opportunity loss for $a = 16$ conditional on a specific state of nature occurring. Column (3) exhibits the probability of the occurrence of each of these possible states of nature. These probabilities were shown previously in Table 12.8. Column (4) is obtained by multiplying each conditional opportunity loss in Column (2) by the corresponding probability in Column (3). The sum of these products is equal to 80 cents, which is the EOL associated with the act of stocking sixteen copies. The EOL for each of the other acts can be computed in the same general manner. The EOL's for all alternative acts are shown in Table 12.12. The reader is urged to verify these figures. As Table 12.12 indicates, the act of stocking nineteen copies yields the minimum EOL of 43.5 cents. Thus, if the news dealer accepts the EOL criterion, he should stock nineteen copies.

Table 12.11

Conditional and Expected Opportunity Losses for the Act of Stocking 16 Copies of LAX Sunday Paper

(1) State of Nature (Number of Copies Demanded): θ	(2) Conditional Opportunity Loss: $\ell(16,\theta)$	(3) Probability of State of Nature: $f(\theta)$	(4) Calculation of EOL $\ell(16,\theta)\cdot f(\theta)$
16	$0	.05	$0
17	.20	.10	.020
18	.40	.12	.048
19	.60	.16	.096
20	.80	.10	.080
21	1.00	.20	.200
22	1.20	.16	.192
23	1.40	.06	.084
24	1.60	.05	.080
Sum		1.00	$ .800 = EOL

Table 12.12

Expected Opportunity Losses of Alternative Actions*

	Action (Number of Copies Stocked)								
	$a = 16$	$a = 17$	$a = 18$	$a = 19$	$a = 20$	$a = 21$	$a = 22$	$a = 23$	$a = 24$
EOL:	.800	.625	.500	.435 ↑ Minimum EOL	.450	.515	.680	.925	1.200

*As an illustration, the EOL of $.800 for $a = 16$ is calculated in Table 12.11.

It is important to note that both the EMV and EOL criteria led to the same optimal act of stocking nineteen copies, since this act has both the maximum EMV and minimum EOL. This may be seen in Table 12.13, which compares the EMV and EOL of each of the alternative acts. As inspection of this table reveals, the sum of the EMV and EOL for each and every act is a constant of $4.00. Since this sum is a constant for any act, the act that has the maximum EMV must have the minimum EOL. This relationship between the EMV and EOL is *always* true, so that the EMV and EOL criteria always lead to the same decision.

Table 12.13

EMV and EOL for LAX Sunday Paper

	Action (Number of Copies Stocked)								
	$a = 16$	$a = 17$	$a = 18$	$a = 19$	$a = 20$	$a = 21$	$a = 22$	$a = 23$	$a = 24$
EMV:	$3.200	$3.375	$3.500	$3.565	$3.550	$3.485	$3.320	$3.075	$2.800
EOL:	.800	.625	.500	.435	.450	.515	.680	.925	1.200
	$4.000	$4.000	$4.000	$4.000	$4.000	$4.000	$4.000	$4.000	$4.000

Note: The EMV figures in this table are taken from Table 12.11, and the EOL figures are taken from Table 12.12.

12.4 INCREMENTAL ANALYSIS: AN APPLICATION OF THE EMV CRITERION

Using the EMV criterion, the optimal act for the decision maker is the act that has the maximum EMV. As we have seen, one way of finding this optimal act is to compute the EMV of each alternative act and then to compare these EMV's. Although this procedure is straightforward, it is laborious, particularly if there are many alternative acts. The inventory-stocking problem, such as the one presented in this chapter, is a case in point. In a realistic inventory problem, the random variable representing demand usually has numerous possible values (numerous possible demand levels), and hence numerous alternative courses of action (numerous stock levels) are available. For such a case, it is not practical to compute the EMV's for all the alternative acts. Fortunately, a simple analytical method, called *incremental analysis*, has been developed for finding the optimal stock level without having to compute the EMV's.

Since the EMV and EOL are equivalent criteria, the formulas for incremental analysis may be stated in terms of monetary values or opportunity losses. In the following discussion, we will present incremental analysis in terms of loss figures. To gain insight into this method, let us return to Table 12.7, which shows the opportunity losses for the LAX Sunday paper. Inspection of this table indicates that the opportunity loss is zero whenever the number of copies stocked equals the number of copies demanded. In other words, there is no opportunity loss if the stock is exactly equal to the quantity demanded (if $a = \theta$). An opportunity loss occurs either when stock exceeds demand ($a > \theta$) or when demand exceeds stock ($\theta > a$). In the case of *overstock* ($a > \theta$), the opportunity loss is 30 cents per copy, which is equal to the purchase cost of a copy. For instance, if the news dealer stocks twenty-two copies ($a = 22$) but demand is only eighteen copies ($\theta = 18$), then there will be an overstock of $(22 - 18) = 4$ copies. The opportunity loss associated with this situation is $(\$.30 \times 4) = \$1.20$, which is exactly the same as that shown in Table 12.7 for $a = 22$ and $\theta = 18$. Similarly, in the case of *understock* ($\theta > a$), the opportunity loss per copy is 20 cents, which represents the additional profit that the news dealer could have made if he had one more copy in stock. For example, if he stocks twenty-two copies ($a = 22$) but demand is twenty-four copies ($\theta = 24$), then there will be an understock of $(24 - 22) = 2$ copies, with a resulting opportunity loss of $\$.20 \times 2 = \$.40$. This is precisely what is shown in Table 12.7 for $a = 22$ and $\theta = 24$.

To generalize the above observations, let:

$$k_o = \text{the opportunity loss per unit of overstock}$$

$$k_u = \text{the opportunity loss per unit of understock}$$

Then the possible opportunity losses associated with an inventory-stocking problem can be calculated from the following function:

$$\ell(a,\theta) = \begin{cases} k_o(a - \theta) & \text{if } a > \theta \\ k_u(\theta - a) & \text{if } \theta > a \end{cases}$$

In words, if stock exceeds demand, the total opportunity loss of overstock is equal to k_o times the number of units of overstock. However, if demand exceeds stock, the total opportunity loss of understock is equal to k_u times the number of units of understock. From this opportunity loss function, a simple analytical solution for optimal stock can be obtained. Specifically, if the state of nature Θ, denoting the demand for a perishable product, is a discrete random variable, the optimal act (optimal stock level) is the smallest value of a that satisfies the following *inequality:*

$$P(\Theta \leq a) \geq \frac{k_u}{k_u + k_o} \tag{12.8}$$

In this formula, $P(\Theta \leq a)$ represents the probability that demand level will be less than or equal to the stock level. Thus this probability, $P(\Theta \leq a)$, can be obtained from the cumulative mass function $F(\theta)$ by setting θ equal to a. That is, since $\theta = a$, we may write $P(\Theta \leq a) = P(\Theta \leq \theta) = F(\theta)$. If Θ is a continuous random variable, the optimal act is the value of a that satisfies the following *equality:*

$$P(\Theta \leq a) = \frac{k_u}{k_u + k_o} \tag{12.9}$$

As an example, let us use incremental analysis to obtain the optimal stock for the LAX Sunday paper. Since demand in this problem is a discrete random variable, we will use Formula (12.8). To apply this formula, we proceed as follows:

Step 1: Determine k_o and k_u. In this example, $k_o = \$.30$, which is the cost of an unsold copy; $k_u = \$.50 - \$.30 = \$.20$, which is the profit that the news dealer forgoes if he fails to stock one unit that could have been sold.

Step 2: Calculate the ratio $k = \dfrac{k_u}{k_u + k_o}$ · For our example, we obtain:

$$k = \frac{k_u}{k_u + k_o} = \frac{.20}{.20 + .30} = .40$$

Step 3: Find the smallest value of a such that $P(\Theta \leq a) \geq k$. To do this, we prepare Table 12.14. The probability mass function, $f(\theta)$, in Column (2), is taken from Table 12.8. The cumulative mass function, $F(\theta)$,

Table 12.14

Cumulative Probabilities for LAX Sunday Paper

(1)	*(2)*	*(3)*	*(4)*
θ or a	*$f(\theta)$*	*$F(\theta)$*	*$P(\Theta \leq a)$*
16	.05	.05	.05
17	.10	.15	.15
18	.12	.27	.27
19	.16	.43	.43
20	.10	.53	.53
21	.20	.73	.73
22	.16	.89	.89
23	.06	.95	.95
24	.05	1.00	1.00
Sum	1.00		

in Column (3), is obtained by cumulating the terms in Column (2). The cumulative function is then re-expressed in terms of the cumulative probability $P(\Theta \leq a)$ in Column (4). From this column, we see that the smallest cumulative probability greater than .40 is equal to .43. That is, $P(\Theta \leq 19) = .43 > .40$. Thus, 19 is the smallest value of a such that $P(\Theta \leq a) > .40$. Consequently, the optimal stock is nineteen copies, which is identical to the result that was obtained previously by computing and comparing the EMV's (or EOL's) of all alternative acts.

PROBLEMS

12.1 Suppose that you have a sum of money that you wish to invest for a 1-year period. You have narrowed your choices to 3 alternatives:

a_1: Invest in a construction firm
a_2: Deposit the money in a savings and loan association
a_3: Invest in an industrial consulting firm.

You are concerned with 3 possible states of nature that might affect the profit you will realize on your investment:

θ_1: Housing starts will increase next year
θ_2: Housing starts will remain at the same level
θ_3: Housing starts will decline next year.

You have determined the possible profits (in thousands of dollars) associated with various action-state pairs as shown in the following payoff table:

State of Nature	*Action*		
	a_1	a_2	a_3
θ_1	9	6	8
θ_2	7	6	7
θ_3	3	6	5

(a) If you adopt the maximin criterion, which course of action should you take?
(b) If you adopt the maximax criterion, which course of action should you take?
(c) If you adopt the Hurwicz criterion with $c = .5$, which course of action should you take?
(d) If you adopt the minimax regret criterion, which course of action should you take?

12.2 Hi-Test Tool Company, Inc., and Ever-Rite Machinery Company, Inc., are direct competitors in several product lines. Currently, each is considering the feasibility of constructing a newly designed automated Capston lathe-manufacturing plant. Hi-Test reported net earnings of $20 million for the preceding year. Hi-Test's management estimates that if it builds the plant and Ever-Rite does not, the *present value* of the *increase* in net-income flow will be $15 million. However, it has concluded that if Hi-Test does not build the plant but its competitor does, the present value of the *decrease* in net-income flow will be $12 million. If both companies construct new plants, Hi-Test's *decrease* in the present value of net-income flow will be $15 million.

(a) Should Hi-Test construct the new plant if its decision criterion is maximin?
(b) Should Hi-Test build the plant if its decision criterion is maximax?
(c) If Hi-Test uses the Hurwicz criterion with the pessimism-optimism coefficient $c = .9$, should it build the plant?
(d) What should Hi-Test do if it uses the minimax regret criterion?

12.3 The Ever-Rite Company in Problem 12.2 had net income of $30 million in the preceding year. The company has arrived at the following estimates:

> If Ever-Rite builds the new plant, but Hi-Test does not, the present value of the *increase* in Ever-Rite's net-income flow will be $27 million. However, if Ever-Rite does not build, but Hi-Test does, the present value of the *decrease* in net-income flow will be $12 million. If both companies build new plants, the present value of the *decrease* in Ever-Rite's net-income flow is estimated to be $15 million.

Find the optimal course of action for Ever-Rite under each of the following decision criteria:

(a) Maximin criterion
(b) Maximax criterion
(c) Hurwicz criterion with $c = .9$
(d) Minimax regret criterion.

12.4 A specialty grocer stocks a type of exotic melon that is flown in daily from Pago Pago. From past experience, he has determined the following probabilities for the daily demand for this type of melon.

Daily Demand	*Probability*
1 melon	.10
2 melons	.40
3 melons	.30
4 melons	.20

The grocer buys melons at $5.00 each, and sells them for $7.00 each. The melon is a highly perishable item. Any melon that is not sold by the end of the day will be spoiled and will have no value on the next day. Thus, the grocer is very much concerned with the number of melons to stock daily.

(a) Construct the conditional profit table for the grocer's stocking problem.
(b) If the grocer adopts the maximum likelihood criterion, how many melons should he stock?
(c) If the grocer adopts the expected state of nature criterion, how many melons should he stock?

12.5 A retailer stocks a perishable product that costs him $3.00 per unit. He buys the product in the morning and sells it at $5.00 per unit during the day. Any unit not sold by the end of the day must be thrown away at a total loss. Let X be the random variable denoting the demand for the fresh product daily. The retailer has found that the probability mass function of X is given by:

$$f(x) = \begin{cases} \dfrac{x-4}{10} & \text{if } x = 5, 6, 7, 8 \\ 0 & \text{otherwise} \end{cases}$$

(a) Construct the conditional payoff table for the retailer's stocking problem.
(b) If the retailer wishes to use the maximum likelihood criterion, how many units should he stock?
(c) If the retailer wishes to use the expected state of nature criterion, how many units should he stock?
(d) Compute the expected monetary value of each of the alternative acts. Using the EMV criterion, how many units should he stock?

12.6 A trading post operator has limited capital and must decide whether he should stock snow shoes or tennis shoes on next month's order. If he stocks snow shoes and it snows, he will make a profit of $100, but if it does not snow he will suffer a $20 loss. If he stocks tennis shoes and it snows he will suffer a $10 loss, but if it does not snow he will make a profit of $90. He estimates a .60 probability that it will snow.
(a) Construct the conditional payoff table for the trading post operator.
(b) If the operator uses the EMV criterion, should he stock snow shoes or tennis shoes?
(c) Construct the conditional opportunity loss table for the operator.
(d) If the operator uses the EOL criterion, should he stock snow shoes or tennis shoes?
(e) Do the EMV and EOL criteria always lead to the same decision? Explain why.

12.7 One of the publications carried by the AXY newsstand is *Nouvelles de Provence*, a French weekly magazine. The dealer pays 70¢ per copy and sells it for $1.00 per copy. Copies that are unsold after a week's time are non-returnable and have no value. The probability distribution of demand is shown below:

Demand (Number of Magazines Per Week)	*Probability*
10	.05
11	.05
12	.10
13	.15
14	.20
15	.25
16	.15
17	.05

Use each of the following decision criteria to determine the optimal number of magazines to stock:
(a) Maximin criterion
(b) Maximax criterion
(c) Hurwicz criterion with $c = .7$
(d) Minimax regret criterion
(e) Maximum likelihood criterion
(f) Expected state of nature criterion
(g) EMV criterion by computing the EMV's of the alternative acts
(h) EOL criterion
  (i) by computing the EOL's of the alternative acts
  (ii) by using incremental analysis

12.8 The manager of a small department store must place his order for an expensive line of Christmas cards for the Christmas season. Each box of cards costs $3, sells at $5 during the season, and the price will be reduced to $2 after the season. He feels that any boxes of cards remaining at the end of the season can be sold at this reduced price. The demand for boxes of cards during the season has been estimated as shown below:

Demand	*Probability*
25	.10
26	.15
27	.30
28	.20
29	.15
30	.10

Use each of the following decision criteria to determine the optimal number of boxes of cards to stock:

(a) Maximin criterion
(b) Maximax criterion
(c) Hurwicz criterion with $c = .4$
(d) Minimax regret criterion
(e) Maximum likelihood criterion
(f) Expected state of nature criterion
(g) EMV criterion by computing the EMV's of the alternative acts
(h) EOL criterion
  (i) by computing the EOL's of the alternative acts
  (ii) by using incremental analysis

12.9 You have received an offer to engage in a speculative real estate venture for an investment of $4,000. If a bond issue passes at the next election, you will receive a net return of $16,000, but if the bond issue fails your investment will be a total loss. What is the minimum probability you would require for the passage of the issue in order for the investment to be a desirable one if you use the EMV criterion?

12.10 The Copycat Corporation is considering installing a Zorex copying machine in its new branch. The Zorex distributor has offered to provide the machine either (i) as an outright sale at a $5,600 price, or (ii) on a rental basis. Either alternative would provide Copycat with the same services. If the branch operates successfully, the present worth of future rentals will be $8,000. If the branch fails, the present worth of future rentals will be only $2,000. On the basis of the EMV criterion, what is the minimum probability of success for the new branch that would make purchase of the machine preferable to rental?

12.11 The conditional opportunity loss table for a particular decision problem is given as follows:

State of Nature	*COL* a_1	a_2
θ_1	$ 0	$100
θ_2	50	0

The expected opportunity loss of a_1 is $35. The expected payoff (or EMV) of a_1 is $225, and the conditional payoff of a_1 given θ_1 is $400.

(a) According to the minimax regret criterion, which course of action is preferable?
(b) Find the probabilities for θ_1 and θ_2.
(c) According to the EOL criterion, which course of action is preferable?
(d) Find the expected payoff of a_2.
(e) Construct the conditional payoff table.

12.12 Jay Adgel believes he can make some money for Christmas by selling Christmas trees. He has decided to sell on weekends, since he can use a parking lot only during that time period. Thus, he must stock the lot with the optimal number for a weekend sales because any trees not sold on a given weekend are worthless. Jay has decided to sell silver-tip trees only, as his profit potential is greater with high-quality than with less expensive trees. After some investigation, Jay concludes that the demand for trees is Poisson with an average of 20. If the cost per tree is \$9.00 and the selling price per tree is \$15, how many trees should he stock per weekend?

12.13 The Empire Department Store has decided to stock a unique design of next year's calendars during the coming fall. The calendars will cost \$1.50 each. The sale of calendars will last for 10 weeks. The store will sell them for \$2.50 each during this period. Any calendars that are not sold by the end of this period will have no value. Based on past experience, the marketing manager estimates that the average demand per week is 90 calendars. Furthermore, he indicates that the weekly demand follows a Poisson process. Determine the optimal number of calendars to stock. [Hint: Use normal approximation to the Poisson, $F_P(x|\lambda,t) \approx F_n(x + .5|\mu,\sigma^2)$. The mechanics are analogous to the normal approximation to the binomial.]

12.14 Weekly demand for a particular perishable product is normally distributed with a mean of 500 kilograms and a standard deviation of 100 kilograms. The product costs the retailer \$6.70 per kilogram and sells for \$10.00 per kilogram. Stock left unsold at the end of the week has no salvage value. Determine the optimal quantity of this product for the retailer to stock each week.

12.15 A flight kitchen manager for Golden Bird Airlines must decide how many meals to prepare for Flight 534. The cost of preparing each meal is \$1.50. Any meals that are not used by passengers will be discarded after landing. If there are not enough meals, however, a meal ticket is issued to each passenger who cannot obtain the meal on the flight. A meal ticket, which costs the airline \$6.00, entitles the passenger to a free meal in the airport restaurant. The manager feels that the demand for meals on this particular flight can be approximated by a normal distribution, with mean equal to 200 and standard deviation equal to 30.

(a) What is the optimal number of meals to prepare for Flight 534?

(b) If the manager orders the number obtained in (a) above, what is the most likely number of meals that will be discarded after landing?

12.16 Funny Sisters, Inc., specializes in designing, making, and selling clothing of original styles for young ladies. The management has just decided to introduce Mini-Blouses for the coming summer. The question yet to be resolved is the size of the production order. Once the production for this order begins, it will not be feasible to revise the order since the facility has been scheduled for other styles. The marketing manager feels that the demand for Mini-Blouses during the season can be approximated by a normal probability distribution. The mean and the standard deviation of this distribution have been estimated as $\mu = 1{,}000$ and $\sigma = 200$. The variable cost per blouse has been estimated as \$5.99. The selling price will be \$11.99 per blouse during the season, and the price will be reduced to \$1.99 after the season. It is felt that any blouses remaining at the end of the season can be sold at this reduced price. The company's policy is to maintain the reputation of carrying new-style clothing. Thus, the company will sell the left-over blouses at a substantial loss rather than store them for next year.

(a) Determine the optimal number of Mini-Blouses to schedule for production.

(b) Suppose that the management feels that the company will suffer a goodwill loss if they run out of stock. If the management estimates that the goodwill loss is \$3.33 per stockout, find the optimal number of Mini-Blouses to schedule for production.

12.17 Suppose you own the Fair-Price Gasoline Station at the corner of Federal Street and State Avenue. You have found that your gasoline sales are fairly constant over time so that you can predict reasonably well when you need the next delivery. Unfortunately, the delivery service is poor and unreliable. Your experience has shown that the lead time (the time between placing a phone order and receiving truck delivery) varies between 2 and 6 hours. If you do not allow enough lead time for delivery, you face the risk of running out of gasoline. When a stockout happens, you will lose your sales and suffer some goodwill loss. On the other hand, if the delivery arrives too early, the truck has to wait until the gasoline in your tank has reached the minimum level, which will cost you a waiting fee. Thus, you are facing the problem of when to place your order. You have quantified your judgment about the uncertain lead time, T (in hours), in terms of the following probability density function:

$$f(t) = \begin{cases} \dfrac{t}{4} - \dfrac{1}{2} & \text{for } 2 \le t < 4 \\ \dfrac{3}{2} - \dfrac{t}{4} & \text{for } 4 \le t \le 6 \\ 0 & \text{otherwise} \end{cases}$$

(a) If you place your order at noon, what is your probability that the truck will arrive no later than 3:00 P.M.?

(b) By analyzing the losses due to early and late delivery, you estimate that the loss due to late delivery is 7 times the loss due to early delivery. At what time should you place your phone order if you need your next delivery at 3:00 P.M.?

12.18 The Chemtex Corporation sells chemicals to pharmaceutical manufacturers. One of the chemicals for which Chemtex receives frequent orders is VIP-27, a compound that is highly perishable in its raw form. Because of its limited market, there is only one manufacturer of VIP-27. This manufacturer is an independent producer who makes the compound available only in 300-pound paper bags, at a price of \$3.00 per pound. Thus, Chemtex can purchase VIP-27 only in multiples of 300 pounds. Furthermore, since the compound perishes rapidly, Chemtex must purchase VIP-27 weekly. When a fresh supply arrives each week, any remainder from the preceding week's supply must be destroyed at a total loss. Because of the perishability of VIP-27, the customers of Chemtex are willing to buy the compound only in the exact amounts that will meet their immediate needs. Over a period of time, total weekly demand has varied between 200 and 1,000 pounds. From analysis of past sales, Chemtex has determined that this total weekly demand for VIP-27 (in hundreds of pounds) can be approximated by the following probability density function:

$$f(x) = \begin{cases} \dfrac{x}{16} - \dfrac{1}{8} & \text{if } 2 \le x \le 6 \\ \dfrac{5}{8} - \dfrac{x}{16} & \text{if } 6 \le x \le 10 \\ 0 & \text{otherwise} \end{cases}$$

As a result of the perishability of the product and the variability of demand, Chemtex charges its customers \$10.00 per pound for VIP-27.

(a) Determine the expected weekly demand for VIP-27 in pounds.

(b) Determine Cehmtex's weekly profit function for sales of VIP-27 if the firm stocks:

  (i) 1 bag each week

  (ii) 2 bags each week

  (iii) 3 bags each week

(c) Using the expected state of nature criterion, how many bags of VIP-27 should Chemtex stock each week?

(d) Compute the EMV for each of the alternative weekly stock levels indicated in (b) above.

(e) Using the EMV criterion, how many bags of VIP-27 should Chemtex stock each week?

Chapter 13
The Value of Information in Decision Making

In most situations wherein decisions are made under uncertainty, the EMV (or equivalently the EOL) is usually considered preferable to other criteria that fail to explicitly combine the monetary consequences of the alternative acts with the probabilities of the possible states of nature. However, even with the EMV criterion, the decision maker does not know for sure what the exact monetary value of his action will be, since he does not know specifically which of the possible states of nature actually will prevail. Then we might ask: If a decision maker may purchase additional information that will enable him to eliminate or reduce his uncertainty, how much would that information be worth? The purpose of the present chapter is to answer this question.

13.1 EMV OF THE DECISION WITH AVAILABLE INFORMATION

Consider the case of Paul Smith, who wishes to invest a sum of $14,000 for a 1-year period. After some investigation, he has found that two alternatives are superior to all others. These alternatives are: (1) to deposit the money in the Westside Bank, and (2) to purchase a parcel of land for the same amount. The land being considered is near one of the two alternative sites that have been proposed for construction of a new airport. One of the proposed airport sites is in the west side of the county, and the other site is in the east side. Paul estimates that the value of this land will appreciate 10 per cent in one year if the west site is accepted for the new airport, and will depreciate 1 per cent otherwise. Furthermore, he learns that the Westside Bank plans to pay an annual simple interest rate of $5\frac{1}{2}$ per cent on savings accounts if the west site is selected for the new airport, but only 5 per cent otherwise.

The decision problem faced by Smith involves a choice between two alternative courses of action:

a_1: Deposit $14,000 in Westside Bank

a_2: Purchase the land for $14,000

The monetary value resulting from either course of action depends on which of the following two possible states of nature happens to occur:

θ_1: West site is selected for airport

θ_2: West site is not selected for airport

Therefore, the actual monetary value of Paul's decision depends not only on which of the two alternative courses of action he chooses, but also on which of the two possible states of nature might occur. Thus, there are four possible monetary values, each associated with one of the four possible action-state pairs. These values are as follows:

$$v(a_1,\theta_1) = \$14{,}000 \times 5\tfrac{1}{2}\,\% = \$770.00$$
$$v(a_1,\theta_2) = \$14{,}000 \times 5\% = \$700$$
$$v(a_2,\theta_1) = \$14{,}000 \times 10\% = \$1{,}400$$
$$v(a_2,\theta_2) = \$14{,}000 \times (-1\%) = -\$140$$

Notice that $v(a_2,\theta_2)$ represents a negative payoff, since the land value will depreciate by 1 per cent if a_2 is chosen and θ_2 occurs. The four conditional monetary values listed above are displayed in the left part of Table 13.1.

As he ponders his decision, Paul Smith is in a state of uncertainty, since he does not know for sure where the airport will be located (whether the true state of nature will be θ_1 or θ_2). However, Paul is not making his decision in a state of total ignorance, since the location of the airport has been a prominently debated public issue that has received considerable study and publicity. On the basis of his personal analysis of the *available information*, Paul estimates the odds to be 3 to 1 that the airport will be located at the west site. In other words, in Paul's judgement, the probabilities of the two possible states of nature are:

$$P(\theta_1) = \tfrac{3}{4} = .75$$
$$P(\theta_2) = \tfrac{1}{4} = .25$$

These subjective probabilities appear in the center column of Table 13.1.

In analyzing his decision problem, Paul Smith has been able to specify the relevant conditional monetary values and probabilities. This enables him to compute the expected monetary value of each of his alternative courses of action. Applying Formula (12.4), the calculation of these EMV's is shown on the right side of Table 13.1. As can be seen from this table, the EMV's of the alternative courses of action are:

$$\text{EMV}(a_1) = \$752.50$$
$$\text{EMV}(a_2) = \$1{,}015.00$$

Table 13.1

Conditional and Expected Monetary Values for Paul Smith

State of Nature	*Conditional Monetary Values*		*Probability of State of Nature*	*Calculation of EMV's*	
	Deposit in Bank a_1	*Purchase Land* a_2		*Deposit in Bank* a_1	*Purchase Land* a_2
West Site Selected θ_1	\$770	\$1,400	.75	$770 \times .75 = 577.50$	$1400 \times .75 = 1{,}050.00$
West Site Not Selected θ_2	700	−140	.25	$700 \times .25 = 175.00$	$-140 \times .25 = -35.00$
				EMV of $a_1 = 752.50$	EMV of $a_2 = 1{,}015.00$

In other words, the act of depositing the funds in the bank has an EMV of \$752.50, whereas the act of purchasing the land has an EMV of \$1,015.00. Thus, using the EMV criterion, Paul should choose to purchase the land, which is the act having the *maximum* EMV. Since this decision is derived from the analysis based on all information currently available, this maximum EMV will be referred to as the *EMV of the decision with available information.*

13.2 EMV OF THE DECISION WITH PERFECT INFORMATION

As we have observed, the analysis of Smith's decision problem based on available information indicates that the optimal course of action is to purchase the land. Nevertheless, even though he knows the *expected* monetary value of this action, he is still uncertain about the *actual* monetary value since he does not know for sure which of the two possible states of nature will occur. However, during the course of contemplating this uncertainty, Smith has the good fortune to learn about a most amazing real estate expert who has the uncanny ability to predict the airport location with 100 per cent accuracy. In other words, if this expert were to predict that the west site will be selected, then Smith could be absolutely sure that the airport will be built on that site. Likewise, if the expert were to predict that the west site will not be selected, Smith could be absolutely sure that this site will not be used for the airport. Smith realizes that such a *perfect prediction,* if he could manage to obtain it from the expert, would *totally eliminate* his uncertainty. That is, since the expert can foretell exactly what the state of nature will be, the cause of Smith's uncertainty would be removed by the expert's prediction. However, Smith is not so naive as to think that the expert would be willing to release his information free of charge. Therefore, before approaching the expert to seek his services, Smith pauses to consider whether he would be better off to purchase the expert's information or to base his decision on the information that is already available.

Smith begins to reason as follows: "If I were to make an agreement with the expert to obtain his prediction, how would I proceed to make my investment decision? Naturally, I would not decide which one of the two alternative courses of action to select until I received his prediction. If the prediction indicates that the west site will be selected (the true state of nature is θ_1), then I will make a profit of either \$770 or \$1,400, depending on which course of action I take. In this case, I will purchase the land, since that is the action that will yield the maximum profit (\$1,400) if the west site is selected for the airport. However, if the prediction indicates that the west site will not be selected for the airport (the true state of nature is θ_2), then I will either make a profit of \$700 or lose \$140, depending on which course of action I take. In that case, I will deposit my money in the bank, since that is the action that will bring me the maximum profit (\$700) if the west site is not selected for the airport."

Up to this point in his reasoning, Smith has determined which course of action he will follow depending on whether the expert predicts θ_1 or θ_2. That is, he will choose a_2 (with a resulting monetary value of \$1,400) if θ_1 is predicted, and he will choose a_1 (with a resulting monetary value of \$700) if θ_2 is predicted. Now Smith continues his reasoning in the following manner: "Since the expert is a perfect predictor, he will predict θ_1 if and only if the true state of nature is in fact θ_1, and he will predict θ_2 if and only if the true state of nature is in fact θ_2. Thus, the probability that he will predict θ_1 is identical to the probability that the true state of nature actually will be θ_1, and the probability that he will predict θ_2 is identical to the probability that the true state of nature actually will be θ_2. Since my estimated probabilities of θ_1 and θ_2 are .75 and .25 respectively, it follows that there is a .75 probability that the expert will predict θ_1 and a .25 probability that he will predict θ_2. Furthermore, since my actual profit is certain to be \$1,400 if he predicts θ_1 and \$700 if he predicts θ_2, there is a .75 probability that I will make \$1,400, and a .25 probability that I will make \$700. Therefore, if I were to use the perfect predictor, the expected profit of my decision would be: \$1,400 (.75) + \$700 (.25) = \$1,225."

The principal steps in Paul Smith's reasoning are summarized in Table 13.2. Column (1) shows the two possible predictions (θ_1 and θ_2) that the expert might make. Column (2) displays the optimal action that would be taken, depending on which prediction is made. Column (3) indicates the monetary values of the corresponding optimal actions. Column (4) gives the probabilities for the possible predictions. Finally, the expected monetary value is calculated in Column (5). The resulting figure of \$1,225 represents the expected monetary value of Smith's decision if he resolves to act on the basis of the additional information that can be obtained from the perfect predictor. This value of \$1,225 may be regarded as the *EMV of the decision with perfect information.*

Table 13.2

Conditional and Expected Monetary Values with Perfect Information for Paul Smith

(1) *Possible Prediction*	(2) *Optimal Action Conditional on* θ_i	(3) *Monetary Value for Optimal Action*	(4) *Probability of Prediction*	(5) *Calculation of EMV*
θ_1	a_2	\$1,400	.75	1,400 × .75 = 1,050
θ_2	a_1	700	.25	700 × .25 = 175
Sum			1.00	\$1,225

The value of \$1,225 may be compared to the \$1,015 that was previously obtained for the EMV of the decision with available information before the possibility of a perfect predictor was considered. Thus, there appears to be some advantage in purchasing the services of the perfect predictor, if a reasonable fee can be negotiated.

More generally, the EMV of the decision with perfect information is computed by (1) finding the optimal action for each possible prediction under the temporary assumption that the decision maker already has obtained the information from the perfect predictor, (2) multiplying the monetary value of each of these optimal actions by the probability of the corresponding prediction, and (3) summing these products. This procedure reflects the real-world situation. If a perfect predictor is used, the decision maker will make his decisions after having obtained the information from the predictor. The decision maker, however, must first evaluate the EMV with perfect information for purposes of determining the value of the perfect predictor. In other words, this EMV is computed at a point in time when the decision maker actually is still *uncertain* about the true state of nature since he has not really received the perfect information.

13.3 EXPECTED VALUE OF PERFECT INFORMATION (EVPI)

For the decision problem faced by Paul Smith, we have pointed out that purchasing perfect information from the expert appears to have some advantage. At this point, Smith needs to know precisely how great this advantage is, so that he can determine the maximum amount that would be reasonable to pay the expert. That is, Smith needs to know the value of the information that can be provided by the perfect predictor.

13.3.1 EVPI in Terms of EMV

The use of the EMV criterion makes the evaluation of perfect information a simple matter. As we have seen, the EMV of Paul Smith's decision with *perfect* information is $1,225, whereas the EMV of the decision with *available* information is $1,015. The difference between these two figures is $210. This amount accounts for the contribution, in expected value, of the perfect information over the available information. Thus, it is the *maximum* amount Smith should pay the perfect predictor. If Smith can negotiate a fee less than this amount, it would be reasonable for him to purchase the perfect information from the expert.

The difference between (1) the EMV of the decision with *perfect* information and (2) the EMV of the decision with *available* information is called the *Expected Value of Perfect Information* (EVPI). The practical significance of EVPI is that it puts an *upper limit* on what a decision maker should pay for additional information, since *any* information, no matter how accurate, would be worth no more than the value of *perfect information.* In real life, of course, the availability of a perfect predictor is rare. However, regardless of whether or not a perfect predictor actually exists, EVPI still puts an upper limit on the worth of additional information. Thus, the usefulness of the concept of EVPI remains undiminished even if a perfect predictor does not actually exist.

13.3.2 EVPI in Terms of EOL

In the preceding section, we defined EVPI in terms of the difference between two EMV's. However, due to the relationship between monetary values and opportunity losses, it is possible to formulate an equivalent definition of EVPI in terms of EOL. For this purpose, let us continue our analysis of Paul Smith's decision problem. From the conditional monetary values shown in Table 13.1, we obtain the corresponding conditional opportunity losses as follows:

$$\ell(a_1,\theta_1) = 1{,}400 - 770 = \$630.$$
$$\ell(a_2,\theta_1) = 1{,}400 - 1{,}400 = \$0.$$
$$\ell(a_1,\theta_2) = 700 - 700 = \$0.$$
$$\ell(a_2,\theta_2) = 700 - (-140) = \$840.$$

These conditional opportunity losses are displayed on the left side of Table 13.3. Using these loss figures, the EOL of each alternative course of action is computed on the right side of the table. As this table indicates, the EOL's of the alternative acts are:

$$\text{EOL}(a_1) = \$472.50$$
$$\text{EOL}(a_2) = \$210.00$$

Hence, using the EOL criterion, Paul should choose to purchase the land, since this is the act that has the minimum EOL. Notice that this is the same choice dictated by the EMV criterion. This should not surprise us since, as we saw in Section 12.3.3, the act having the maximum EMV is always the act having the minimum EOL.

We may now make the important observation that the minimum EOL of \$210 is identical to the EVPI obtained earlier. This is not a mere coincidence. Rather, it is always true that the EVPI is equal to the minimum EOL. To understand this relationship, we may focus on the following figures, which are extracted from Tables 13.1 and 13.3:

	Alternative Acts	
	a_1	a_2
EMV (from Table 13.1)	\$ 752.50	\$1,015.00
EOL (from Table 13.3)	472.50	210.00
	\$1,225.00	\$1,225.00

Table 13.3

Conditional and Expected Opportunity Losses for Paul Smith

State of Nature	*Conditional Opportunity Losses*		*Probability of State of Nature*	*Calculation of EOL's*	
	Deposit in Bank a_1	*Purchase Land* a_2		*Deposit in Bank* a_1	*Purchase Land* a_2
West Site Selected θ_1	$630	$ 0	.75	$630 \times .75 = 472.50$	$0 \times .75 = 0.00$
West Site Not Selected θ_2	0	840	.25	$0 \times .25 = 0.00$	$840 \times .25 = 210.00$
				EOL of a_1 = 472.50	EOL of a_2 = 210.00

Here we see again, as we previously observed in Section 12.3.3, that the sum of the EMV and EOL is a constant value for all alternative acts. In the case of Paul Smith's decision problem, this sum is equal to $1,225. The reader may have observed that this amount is equal to the EMV of the decision with perfect information, which was computed in Table 13.2. This relationship may be readily grasped by intuitive reasoning, without need for formal mathematical proof. The EMV of an act represents the expected profit of that act based on available information. Similarly, the EOL of an act represents the expected opportunity loss associated with that act due to lack of perfect information. In other words, the EOL indicates the additional expected profit that the decision maker could realize if he had a perfect predictor. Specifically, if Smith were to choose a_1, his expected profit with available information is $752.50 and the additional expected profit that could be obtained from using the perfect predictor is $472.50. However, if Smith were to choose a_2, his expected profit with available information is $1,015 and the additional expected profit that could be obtained from using the perfect predictor is $210. Regardless of which act Smith chooses, the sum of the EMV and EOL is $1,225, which represents the total expected profit for Smith if he has a perfect predictor. In other words, the sum of the EMV and EOL of any *act* with *available* information is equal to the EMV of the *decision* with *perfect* information.

If Smith were to choose a_1, the perfect predictor would be worth $472.50 since, as we have seen, this amount represents the additional expected profit that could be obtained by using a perfect predictor. Similarly, if Smith were to choose a_2, the perfect predictor would be worth $210. In the event that Smith were forced to make a decision based on available information, he would choose a_2 rather than a_1, since a_2 is the optimal act under the EMV criterion. Hence, the additional expected profit that Smith could gain by using the perfect predictor is only $210 rather than $472.50. Thus, the perfect predictor is worth a maximum amount of $210 to Smith. In other words, the EVPI is $210. This amount is also the EOL of the optimal act. Generalizing this observation, we may define EVPI as the EOL of the optimal act based on available information. Since the EOL of the optimal act is a minimum, it should be clear that EVPI is equal to the minimum EOL.

To gain further understanding of the fact that the EVPI equals the minimum EOL, let us examine the conditional opportunity losses in Table 13.3. We have seen that, without perfect information, Smith's optimal choice is a_2. However, with a perfect predictor, Smith's optimal choice will depend on which state of nature is predicted. On the one hand, if θ_1 is predicted, then the optimal act is a_2. In this case, the optimal act with *perfect* information is the same as the optimal act with *available* information. Thus, Smith will gain nothing if θ_1 is predicted, and hence the value of this specific prediction from the perfect predictor is zero. On the other hand, if θ_2 is predicted, then the optimal act is a_1. In that case, the optimal act with *perfect* information differs from the optimal act with *available* information. Thus, Smith will switch from a_2 to a_1 if θ_2 is predicted. Since the profit for a_1 is $700 and the profit for a_2 is $-\$140$, Smith will gain an additional profit of $840 by switching from a_2 to a_1. Recalling that the probability is .75 that θ_1 will be predicted and .25 that θ_2 will be predicted, Smith's expected gain from using the perfect predictor may be computed as follows:

$$(0 \times .75) + (840 \times .25) = \$210.$$

This result agrees precisely with our earlier calculations of the EVPI.

13.4 EVALUATING IMPERFECT INFORMATION

In the preceding section we observed that the EVPI for Paul Smith's decision problem is $210, which represents the upper limit of the amount he should pay for any additional information. With high hopes, Smith telephones the real estate expert to negotiate a deal to purchase perfect information. To Smith's consternation, the expert adopts a supercilious air and steadfastly refuses to provide his services for less than $500. Since this fee far exceeds the EVPI, Smith gives up in disgust and tells the expert to go jump in the lake.

While calming down from this frustrating experience, Smith happens to recall a speech given at a recent civic meeting by John Davis, a successful real estate consultant. Smith wonders if Davis might be willing to conduct a study and prepare a report predicting which airport site will be selected. Consequently, Smith telephones Davis and explains his problem. Davis indicates that, for a fee of $100, he is willing to prepare the report that Smith desires. Davis states that his prediction will be quite reliable, but he admits that it will not be a perfect prediction by any means. He estimates that his prediction will be 80 per cent accurate. That is, if the west site actually is to be selected for airport construction, there is an 80 per cent chance that he will be able to predict this fact. If, however, the west site is not to be selected, the chances that he will be able to predict this fact are also 80 per cent.

Since Smith can obtain the prediction from Davis for an amount less than the EVPI, he realizes that it might be worthwhile to purchase the prediction. However, the decision whether or not to purchase Davis' report cannot be answered by simply comparing the cost of the report with the EVPI, because the report would not provide a perfect prediction. Rather, since Davis can provide only *imperfect information*, Smith must compare the cost of the report with the expected value of the imperfect information provided by the report in order to determine whether he should purchase the report.

Intuitively, the reader might rush to the conclusion that the expected value of imperfect information is equal to the EVPI multiplied by the degree of accuracy, and hence assume that the expected value of the above imperfect information would be equal to ($\$210 \times 80\%$) = \$168. Unfortunately, this intuitive conclusion is definitely *wrong*. The correct approach to evaluating imperfect information is a complex process, which we will now examine.

13.4.1 Revision of Probabilities Due To Additional Information

Although the procedure for evaluating imperfect information is more complex than the procedure for evaluating perfect information, the underlying principle is the same for both procedures. This basic principle is that *the introduction of additional information during the process of making a decision generally results in some change in the probabilities of the possible states of nature.*

To illustrate this principle, let us begin by recalling that Paul Smith has a certain amount of information available to him. On the basis of his analysis of that information, Smith feels it is more likely than not that the west site will be selected for the airport. Consequently, he has assigned the following probabilities to the possible states of nature:

$$P(\theta_1) = .75$$
$$P(\theta_2) = .25$$

It is important to note that these probabilities reflect the amount of information that is available to Smith *prior* to obtaining any additional information. As we observed in our discussion of Bayes' Theorem in Chapter Five, such probabilities are called *prior probabilities.* It is also important that this pair of probabilities constitutes a *probability distribution,* since the two possible states of nature, θ_1 and θ_2, are mutually exclusive and collectively exhaustive. Since this probability distribution consists of prior probabilities, it is called a *prior distribution.*

Let us now consider what the effect would be on Smith's prior distribution if he had been able to obtain perfect information from the expert. Since the effect of the expert's information depends on whether he predicts θ_1 or θ_2, we will examine the effects of these two possible predictions separately. *If the expert were to predict* θ_1, then Smith would be certain that θ_1 will prevail, and also certain that θ_2 will not prevail. This may be expressed in probability notation as follows:

$$P(\theta_1 | \text{expert predicts } \theta_1) = 1$$
$$P(\theta_2 | \text{expert predicts } \theta_1) = 0$$

These two probabilities constitute a *posterior distribution* since this distribution consists of probabilities obtained *posterior to* receiving the perfect information from the expert. Thus, if the expert were to predict θ_1, then his prediction would change the probability distribution of the state of nature

$$\begin{array}{ccc} & \underline{\textit{Prior}} & & \underline{\textit{Posterior}} \\ \text{from} & \begin{cases} P(\theta_1) = .75 \\ P(\theta_2) = .25 \end{cases} & \text{to} & \begin{cases} P(\theta_1|\text{expert predicts } \theta_1) = 1 \\ P(\theta_2|\text{expert predicts } \theta_1) = 0 \end{cases} \end{array}$$

However, if the expert were to predict θ_2, then Smith would be certain that θ_1 will not prevail and also certain that θ_2 will prevail. Using probability notation, we have:

$$P(\theta_1|\text{expert predicts } \theta_2) = 0$$

$$P(\theta_2|\text{expert predicts } \theta_2) = 1$$

These two probabilities constitute the posterior distribution that would result from a prediction of θ_2. That is, if the expert were to predict θ_2, then his prediction would change the probability distribution of the state of nature

$$\begin{array}{ccc} & \underline{\textit{Prior}} & & \underline{\textit{Posterior}} \\ \text{from} & \begin{cases} P(\theta_1) = .75 \\ P(\theta_2) = .25 \end{cases} & \text{to} & \begin{cases} P(\theta_1|\text{expert predicts } \theta_2) = 0 \\ P(\theta_2|\text{expert predicts } \theta_2) = 1 \end{cases} \end{array}$$

Thus there are two possible posterior distributions, each associated with one of the two possible predictions. Furthermore, we see that all the posterior probabilities are either 1 or 0. This reflects the fact that uncertainty would be completely eliminated by the perfect information that the expert can provide.

As we can see from the above discussion, posterior distributions of the possible states of nature resulting from additional information are obtained quite easily when the information is *perfect*. If the information is *imperfect*, however, the derivation of posterior distributions is more laborious, but may be accomplished readily through the use of Bayes' Theorem. To illustrate this process, let us proceed with Smith's decision problem.

Recall that Smith's prior probabilities of the possible states of nature are .75 and .25 for θ_1 and θ_2, respectively. Furthermore, if Smith desires, John Davis can provide him with additional information that is 80 per cent accurate. In other words, if the west site is to be selected for airport construction, the probability that Davis will be able to predict this fact has been estimated as .80. If, however, the west site is not to be selected, the probability that he will be able to predict this fact is also .80. To express these probabilities in formal notation, we will designate the two possible outcomes of the predictor's analysis by:

o_1: The predictor indicates that the west site will be selected.

o_2: The predictor indicates that the west site will *not* be selected.

Then we may express the various probabilities as follows:

(i) the prior probability of θ_1:

$$P(\theta_1) = .75$$

(ii) the prior probability of θ_2:

$$P(\theta_2) = .25$$

(iii) the conditional probability that the prediction will be o_1 if θ_1 is true:

$$P(o_1|\theta_1) = .80$$

(iv) the conditional probability that the prediction will be o_2 if θ_1 is true:

$$P(o_2|\theta_1) = 1 - P(o_1|\theta_1) = 1 - .80 = .20$$

(v) the conditional probability that the prediction will be o_2 if θ_2 is true:

$$P(o_2|\theta_2) = .80$$

(vi) the conditional probability that the prediction will be o_1 if θ_2 is true:

$$P(o_1|\theta_2) = 1 - P(o_2|\theta_2) = 1 - .80 = .20$$

Since there are two possible outcomes of the predictor's analysis (o_1 and o_2), the effect of the prediction on the probabilities of θ_1 and θ_2 depends on which outcome actually occurs. That is, in evaluating the potential benefit of purchasing Davis' information, Smith must consider two possible posterior distributions—the distribution that would result from o_1 and the distribution that would result from o_2. These posterior distributions may be obtained by using Bayes' Theorem, which was discussed in Section 5.5. For this purpose, we will follow the computational procedure illustrated in Table 5.9.

If Davis' prediction were o_1, this information would change the prior probabilities $P(\theta_i)$ to the posterior probabilities $P(\theta_i|o_1)$. The computation of these posterior probabilities is shown in Table 13.4. Column (1) contains the possible states of nature, θ_1 and θ_2. Column (2) shows the prior probabilities for θ_1 and θ_2, respectively. Column (3) shows the conditional probability of o_1 given each of the possible states of nature. Column (4) is obtained by multiplying the probabilities in Column (2) by those in Column (3). Each of the two figures thus obtained in Column (4) is the joint probability of the occurrence of both the particular state of nature and the specific outcome. The sum of the two probabilities in this column is the marginal probability of o_1. This probability, $P(o_1)$, is the probability that the predictor will indicate that the west site will be selected for airport construction, regardless of whether the true state of nature should be θ_1 or θ_2. If, however, the outcome o_1 actually has been reported by the predictor, then the probability of o_1—the Column (4) total—must become 1.00. Consequently, the other figures in Column (4) must be adjusted accordingly. This is done by dividing the individual figures in Column (4) by the column total; the results are shown in Column (5). The probabilities in this column constitute the posterior distribution that would result if the outcome of Davis' analysis should be o_1.

Table 13.4

Calculation of Posterior Probabilities for Paul Smith (Given the Outcome of the Predictor's Analysis is o_1)

(1)	*(2)*	*(3)*	*(4)*	*(5)*
State of Nature θ_i	*Prior Probability* $P(\theta_i)$	*Conditional Probability* $P(o_1 \mid \theta_i)$	*Joint Probability* $P(\theta_i \cap o_1)$	*Posterior Probability* $P(\theta_i \mid o_1)$
θ_1	.75	.80	.60	.9231
θ_2	.25	.20	.05	.0769
Sum	1.00		.65	1.0000
			$P(o_1)$	

As we can see from Column (3) of Table 13.4, $P(o_1 \mid \theta_1) = .80$ and $P(o_1 \mid \theta_2) = .20$. These conditional probabilities indicate that there is a .80 chance that the outcome o_1 would occur if in fact θ_1 were the true state of nature, and a .20 chance that o_1 would occur if in fact the true state of nature were θ_2. Since these probabilities indicate how "likely" it would be to obtain o_1 depending on which state of nature actually is true, they are commonly called *likelihoods*. Notice that the likelihoods in Column (3) represent the probabilities of the *same* outcome (o_1) conditional on the different possible states of nature (θ_1 or θ_2). Since the sum of the probabilities of the same event given different conditions is meaningless, the total of this column is not calculated. The fact that the sum of this column happens to be equal to 1 is merely a coincidence due to the particular figures involved in our example.

If the outcome of Davis' analysis should be o_2, rather than o_1, the posterior distribution would consist of probabilities in the form $P(\theta_i \mid o_2)$ rather than $P(\theta_i \mid o_1)$. The computation of these posterior probabilities is presented in Table 13.5. This table differs from Table 13.4 in that Column (3) shows the conditional probability of o_2, rather than o_1, given each of the possible states of nature. As a result, the figures in Columns (4) and (5) of Table 13.5

also are different from those in the corresponding columns of Table 13.4. However, in both tables these two columns are obtained in exactly the same manner as those in Table 13.4.

Table 13.5

Calculation of Posterior Probabilities for Paul Smith (Given the Outcome of the Predictor's Analysis is o_2)

(1) State of Nature θ_i	(2) Prior Probability $P(\theta_i)$	(3) Conditional Probability $P(o_2\|\theta_i)$	(4) Joint Probability $P(\theta_i \cap o_2)$	(5) Posterior Probability $P(\theta_i\|o_2)$
θ_1	.75	.20	.15	.4286
θ_2	.25	.80	.20	.5714
Sum	1.00		.35 ↑ $P(o_2)$	1.0000

By comparing Column (5) in Table 13.4 with the same column in Table 13.5, we see that the two posterior distributions are markedly different, because the effect of Davis' information on the prior distribution differs according to the specific prediction that is made. That is, if the outcome of Davis' analysis is o_1, then the probability distribution of the state of nature is changed

$$\begin{array}{cc} \textit{Prior} & \textit{Posterior} \\ \text{from} \begin{cases} P(\theta_1) = .75 \\ P(\theta_2) = .25 \end{cases} & \text{to} \begin{cases} P(\theta_1|o_1) = .9231 \\ P(\theta_2|o_1) = .0769 \end{cases} \end{array}$$

However, if the outcome of Davis' analysis is o_2, then the probability distribution of the state of nature is changed

$$\begin{array}{cc} \textit{Prior} & \textit{Posterior} \\ \text{from} \begin{cases} P(\theta_1) = .75 \\ P(\theta_2) = .25 \end{cases} & \text{to} \begin{cases} P(\theta_1|o_2) = .4286 \\ P(\theta_2|o_2) = .5714 \end{cases} \end{array}$$

The reader may recall that, in the case of a perfect predictor, all the posterior probabilities are either 0 or 1. In the case of an imperfect predictor, however, the posterior probabilities are not limited to 0 and 1. Rather, the posterior probabilities resulting from imperfect information can take on any values between 0 and 1. Posterior probabilities other than 0 and 1, such as those obtained above, reflect the fact that uncertainty still exists after imperfect information has been obtained.

13.4.2 EMV of the Decision with Imperfect Information

We have just seen the potential effect of the additional information that Davis can provide on Smith's prior probabilities. We will now examine the implications of this effect to Smith in determining if he should purchase Davis' information. Since the effect of Davis' information on the prior distribution depends on whether his prediction is o_1 or o_2, we will examine each of these two cases separately.

As derived earlier, if the outcome of Davis' analysis is o_1, the posterior probabilities are: $P(\theta_1|o_1) = .9231$ and $P(\theta_2|o_1) = .0769$. Thus, if Davis were to predict that the west site would be selected for airport construction (o_1), it would be highly likely, but not certain, that the west site actually would be used. Since Smith cannot be entirely sure that θ_1 will occur, he must consider the monetary value associated with θ_2 as well as θ_1 in analyzing his investment decision. Based on the EMV criterion, this analysis can be performed by using the posterior probabilities to compute the EMV of each of the two alternative courses of action. The calculation of each of these two EMV's is shown below:

$$\begin{aligned}
\text{EMV}(a_1|o_1) &= v(a_1,\theta_1)\,P(\theta_1|o_1) + v(a_1,\theta_2)\,P(\theta_2|o_1) \\
&= (770)(.9231) + (700)(.0769) \\
&= \$764.62 \\
\text{EMV}(a_2|o_1) &= v(a_2,\theta_1)\,P(\theta_1|o_1) + v(a_2,\theta_2)\,P(\theta_2|o_1) \\
&= (1{,}400)(.9231) + (-140)(.0769) \\
&= \$1{,}281.57
\end{aligned}$$

Since these EMV's are calculated using the posterior probabilities, they are called *posterior* EMV's. Thus, if Davis' prediction were o_1, Smith should choose to purchase the land (a_2) since this action has a higher posterior EMV.

If, however, the outcome of Davis' analysis is o_2 rather than o_1, the posterior probabilities are: $P(\theta_1|o_2) = .4286$ and $P(\theta_2|o_2) = .5714$. Hence, if Davis were to predict that the west site would not be selected for airport construction (o_2), it would be only likely, but not certain, that the west site actually would not be used. Since Smith cannot be completely sure that the true state of nature is θ_2, he must consider the monetary values associated with θ_1 as well as θ_2 in evaluating his investment alternatives. This evaluation can be accomplished by comparing the following posterior EMV's:

$$\begin{aligned}
\text{EMV}(a_1|o_2) &= v(a_1,\theta_1)\ P(\theta_1|o_2) + v(a_1,\theta_2)\ P(\theta_2|o_2) \\
&= (770)(.4286) + (700)(.5714) \\
&= \$730.00 \\
\text{EMV}(a_2|o_2) &= v(a_2,\theta_1)\ P(\theta_1|o_2) + v(a_2,\theta_2)\ P(\theta_2|o_2) \\
&= (1{,}400)(.4286) + (-140)(.5714) \\
&= \$520.04
\end{aligned}$$

Thus, if Davis' prediction were o_2, Smith should choose to deposit his money in the bank (a_1) since this action has a higher posterior EMV.

So far we have been able to determine the optimal action for Smith if a specific prediction were made by Davis. That is, if Davis were to predict o_1, Smith's optimal action would be a_2, which has a posterior EMV of \$1,281.57. If, however, Davis were to predict o_2, Smith's optimal action would be a_1, which has a posterior EMV of \$730.00. Naturally, if Smith were to make an agreement to purchase Davis' prediction, he would not decide which one of the two alternative courses of action to select until he had actually received the prediction. Under such conditions, the EMV of Smith's decision would be either \$1,281.57 or \$730.00, depending on whether o_1 or o_2 were predicted. From Tables 13.4 and 13.5, we see that $P(o_1) = .65$ and $P(o_2) = .35$. That is, there is a .65 probability that Davis' prediction will be o_1 and a .35 probability that his prediction will be o_2. It then follows that there is a .65 probability that Smith's EMV would be \$1,281.57 and a .35 probability that his EMV would be \$730.00. Consequently, Smith's expected profit can be calculated as follows:

$$(\$1{,}281.57)(.65) + (\$730.00)(.35) = \$1{,}088.52$$

The resulting figure of \$1,088.52 represents the expected monetary value of Smith's decision if he resolves to act on the basis of the additional information that can be obtained from the imperfect predictor. Hence, this amount will be referred to as the *EMV of the decision with imperfect information.*

Table 13.6 summarizes the major steps used in deriving the EMV of Smith's decision with imperfect information. Column (1) displays the two possible outcomes of the predictor's analysis; i.e., Davis' two possible predictions (o_1 and o_2). Column (2) gives Smith's optimal action for each possible prediction under the *temporary assumption* that he has already obtained the prediction. Column (3) presents the posterior EMV of each of the optimal actions. Column (4) indicates the probability of the occurrence of each of the possible predictions. Column (5) is obtained by multiplying the posterior

EMV of each of the optimal actions by the probability of the corresponding prediction. Summing these products, we obtain \$1,088.52, which is the EMV of Smith's decision with imperfect information.

Table 13.6

EMV of Smith's Decision with Imperfect Information

(1) Outcome of Predictor's Analysis o_i	(2) Optimal Action Conditional on o_i	(3) Posterior EMV	(4) Probability of Outcome $P(o_i)$	(5) Calculation of Prior Expectation of Posterior EMV's (3) × (4)
o_1	a_2	\$1,281.57	.65	\$833.02
o_2	a_1	730.00	.35	255.50
Sum				\$1,088.52

13.4.3 Expected Value of Imperfect Information (EVII)

In Section 13.1 we considered how Paul Smith might make his investment decision with available information. We observed that, without any additional information that might be provided by a predictor, his decision should be to select a_2 (purchase the land), which has an EMV of \$1,015.00. In Section 13.4.2, we examined how Smith might make his investment decision with additional information that could be provided by the imperfect predictor. We saw that, if Smith were to act on the basis of Davis' additional information, his optimal action would depend on the specific prediction made by Davis. Prior to actually receiving the prediction, the EMV of Smith's decision with imperfect information is \$1,088.52. Thus, the excess of the EMV with the predictor (\$1,088.52) over the EMV without any predictor (\$1,015.00) is equal to \$73.52. This amount is the contribution, in expected value, of the imperfect predictor. In other words, this is the *Expected Value of Imperfect Information* (EVII). Since the cost of obtaining the additional information from Davis (\$100) exceeds its value (\$73.52), Smith should not purchase Davis' information.

More generally, if a decision maker is to purchase imperfect information from a predictor, he will make his decision concerning the choice of alternative courses of action after he has actually obtained such information. The real-world situation, however, is that the additional information will not be made available to the decision maker until he has agreed to pay a price for such information. Consequently, if the decision maker wishes to determine if the additional information can sufficiently increase the EMV of his decision to justify paying that price, he must do so prior to the receipt of this information. Therefore, we may summarize the procedure for evaluating the imperfect information as follows:

1. Examine the effect of the potential information on the prior distribution of the state of nature; i.e., derive the posterior distribution that would result from each of the possible predictions.
2. Compute the posterior EMV of each alternative course of action and identify the optimal action given each possible prediction. This is done under the *temporary assumption* that the decision maker has already obtained the information.
3. Determine the EMV of the decision with imperfect information. This is accomplished by (a) multiplying the posterior EMV of each of the optimal actions by the probability of the corresponding prediction, and (b) summing these products. This computation is necessary because at this point in time the decision maker has not really received the actual prediction.
4. Calculate the Expected Value of Imperfect Information by finding the difference between (a) the EMV of the decision with *imperfect* information, and (b) the EMV of the decision with *available* information.

PROBLEMS

13.1 For Problem 12.4 in Chapter Twelve, do the following:
   (a) Determine the EMV of the grocer's decision with available information.
   (b) Determine the EMV of the grocer's decision with perfect information.
   (c) Find the EVPI by using the results in (a) and (b) above.
   (d) Construct the opportunity loss table for the grocer. According to the EOL criterion, how many melons should he stock?
   (e) Is the EOL associated with the optimal stock in (d) equal to the EVPI obtained in (c)? Is this a general phenomenon or a mere coincidence? Why?
   (f) Find the sum of the EMV and the EOL for each alternative act. Is this sum constant for all alternative acts?
   (g) How is the sum in (f) related to the EMV of the grocer's decision with perfect information obtained in (b) above?

13.2 For Problem 12.5 in Chapter Twelve, do the following:
   (a) Determine the EMV of the retailer's decision with perfect information.
   (b) Find the EVPI by comparing (i) the EMV of the retailer's decision with perfect information and (ii) the EMV of his decision with available information.
   (c) Construct the conditional opportunity loss table for the retailer. Compute the EOL of each of the alternative acts. Using the EOL criterion, how many units should he stock?
   (d) Is the EOL associated with the optimal act in (c) equal to the EVPI obtained in (*b*)? Is this a mere coincidence? Explain.

13.3 Suppose that the trading post operator in Problem 12.6 has found an extremely capable weather forecaster. This particular forecaster can foretell *for sure* whether or not it is going to snow. What is the EMV for the operator if he decides to purchase the forecaster's perfect prediction? What is the maximum amount that the operator should pay for the perfect prediction?

13.4 Suppose that the AXY newsstand in Problem 12.7 has received an offer to consider an option of consignment by the magazine publisher. The option allows the dealer to return all the unsold copies of *Nouvelles de Provence* to the publisher, and pay only for the copies sold.
(a) Under the consignment option, what is the dealer's EMV?
(b) What is the maximum amount that the dealer should pay the publisher for the right to use the consignment option?

13.5 Suppose that the department store in Problem 12.8 found that the Christmas card printer does sell on consignment, with prices of the consignment option subject to negotiation.
(a) If the store should decide to use the consignment option, what would be its EMV from selling Christmas cards?
(b) What is the maximum amount that the store should pay for the option?

13.6 Jack Willis, marketing manager of Stratford Cosmetics Company, is considering whether or not to introduce the firm's newly developed Cheshire Kitten vanishing cream into a particular marketing area. On the basis of the sales of other of the firm's products in this market, Jack estimates the following probabilities of demand during the first six months:

Demand Level	*Probability*
θ_1: High	.30
θ_2: Moderate	.50
θ_3: Low	.20

Jack estimates that sales of this product will result in a \$60,000 profit if sales are high, and a \$20,000 profit if sales are moderate. However, if sales are low, he estimates a \$70,000 loss.

At this point, Jack is faced with the decision of choosing one of the two alternative acts:

a_1: Introduce Cheshire Kitten into the market
a_2: Do not introduce Cheshire Kitten into the market

(a) Prepare a payoff table of conditional monetary values for Jack's decision problem.
(b) Determine the EMV of each of Jack's alternatives. On the basis of the EMV criterion, which act should Jack choose?
(c) At this point, what is the amount of Jack's EVPI? How do you interpret this figure?

Reluctant to make his decision hastily, Jack considers the possibility of requesting his marketing research group to conduct a survey in the proposed target area. The survey will cost \$1,500. The group's report would consist essentially of one of the following predictions of the demand for Cheshire Kitten:

o_1: Demand will be high
o_2: Demand will be moderate
o_b: Demand will be low

From past performance of the research group, Jack prepares the following table of conditional probabilities which reflect the reliability of the survey procedure:

If Actual Demand is	Probability that Survey Prediction Will Be		
	High	Moderate	Low
High	.60	.20	.20
Moderate	.10	.80	.10
Low	.10	.20	.70

(d) Calculate the posterior probability of each demand level if the research group should predict high demand.
(e) Calculate the posterior probability of each demand level if the research group should predict moderate demand.
(f) Calculate the posterior probability of each demand level if the research group should predict low demand.
(g) Calculate the posterior EMV of each act if the research group should predict high demand.
(h) Calculate the posterior EMV of each act if the research group should predict moderate demand.
(i) Calculate the posterior EMV of each act if the research group should predict low demand.
(j) If Jack should follow through on his plan to have the survey conducted, what would be the EMV of his decision with the imperfect information provided by the survey prediction?
(k) Should Jack request that the survey be conducted? Explain.

13.7 The Ample Money Credit Card Company (AMCCC) issues general purpose credit cards. Mr. Eagle, credit manager for AMCCC, must accept or reject each credit-card application. He may make his decision either (i) on the basis of his subjective evaluation of the information contained in the application, or (ii) on the basis of his subjective evaluation plus a credit rating that he can purchase from the Better Credit Rating Center (BCRC). Tammy White has just applied for a credit card. After his usual evaluation procedure, Mr. Eagle assigns a prior probability of 0.6 that Tammy is a good risk. From past experience, the credit manager estimates that a good credit risk will produce a profit of $500 whereas a bad credit risk will result in a loss of $190.

(a) Let a_1 and a_2 represent the two alternative courses of action that are available to Mr. Eagle—issue or not issue a credit card to Tammy. Let θ_1 and θ_2 denote the two possible states of nature—Tammy is a good or bad risk. Prepare the conditional payoff table for Mr. Eagle. According to the EMV criterion, should Mr. Eagle issue a credit card to Tammy?
(b) Construct the opportunity loss table corresponding to the payoff table in (a) above. According to the EOL criterion, should Mr. Eagle issue the credit card?
(c) Is the decision found in (b) the same as that found in (a)? Do both the EMV and the EOL criteria always lead to the same optimal decision? Why?

Mr. Eagle wishes to consider using BCRC's service. By examining his applicants' file for the past year, he has found that among those applicants investigated by BCRC, in 70 per cent of those cases in which the applicant actually was a good risk, BCRC had given a "favorable" report. In 90 per cent of those cases in which an applicant was in fact a bad risk, BCRC had given an "unfavorable" report.

(d) What would be the posterior probabilities of the possible states of nature if BCRC's report concerning Tammy should be:
  (i) favorable
  (ii) unfavorable

(e) What would be the posterior EMV of each of Mr. Eagle's alternative acts if BCRC's report should be:
  (i) favorable
  (ii) unfavorable

(f) Determine the EMV of Mr. Eagle's decision if he should purchase BCRC's service.

(g) Would it be reasonable for Mr. Eagle to pay $40 for BCRC's service? Why?

(h) Suppose that another credit-rating agency, called Sigma, can provide perfect prediction concerning Tammy. That is, Sigma can foretell for sure whether or not Tammy is a good risk. What is the maximum amount that Mr. Eagle should be willing to pay Sigma for its service?

13.8 Assume that you are the president of a large engineering research and development firm that is nearing completion of a major project and you must decide what the company's next project will be. You have been considering two speculative ventures:

a_3: design a revolutionary atomic power-generating system

a_2: design a rotary engine for buses and trucks.

In addition to these speculative projects, you have a third alternative:

a_3: accept an offer to design a tunnel under the Mississippi River.

In considering your alternatives, you have been particularly concerned with the eventual fate of two environmental control bills—the Mossback Bill and the Goodheart Bill. The provisions of these two bills are such that passage of either would preclude passage of the other. Thus you are concerned with three possibilities:

θ_1: The Mossback Bill will be passed

θ_2: The Goodheart Bill will be passed

θ_3: Neither bill will be passed

On the basis of the information that is currently available, you estimate the probabilities of these possible states of nature as follows:

$$P(\theta_1) = .30 \qquad P(\theta_2) = .50 \qquad P(\theta_3) = .20$$

By proceeding with the atomic power project, you estimate that you would lose $7 million if the Mossback Bill passes, make a profit of $4 million if the Goodheart Bill passes, and make a profit of $8 million if neither bill passes. By undertaking the rotary engine project, you figure that you would make a profit of $6 million if the Mossback Bill passes, suffer a loss of $3 million if the Goodheart Bill passes, and make a profit of $5 million if neither bill passes. Regardless of the fate of either bill, undertaking the tunnel project would yield a sure profit of $2 million.

(a) Using the EMV criterion, if you make your decision on the basis of the information currently available to you, which project would you choose?

(b) What is the EMV of the decision with available information?

(c) What is the EVPI? How do you interpret this figure?

Due to the impact of your decision, you feel that you would like to obtain additional information regarding the possible states of nature before you commit yourself. You therefore consider the possibility of seeking the consulting services of Dr. Stanley Livingstone, an eminent political scientist and ecologist. If you should enlist Dr. Livingstone's services, you would ask him to fly to Washington, interview key congressmen and administration officials, and report his conclusions to you. His report would be one of the following:

o_1: Passage of the Mossback Bill appears likely
o_2: Passage of the Goodheart Bill appears likely
o_3: Most likely neither bill will pass

Since Dr. Livingstone has provided his services to your company many times during the past 20 years, you have a reasonable background of experience on which to evaluate his predictions. You therefore estimate the following conditional probabilities of receiving each possible report given each possible state of nature:

	State of Nature		
Report	θ_1	θ_2	θ_3
o_1	.7	.1	.2
o_2	.1	.7	.2
o_3	.2	.2	.6

(d) What would be the posterior probabilities of the possible states of nature if Dr. Livingstone's report should be:
  (i) o_1?
  (ii) o_2?
  (iii) o_3?
(e) What would be the EMV of your decision if you proceed to retain Dr. Livingstone and base your decision on the imperfect information contained in his report?
(f) What is the maximum amount you should be willing to pay Dr. Livingstone for his services?

Chapter 14

Utility In Decision Making

In Chapter Twelve, the EMV criterion was applied to the news dealer's problem of deciding how many copies of the LAX Sunday paper he should stock each week. For such an application, in which the same decision can be repeated time after time, the act that maximizes the EMV will yield the maximum average profit in the long run. Generally speaking, as the number of repetitions becomes large, the *actual* average profit will approach the theoretical *expected* profit. Consequently, the EMV criterion is particularly useful for such decison makers as insurance underwriters, loan officers, and purveyors of perishable goods, since they characteristically engage in making a large volume of decisions that are very similar in nature from one occasion to the next.

In Chapter Thirteen, the EMV criterion was used to analyze Paul Smith's problem of deciding whether to deposit his money in the bank or to purchase the land. In contrast to the news dealer's decision problem, which is repetitive in nature, Smith's problem is a one-time decision. In fact, many of the most important personal and business decisions are made in unique situations. For such a one-time decision, there can be no long-run average profit because a long-run series of repeated decisions does not occur. Since a long-run average does not exist, the EMV becomes a theoretical concept that is difficult to interpret. In such a situation, the decision maker may not wish to use the EMV criterion for decision making. This chapter will present an alternative approach.

14.1 THE DECISION MAKER'S ATTITUDE TOWARD RISK

As we have just observed, the EMV criterion may be undesirable for one-time decisions in which the concept of a long-run average is invalid. However, this is not the only reason why a decision maker may shy away from the EMV criterion. Even for situations in which the idea of long-run average profit is valid, the decision maker's personal attitude toward risk may lead him to adopt some approach that will allow him to exercise a greater amount of judgment than the EMV criterion permits. At this point, let us consider several examples that demonstrate why the EMV criterion may not be a valid guide for some decision making, and how a decision maker's attitude may influence his decision making.

As a first example, consider an individual who has to make a choice between the two alternatives:

1. receive a \$2 million gift for certain, and
2. a 50-50 chance of receiving either \$6 million or nothing.

The EMV of alternative (2) is:

$$(\$6{,}000{,}000)\left(\frac{1}{2}\right) + (\$0)\left(\frac{1}{2}\right) = \$3{,}000{,}000$$

Clearly, the EMV of act (2) is greater than the payoff of act (1). In spite of this fact, most people would choose act (1). One might argue that he would like to have the \$2 million for certain to enjoy a quite comfortable life rather than to play a game in which, on the flip of a fair coin, he might receive nothing at all. It would be even nicer, of course, to have \$6 million, but one might feel that \$6 million is not so much more preferable to \$2 million that it is worth a risk of winding up with nothing. In other words, he might feel that \$6 million is not worth three times as much as \$2 million.

As a second example, consider the case of two management consultants, John Jones and Kent Kemp, who have agreed to prepare a prospectus for the organizers of a new company. When the prospectus is completed, each of the two consultants will receive an amount of \$9,000 in the form of either cash or stock at his own choice. This choice, however, has to be made before they start to work. In view of the potential products and the key personnel of the company, John and Kent estimate that the odds are 2 to 1 that the company will operate successfully. The price of the stock will be doubled if the company succeeds, but the price will drop to 30 per cent of its initial value otherwise. Thus, the \$9,000 in stock will be worth either $(\$9{,}000 \times 200\%) = \$18{,}000$ or $(\$9{,}000 \times 30\%) = \$2{,}700$, depending on whether or not the company succeeds. Consequently, the EMV of the stock is

$$(\$18{,}000)\left(\frac{2}{3}\right) + (\$2{,}700)\left(\frac{1}{3}\right) = \$12{,}900$$

Now, John and Kent start arguing with each other about the choice between the cash and the stock.

John: *Look! The EMV of the stock is substantially higher than the cash amount. Of course, I prefer the stock.*

Kent: *Be careful! This is a one-time decision. If you choose the stock, your consulting fee will be either \$18,000 or \$2,700, but* not *the EMV of \$12,900. Obviously, the sure amount of \$9,000 is preferable to a risky stock.*

John: *I know the EMV is not the actual amount. However, think about the great chance of receiving that large amount of $18,000. I personally believe that you are overly cautious. Have you thought hard enough about what you could do with that amount?*

Kent: *Certainly, I have. The trouble is that you have not taken the whole thing seriously enough. How can you even consider giving up a sure amount of $9,000 for such a wild gamble?*

John: *The worst thing that could happen is to receive $2,700. After all, it isn't too bad.*

Kent: *What do you mean by that? It drops from $9,000 to $2,700. That is bad enough for me. I have just decided to take the cash. I hope you will too.*

John: *In spite of your decision, I am going to take the stock. I hope you will change your mind.*

Kent: *I will never change my mind. But I wish you good luck.*

Clearly, John and Kent have come to opposite decisions even though they both had the same alternatives and the same information available to them. However, these two decision makers had different attitudes toward the same risk, and each of them based his decision on his own attitude toward risk.

A decision maker's attitude toward risk may depend on a combination of factors such as his financial condition, his propensity for gambling, and his temperamental predisposition. Consequently, not only may two decision makers have different attitudes toward risk, but each of them also may have different attitudes toward risk at different points in time.

To illustrate the above point, let us consider a third example. An architectural and engineering (A & E) firm has been asked by a client to conduct a feasibility study for a construction project. The study is to determine whether or not it is economically feasible to build a new manufacturing plant in a specific industrial area. The client will pay a nominal fee of $1,000, although it will cost the A & E firm $11,000 to conduct the study. However, should the study convince the client that the plant construction is feasible, the client will award the firm a design contract for preparing all the construction documents. The A & E firm estimates that this design contract, which is on a "cost plus" basis, will yield a profit of $30,000. Consequently, the firm would make a net profit of ($1,000 − $11,000) + ($30,000) = $20,000 if the plant construction is feasible, and would suffer a loss of ($11,000 − $1,000) = $10,000 otherwise. The firm further estimates that the odds are 3 to 2 that the plant construction will be feasible. Therefore, the EMV of agreeing to undertake the feasibility study is: ($20,000)$(\frac{3}{5})$ + (−$10,000)$(\frac{2}{5})$ = $8,000, and the corresponding figure of declining the offer obviously is zero. According to the EMV criterion, the A & E firm clearly should accept the client's offer. Since this firm engages in many feasibility studies and building designs for clients, the use of the EMV criterion for selecting various offers would lead the firm to the achievement of maximum average profit in the long run. Thus, the EMV criterion is used by this firm under normal circumstances. However, the A & E firm now has inadequate working capital and is extremely hard pressed for cash. The manager figures that should the loss of $10,000 be incurred, his firm would undergo even more serious financial difficulties. As a result, the manager has decided to decline the client's offer. Clearly, the manager is more averse to risk at this time than he usually is.

Finally, let us consider an example in which the EMV's of the alternatives are negative. Specifically, take the case of a manufacturer who must decide whether to insure his $10 million plant against fire for an annual premium of $2,000. The manufacturer believes the chance that his plant will be destroyed by fire during the next year is 1 in 10,000. Thus the EMV of the act of not buying the insurance is:

$$(-\$10{,}000{,}000)\left(\frac{1}{10{,}000}\right) + (\$0)\left(\frac{9{,}999}{10{,}000}\right) = -\$1{,}000$$

In other words, the expected loss is $1,000. However, the manufacturer realizes that should the plant burn, the *actual* loss would be $10,000,000 rather than the theoretical expected loss of $1,000. Such a disastrous occurrence would force the manufacturer to go out of business. Consequently, he is willing to pay an insurance premium of $2,000, which is higher than the expected loss, to insure his plant against fire. This phenomenon is fairly common among many firms and individuals. Since the insurance company must pay its expenses and make a profit in addition to covering the risk, the insurance premium naturally is higher than the expected loss. From the insured's viewpoint, the expected monetary value of the benefit is smaller than the payment of the insurance premium. However, many firms and individuals are willing to pay such an insurance premium to guard against a possible heavy loss (disastrous occurrence), even though the chance of a disaster's occurring is quite small.

14.2 UTILITY FUNCTIONS

The examples in the preceding section illustrate various types of situations in which a decision maker would shy away from the EMV criterion. The fundamental reason for the inadequacy of the EMV criterion in such situations is that the calculation of EMV implicitly assumes that the *value* of money is a *linear function* of the *amount* of money. This means that the value of money to any decision maker is directly proportional to the amount of money. For instance, this assumption says that the value of $6 million to a decision maker is three times the value of $2 million, or six times the value of $1 million, and so on. However, as we have seen, this is not necessarily so. If the assumption of linearity is not satisfied, then the monetary amounts will not accurately reflect the *relative values* that the decision maker attaches to various monetary consequences. As a result, the EMV's that are computed from such monetary consequences will fail to represent the relative desirability among the alternative courses of action. One way of resolving this difficulty is to adopt the utility approach developed by Von Neumann and Morgenstern.

In decision theory, *utility* is a measure of the relative value of different amounts of money to the decision maker. The basic idea of the Von Neumann-Morgenstern approach is to express the possible consequences of the alternative acts in terms of utility rather than monetary amount. For this purpose, utility measures are obtained in such a way that they reflect the relative preferences of the various possible consequences to the decision maker.

To apply the Von Neumann-Morgenstern approach, it is necessary to obtain a utility corresponding to each of the possible monetary consequences involved in the decision problem. One way of doing this is to ask the decision maker to indicate the utility measure that he would assign to each of these monetary consequences. However, experience has shown that this direct assignment of utilities is often difficult and tedious, particularly if a decision problem involves many monetary consequences. Therefore, it is generally more practical to establish a *utility function* for the decision maker.

The procedure used to establish the utility function for a specific decision maker involves several steps. To illustrate these procedural steps, consider again the A & E firm example. Recall that this firm was considering a contract to conduct a feasibility study for a client. For this proposed contract there was a .60 probability of making a net profit of \$20,000 and a .40 probability of suffering a loss of \$10,000.

The *first* step for constructing a utility function is to select two monetary values that are sufficiently wide apart that they encompass all the possible monetary consequences associated with the decision problem concerned. Specifically, one monetary value denoted as x_b, is selected so that it is at least as large as the best possible monetary consequence in the decision problem. The other monetary value, denoted as x_w, is selected so that it is at most equal to the worst possible monetary consequence. For our example, x_b should be equal to or greater than \$20,000 and x_w should be equal to or less than $-\$10{,}000$. Suppose that the A & E firm selects $x_b = \$20{,}000$ and $x_w = -\$10{,}000$.

The *second* step is to assign two arbitrary real numbers: one to x_b and one to x_w such that the number assigned to x_b exceeds that assigned to x_w. These two numbers are the utilities of the two reference monetary consequences. Suppose that the A & E firm assigns:

$$u(\$20{,}000) = 1$$

$$u(-\$10{,}000) = 0$$

where the symbol u denotes "utility." For instance, $u(\$20{,}000) = 1$ is read as "the utility of \$20,000 is equal to 1." It is important to bear in mind that the assignment of the numbers 1 and 0 is arbitrary. Any other numbers could have been assigned, just so the utility assigned to the most desirable monetary consequence is greater than that assigned to the least desirable monetary consequence.

The *third* step is to determine the utilities corresponding to several selected monetary amounts between x_b and x_w. This is accomplished by obtaining responses from the decision maker in a series of simple, imaginary situations. In each of these situations, the decision maker is asked to imagine that he is committed to a contract and does not know for sure what its consequence will be. If the decision maker regards the contract as undesirable to him, he is asked to specify the maximum amount that he would be willing to pay in order to be released from the contract. However, if he regards the contract as desirable, he is asked to determine the minimum amount for which he would be willing to sell the contract. In either case, the amount specified by the decision maker is called the *certainty equivalent* of the contract.

To make the decision maker's judgment task as easy as possible, each imaginary contract that is presented to him has precisely two possible consequences. In describing a contract to the decision maker, both the specific monetary amount and the probability of each of these consequences are specified. Such an imaginary contract is called a *reference contract.* The probabilities assigned to the two monetary consequences of a reference contract can be arbitrarily selected, although it is common practice to assign each consequence a probability of $\frac{1}{2}$.

In using a reference contract to obtain the utility of a monetary amount, it is necessary that the monetary amounts of the two consequences specified in the contract have known utilities. In Step 2, the utilities of x_b and x_w were specified, and at the outset of Step 3 these are the only two monetary amounts whose utilities have been established. Consequently, we must begin the series of imaginary situations with a reference contract that has x_b and x_w as its two monetary consequences. Thus, following this procedure, the A & E firm is asked to imagine that it is committed to a contract that has a $\frac{1}{2}$ probability of a \$20,000 profit and a $\frac{1}{2}$ probability of incurring a \$10,000 loss. In responding to this imaginary situation, the firm indicates that, in its current financial condition, a loss of \$10,000 would be disastrous, and a contract that has a 50 per cent chance of incurring such a loss would therefore be undesirable. Under such conditions, management makes the subjective estimate that it would be willing to pay up to \$4,000 in order to be released from this contract. However, it would keep the contract rather than pay a sum in excess of \$4,000. In other words, this firm is *indifferent* between the contract and an immediate payment of \$4,000. This is equivalent to saying that the utility of $-\$4{,}000$ is equal to the utility of the contract. Since the utilities of the two possible monetary consequences already have been specified, the utility of the reference contract may be computed as follows:

$$\frac{1}{2}[u(\$20{,}000)] + \frac{1}{2}[u(-\$10{,}000)] = \left(\frac{1}{2}\right)(1) + \left(\frac{1}{2}\right)(0) = .50$$

Hence, the certainty equivalent of the contract is $-\$4{,}000$ and the utility of this amount is .50.

Now, since the utility of $-\$4{,}000$ has been determined, this monetary value may be used as one of the consequences of a new reference contract, together with either x_b or x_w as the other consequence. Thus we now are able to ask the A & E firm to imagine another reference contract with a $\frac{1}{2}$ probability of incurring a \$4,000 loss and a $\frac{1}{2}$ probability of making a \$20,000 profit. Through questioning, it is determined that the A & E firm would regard such a contract as somewhat desirable, but would be willing to sell it if they could obtain a price of at least \$2,600. This implies that the utility of \$2,600 is equal to the utility of the contract. Hence, we obtain:

$$\begin{aligned} u(\$2{,}600) &= \frac{1}{2}[u(-\$4{,}000)] + \frac{1}{2}[u(\$20{,}000)] \\ &= \frac{1}{2}(.50) + \frac{1}{2}(1) = .75 \end{aligned}$$

Thus, the utility attached to \$2,600 is .75.

In a similar manner, the A & E firm is asked to consider a third reference contract with a $\frac{1}{2}$ probability at $-\$4{,}000$ and a $\frac{1}{2}$ probability at $-\$10{,}000$. For such an undesirable contract, the A & E firm figures that it would be willing to pay up to \$7,500 in order to be released from this contract. This indicates that the utility of $-\$7{,}500$ is equal to the utility of the contract. That is:

$$\begin{aligned} u(-\$7{,}500) &= \frac{1}{2}[u(-\$4{,}000)] + \frac{1}{2}[u(-\$10{,}000)] \\ &= \frac{1}{2}(.50) + \frac{1}{2}(0) = .25 \end{aligned}$$

Hence, the utility for $-\$7{,}500$ is .25.

Thus far, the A & E firm has determined five money-utility pairs. These pairs are displayed in Table 14.1. If the firm wishes, it may follow the same procedure to determine the utilities for other monetary consequences. That is, specify a reference contract that has two possible monetary consequences, x_1 and x_2, each of which has a known utility and a $\frac{1}{2}$ probability of occurring. Then, ask the decision maker to specify his certainty equivalent for the contract. Using x_0 to denote the certainty equivalent, the utility for x_0 is obtained from the following equation:

$$u(x_0) = \frac{1}{2}u(x_1) + \frac{1}{2}u(x_2) \tag{14.1}$$

Table 14.1

A & E Firm's Utilities for Selected Monetary Consequences

Monetary Consequence	*Utility*
−$10,000	0.00
−7,500	0.25
−4,000	0.50
2,600	0.75
20,000	1.00

In Formula (14.1), the decision maker's response, x_0, depends on his attitude toward risk. This attitude, in turn, depends on such factors as his asset position and his propensity for risk-taking. Thus, it is imperative that the decision maker's asset position remain constant during the construction of his utility function. This requirement is implicit in Formula (14.1). If we denote the decision maker's present asset position as m_0, then the corresponding *explicit* formula[1] is:

$$u(m_0 + x_0) = \frac{1}{2}u(m_0 + x_1) + \frac{1}{2}u(m_0 + x_2) \tag{14.2}$$

The *fourth* step is to plot all the assessed money-utility pairs on a graph, with the horizontal axis denoting the monetary consequence and the vertical axis showing the utility. It is a common practice to assess only a limited number of pairs, and then use these pairs to approximate the entire utility function by drawing a smooth curve. The utility function for the A & E firm shown in Figure 14.1 is obtained by smoothing the points represented by the money-utility pairs in Table 14.1. Once this utility curve is established, the utility for any monetary consequence on the horizontal axis may be obtained by reading the height of the curve at that point. For instance, the utility for $0 is equal to .67.

[1] For a brief but clear discussion of the significance of a decision maker's asset position and the explicit formula, see Chi-Yuan Lin and Paul D. Berger, "On the Selling Price and Buying Price of a Lottery," *The American Statistician*, December 1969. For a detailed and technical discussion of this subject, see John Pratt, "Risk Aversion in the Small and in the Large," *Econometrica*, Volume 32, No. 1–2, January-April 1964.

The *final* step is to check for consistency, using several new reference contracts. As an illustration, suppose that the A & E firm considers the reference contract with a 50-50 chance of \$2,600 or −\$7,500. The firm determines that it would be indifferent between a payment of \$4,000 and this

Figure 14.1

Utility Function for the A&E Firm

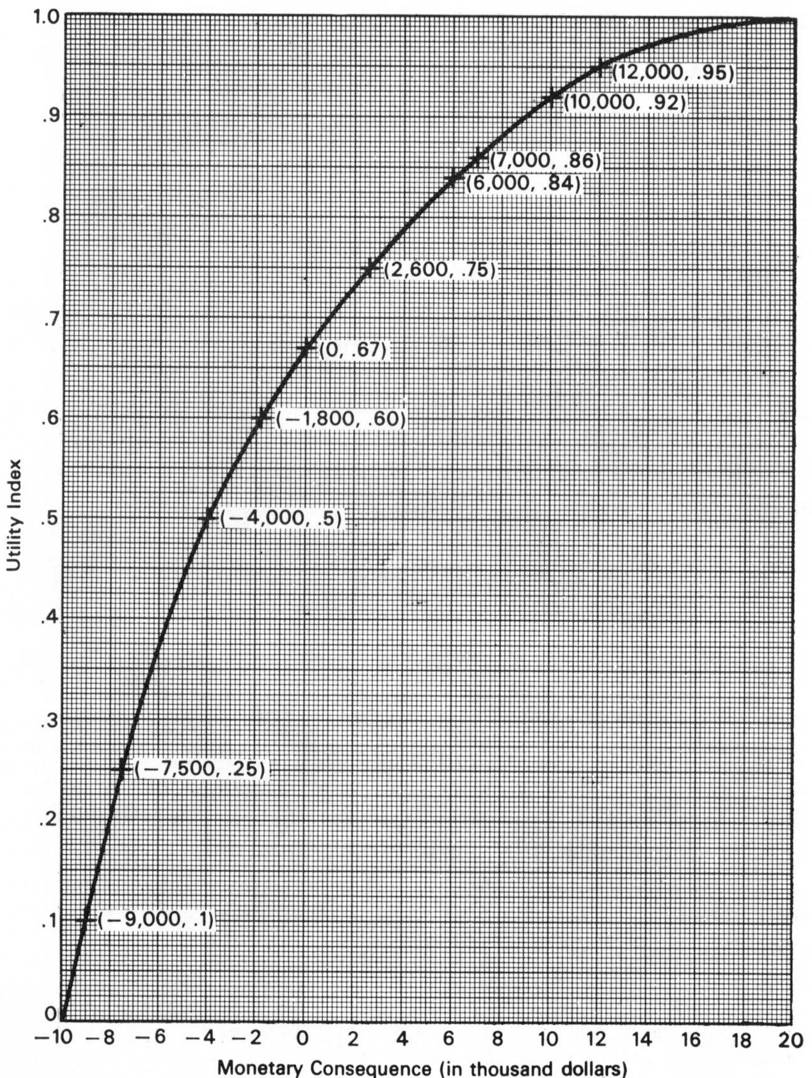

undesirable contract. This means the utility of $-\$4{,}000$ is equal to the utility of the contract:

$$u(-\$4{,}000) = \frac{1}{2}[u(\$2{,}600)] + \frac{1}{2}[u(-\$7{,}500)]$$

$$= \frac{1}{2}(.75) + \frac{1}{2}(.25) = .50$$

Thus, the utility of $-\$4{,}000$ is .50. Since this is identical to that obtained previously, no inconsistency is revealed. If inconsistencies are found, appropriate revisions should be made so that the decision maker is confident that the function properly expresses his utilities for monetary consequences.

14.3 EXPECTED UTILITY CRITERION

Once the utility function of a decision maker has been established, the monetary consequences of a decision problem may be re-expressed in terms of utilities rather than monetary values. This makes it possible to replace the EMV criterion with the *Expected Utility* (EU) criterion. Application of the EU criterion involves (1) computing the expected utility of each of the alternative acts, and (2) selecting that act that has the maximum expected utility. The computation of the EU of an act is identical to that of the EMV of an act, except that the original monetary values are replaced by the corresponding utilities.

As a simple illustration, let us return to the A & E firm's problem of deciding whether or not to accept the offer to perform a feasibility study. Recall that, if the offer is accepted, the firm has a .60 probability of making an eventual profit of \$20,000 and a .40 probability of suffering a \$10,000 loss. Thus, the EMV for the act of accepting the offer is \$8,000, whereas the EMV for the act of declining the offer is \$0. Hence, according to the EMV criterion, the firm should accept the offer. However, as we have seen, the firm has decided to decline the offer in defiance of the EMV criterion.

Let us now see whether the A & E firm's decision to decline the offer can be justified by the EU criterion. From the firm's utility function in Figure 14.1, we see that the utilities corresponding to \$20,000 and $-\$10{,}000$ are 1 and 0, respectively. Therefore, we obtain the expected utility for the act of accepting the offer as follows:

$$1(.60) + 0(.40) = .60$$

For the alternative act of refusing the offer, which has a monetary value of $0, Figure 14.1 yields a corresponding utility of .67. Since the EU of declining the offer (.67) is greater than the EU of accepting the offer (.60), the preferable act under the EU criterion is to decline the offer. Hence, the A & E firm's decision is justified by the EU criterion.

It should be pointed out that once the expected utility of an act is computed, the certainty equivalent of the act may readily be obtained from the utility function. For example, to obtain A & E firm's certainty equivalent for the act of accepting the feasibility study, we first note that he expected utility of this act is .60. From Figure 14.1, it may be seen that this utility has a corresponding monetary value of −$1,800. This amount of −$1,800 is the certainty equivalent of that act.

In evaluating alternative acts, a utility function is particularly useful if each alternative involves several possible monetary consequences. To illustrate this point, suppose that the A & E firm has been offered two contracts, *A* and *B*, by different clients. Due to its limited resources (particularly technical manpower), the firm can accept only *one* of these contracts, even though both of them seem profitable. The A & E firm has estimated the monetary consequences and the probabilities associated with each of these contracts as shown below:

Contract A		*Contract B*	
Probability	*Monetary Consequences*	*Probability*	*Monetary Consequences*
.3	$12,000	.3	$10,000
.5	6,000	.4	7,000
.2	−9,000	.3	−4,000

The reader can verify that the EMV of contract *A* is $4,800, whereas the EMV of contract *B* is $4,600. According to the EMV criterion, the A & E firm should choose contract *A*. This criterion, however, fails to consider the firm's *present attitude toward risk* as reflected in the utility function shown in Figure 14.1. To consider its attitude toward risk in evaluating the contracts, the firm first reads off the utility for each of the six monetary consequences associated with these contracts. These utilities, together with the initial estimates of probabilities and monetary consequences, are shown below:

Contract A			Contract B		
Probability	*Monetary Consequence*	*Utility*	*Probability*	*Monetary Consequence*	*Utility*
.3	$12,000	.95	.3	$10,000	.92
.5	6,000	.84	.4	7,000	.86
.2	−9,000	.10	.3	−4,000	.50

Using the above information, the expected utility of contract A is computed as follows:

$$\text{EU(contract } A) = (.3)(.95) + (.5)(.84) + (.2)(.10) = .725$$

Similarly, the expected utility of contract B is computed as shown below:

$$\text{EU(contract } B) = (.3)(.92) + (.4)(.86) + (.3)(.50) = .770$$

According to the EU criterion, the A & E firm should choose contract B since it has the higher expected utility. It is interesting that the expected utility of either contract is greater than .67, which is the utility of $0. Thus, without the resources constraint, the acceptance of either contract would be preferable to doing nothing.

As pointed out in the preceding example, once the expected utilities are computed, the certainty equivalents can be obtained from the utility function. As the reader may verify from Figure 14.1, the certainty equivalent of contract A (which has an EU of .725) is $1,700. Similarly, the certainty equivalent of contract B (which has an EU of .770) is $3,200. Thus, contract B is preferable to contract A, since contract B has the higher certainty equivalent. This is the same act that was selected by maximizing the expected utility. This is not a mere coincidence. Rather, it is always true that the act that has the maximum EU will also have the maximum certainty equivalent.

It should be emphasized that a utility function established for a given decision situation reflects the decision maker's attitude toward risk *at the time when such a function is established.* Thus, in using the utility function to evaluate alternatives associated with subsequent decision problems as illustrated above, extreme caution should be exercised to make certain that the decision maker's attitude toward risk has not changed significantly. Otherwise, such evaluations of alternatives might be in error. If a significant change has occurred, a new utility function must be constructed.

14.4 THREE BASIC SHAPES OF UTILITY FUNCTIONS

The A & E firm's utility function depicted in Figure 14.1 represents only one of numerous possible shapes of utility functions. Three basic shapes of utility functions, however, are of particular interest. To illustrate these three basic shapes, consider three different decision makers—Abe, Bob, and Chuck—whose utility functions are shown in Figure 14.2.

The utility functions shown in Figure 14.2 and Figure 14.1 have a common characteristic in that each has a positive slope over its range. In other words, each utility function is an increasing function of money. Stated

Figure 14.2

Three Basic Shapes of Utility Functions

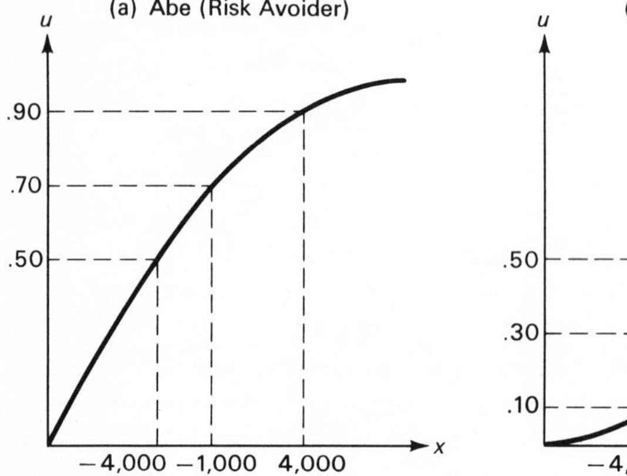

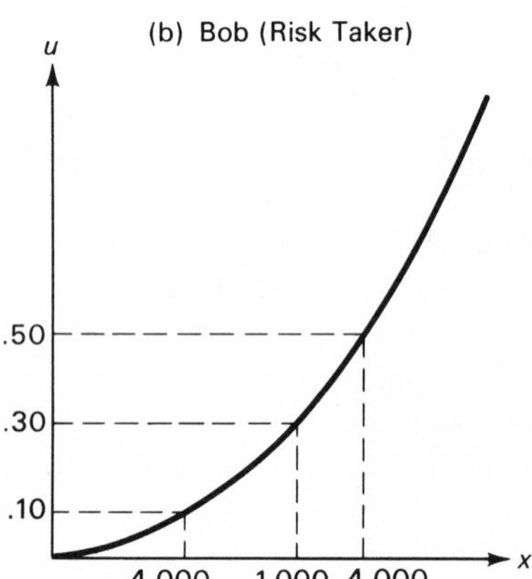

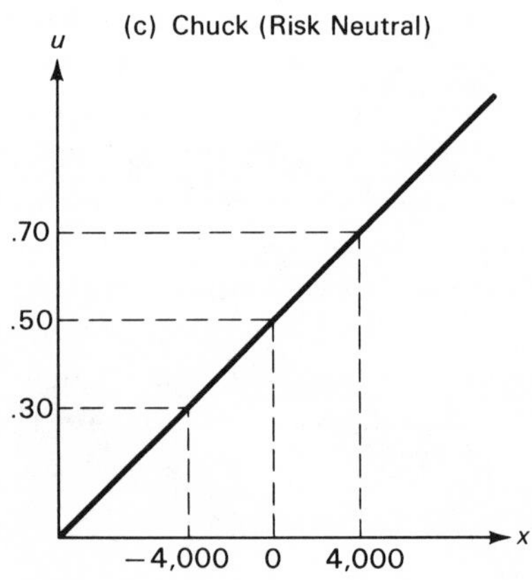

in a less technical manner, each curve rises consistently from the lower-left to the upper-right side of the graph. This simply means that all of these decision makers attach greater utility to a larger amount of money than to a smaller amount. This phenomenon, which often is referred to as "positive marginal utility for money," holds for all but a very few mystical decision makers.

The utility function for Abe, shown in Figure 14.2a, is similar to that shown in Figure 14.1. The concave downward shape of the utility curve indicates that the decision maker's utility for money increases at a decreasing rate. In other words, he has a *diminishing marginal utility for money.* A decision maker with this shape of utility function is averse to risk or he is a risk avoider. In other words, he prefers a *sure* monetary amount to the same amount of *expected* monetary value. As an illustration, suppose that Abe has to decide whether or not to accept a contract with a 50-50 chance of \$4,000 or $-$\$4,000. The EMV of this contract clearly is zero. As shown in Figure 14.2a, the utilities for these two possible monetary consequences are .90 and .50 respectively. Thus, the expected utility of the contract is: $(.90)(\frac{1}{2}) + (.50)(\frac{1}{2}) = .70$. As depicted in the graph, the certainty equivalent of the contract, which is the monetary value corresponding to the utility of .70, is equal to $-$\$1,000. This means that the decision maker is willing to pay up to \$1,000 in order to be released from the contract. Consequently, Abe prefers doing nothing rather than accepting the contract. In the vernacular, this example shows that this kind of decision maker prefers one bird in the hand to a 50-50 chance of two birds in the bush.

The utility function shown in Figure 14.2b is convex in shape, indicating that Bob has an *increasing marginal utility* for money. That is, Bob's utility for money increases at an increasing rate. A decision maker with this shape of utility function is *not* risk averse. Indeed, he is a risk lover or risk taker. He prefers to take a risk with a given expected payoff rather than accept an equal amount for certain. Suppose that Bob considers the same contract presented to Abe. For Bob, the EU of this contract is $.50(\frac{1}{2}) + .10(\frac{1}{2}) = .30$. Thus, as indicated in Figure 14.2b, the certainty equivalent of the contract for Bob is \$1,000 since this is the amount corresponding to the utility of .30. Clearly, this shows that he prefers this risk alternative rather than doing nothing. This is because he is much more interested in the possibility of receiving \$4,000 than he is concerned with the possibility of losing \$4,000. The reader may wonder if such decision makers exist. In fact, they do. For example, Grayson[2] has shown that some oil wildcatters are not willing to sell their rights to a venture for a price equal to its EMV, especially if some of the possible, but not very probable, monetary consequences hold out the promise of a "new way of life."

[2] C. J. Grayson, Jr., *Decision Under Uncertainty: Drilling Decisions by Oil and Gas Operators,* Division of Research, Graduate School of Business Administration, Harvard University, Boston, 1965.

The utility function shown in Figure 14.2c is a linear function, indicating that Chuck has a *constant marginal utility* for money. That is, Chuck's utility for money increases at a constant rate. A decision maker with this type of utility function is neutral to risk. He is indifferent between a sure amount and an alternative for which the EMV is equal to the sure amount. Suppose that Chuck considers the same contract presented to Abe and Bob. For Chuck, the EU of this contract is $.70(\frac{1}{2}) + .30(\frac{1}{2}) = .50$. Thus, as indicated in Figure 14.2c, the certainty equivalent of the contract for Chuck is $0, which is the amount corresponding to the utility of .50. Clearly, this shows that he is indifferent between the contract and doing nothing.

It is important to note that, when a utility function is linear, the certainty equivalent of a risky alternative is *always* equal to the expected monetary value of the same alternative. Consequently, if a decision maker's utility function is linear, the optimal act obtained by maximizing the certainty equivalent is always identical to that obtained by maximizing the EMV. Stated in a different way, the criterion of maximizing expected utility is identical to the criterion of maximizing the expected monetary value whenever the decision maker has a linear utility function. Conversely, it is also true that if a decision maker uses the EMV criterion, he is assuming that his utility function for money is linear over the range of monetary consequences concerned.

PROBLEMS

14.1 Lance Buck, the famous industrialist and jetset playboy, has retained the noted psychologist, Dr. Sigmund Wise, as a member of his personal staff. To gain insight into Buck's decision-making behavior, Dr. Wise asks him to respond to a series of reference contracts. Buck agrees readily, and the following dialogue ensues:

Wise: Imagine that you are committed to a contract that gives you a 50 per cent chance of a $100,000 profit and a 50 per cent chance of a $20,000 loss. Would this contract be desirable to you?

Buck: Of course!

Wise: If you had a chance to sell the contract, what is the least amount you would accept for it?

Buck: Hmm. . . I suppose I'd be willing to let it go for about $70,000.

Wise: Now suppose that you are committed to another contract with a 50 per cent chance of losing $20,000 and a 50 per cent chance of making $70,000. How do you like this one?

Buck: Sounds good to me.

Wise: What is the lowest price you would accept for this one?

Buck: Well, since it's not as good a deal as the first contract, I guess I'd take $40,000 for it.

Wise: I see. Let's try one more. Suppose you hold a contract that has a 50 per cent chance of yielding a $100,000 profit and a 50 per cent chance of a $70,000 profit. I assume that this contract is not entirely abhorrent to you?

Buck: I can live with it comfortably.

Wise: If you were to sell, how much would you ask?

Buck: I'd have to get at least $90,000 for that one.

At this point the dialogue is interrupted by a transatlantic telephone call from Buck's broker in London.

(a) On the basis of Buck's responses to the reference contracts, prepare a graph of his utility function for money. Assume a utility of 0 for − $20,000 and a utility of 1 for $100,000.

(b) From the graph, what utility measure would be assigned to a monetary amount of $80,000?

(c) Is Buck a risk taker or a risk avoider? Explain.

14.2 Assume that you hold a reference contract that gives you a $\frac{1}{2}$ probability of winning $5,000 and a $\frac{1}{2}$ probability of winning $1,000.

(a) What is the minimum price you would accept if you had an opportunity to sell this contract?

(b) On the basis of your response to (a) above, present yourself with a new reference contract. Continue doing this until you have responded to several contracts. Then prepare a graph of your personal utility function for money.

14.3 Suppose that the A & E firm discussed in this chapter has been asked by another client to conduct a study under circumstances quite similar to those described in the text. The client will pay a nominal fee of $2,500 for the preparation of a feasibility study for the construction of a proposed facility. It will cost A & E $10,000 to conduct the study. However, the client will award the construction contract to the A & E firm if the study convinces the client that construction of the new facility is desirable. A & E estimates that if they are fortunate enough to obtain the construction contract for the new facility, their profit should be $25,000. A & E's best guess of the probability that they will obtain the contract to build the new facility is .75.

(a) Assuming that the utility function shown in Figure 14.1 is pertinent to this new contract situation, determine the expected utility for the act of accepting the feasibility study.

(b) Should A & E accept the study?

14.4 A decision maker has determined the possible utilities associated with the four action-state pairs for a particular decision problem as shown in the following table:

State of Nature	*Utility*	
	a_1	a_2
θ_1	.5	.9
θ_2	.8	.2

He feels, however, that it is extremely difficult to specify the probability for θ_1. Naturally, if $P(\theta_1)$ is specified, then $P(\theta_2)$ is $1 - P(\theta_1)$. Because of the difficulty, he is interested in knowing the minimum value of $P(\theta_1)$ for which he would prefer a_2 over a_1. Find this minimum value.

14.5 As a talented decision analyst, you have helped Mr. S.Q. Root to assess his utility function for money. Because of your mathematical insight, you have discovered that Mr. Root's utility function can be accurately represented by a mathematical function. Specifically, if x is used to denote money in dollars, then the utility of money, which will be designated as $u(x)$, may be expressed by the following function:

$$u(x) = \begin{cases} |\sqrt{x}| & \text{if } x > 1 \\ x & \text{if } x \leq 1 \end{cases}$$

Mr. Root wishes to compare the following two alternative courses of action:

(a) Compute the EMV of each of the two alternative acts. According to the EMV criterion, which is the optimal act? Should Mr. Root adopt the EMV criterion for decision making? Why?
(b) Calculate the EU of each of the two alternative acts. According to the EU criterion, what should Mr. Root choose?
(c) Find the certainty equivalent for each of the two alternative acts. By comparing the certainty equivalents, what should Mr. Root choose?
(d) Is the optimal decision found in (c) the same as that obtained in (b)? Is this a general phenomenon or a mere coincidence?
(e) Is Mr. Root risk averse, risk taking, or risk neutral? Why?

Appendix I
Summaries

1. SUMMARY OF SET NOTATION
2. SUMMARY COMPARISON OF DISCRETE AND CONTINUOUS RANDOM VARIABLES
3. SUMMARY OF COMMON DISCRETE PROBABILITY MODELS
4. SUMMARY OF COMMON CONTINUOUS PROBABILITY MODELS
5. BASIC RULES FOR DERIVATIVES
6. BASIC RULES FOR INTEGRATION

1. Summary of Set Notation

Symbol	*How used*	*How read*	*Meaning*
$\in$	$b \in B$	b is an element of the set B	b belongs to the set B.
$\in$	$b \in B$	b is not an element of the set B	b does not belong to the set B.
$\emptyset$	$S = \emptyset$	S is a null set or an empty set	The set S has no elements.
$\subseteq$	$A \subseteq B$	The set A is a subset of the set B	Each element of A is also an element of B. Remark: $\emptyset$ is a subset of every set; a set is a subset of itself; i.e., $B \subseteq B$.
$\subset$	$A \subset B$	The set A is a proper subset of the set B	A is a subset of B but A is neither empty nor the entire set B.
$\mid$	$D = \{x \mid x$ is a decimal digit$\}$	D is the set of elements x such that x is a decimal digit	$D = \{0,1,2, \ldots ,9\}$
$\mathcal{U}$		Universal set	Set of all elements of discourse.
$'$	A'	The complement of A (with respect to $\mathcal{U}$)	Set of elements of $\mathcal{U}$ that do not belong to A; i.e., $A' = \{x \in \mathcal{U} \mid x \in A\}$.
$\cap$	$A \cap B$	A intersection B or A cap B.	Set of elements that belong to both A and B.
$\cup$	$A \cup B$	A union B or A cup B	Set of elements that belong to A or B or both.

2. Summary Comparison of Discrete and Continuous Random Variables

	Discrete	*Continuous*
Random Variable	X can take on only a countable number of values so that these values can be listed	X can take on any of an infinite number of values along a continuum between specified limits
Probability Function	$f(x)$ is a probability mass function if: $\begin{cases} f(x) \geq 0 \\ \Sigma f(x) = 1 \end{cases}$ If $f(x)$ is a p.m.f., then: $P(X = x) = f(x)$	$f(x)$ is a probability density function if: $\begin{cases} f(x) \geq 0 \\ \int_{-\infty}^{\infty} f(x)\,dx = 1 \end{cases}$ If $f(x)$ is a p.d.f., then: $P(X = x) = 0$; $P(a \leq X \leq b) = \int_a^b f(x)\,dx$
Cumulative Function	Cumulative mass function $F(x) = \sum_{t \leq x} f(t)$	Cumulative density function $F(x) = \int_{-\infty}^{x} f(t)\,dt$
Median	x_{md} is the smallest x-value for which $F(x)$ is at least .50	x_{md} is the x-value for which $F(x)$ equals .50
Mode	x_{mo} is an x-value whose probability is greater than the probabilities of x-values in its immediate neighborhood	If $f(x)$ is twice differentiable, x_{mo} is an x-value such that $\begin{cases} f'(x) = 0 \\ f''(x) < 0 \end{cases}$
Mean	$\mu = E(X) = \Sigma\, x f(x)$	$\mu = E(X) = \int_{-\infty}^{\infty} x f(x)\,dx$
Variance	$\sigma^2 = V(X)$ $= \Sigma(x - \mu)^2 f(x)$ $= \Sigma\, x^2 f(x) - \{\Sigma\, x f(x)\}^2$	$\sigma^2 = V(X)$ $= \int_{-\infty}^{\infty} (x - \mu)^2 f(x)\,dx$ $= \int_{-\infty}^{\infty} x^2 f(x)\,dx - \left\{\int_{-\infty}^{\infty} x f(x)\,dx\right\}^2$

3. Summary of Common Discrete Probability Models

Name	*Probability Mass Function*	*Parameters*	*Mean*	*Variance*
Hyper-geometric	$f_h(x\|N, n, D)$ $= \dfrac{\binom{D}{x}\binom{N-D}{n-x}}{\binom{N}{n}}, \quad x = 0, 1, \ldots, k$ where $k = n$ or D, whichever is smaller	$N = 1, 2, \ldots$ $D = 0, 1, \ldots, N$ $n = 1, 2, \ldots, N$	$n\left(\dfrac{D}{N}\right)$	$n\left(\dfrac{D}{N}\right)\left(\dfrac{N-D}{N}\right)\left(\dfrac{N-n}{N-1}\right)$
Binomial	$f_b(x\|n,p)$ $= \binom{n}{x} p^x(1-p)^{n-x}, \quad x = 0, 1, \ldots, n$	$0 < p < 1$ $n = 1, 2, \ldots$	np	$np(1-p)$
Negative Binomial (Pascal)	$f_{nb}(n\|x,p)$ $= \binom{n-1}{x-1} p^x(1-p)^{n-x}, \quad n = x, x+1, \ldots$	$0 < p < 1$ $x = 1, 2, \ldots$	$\dfrac{x}{p}$	$\dfrac{x(1-p)}{p^2}$
Geometric	$f_g(n\|p)$ $= p(1-p)^{n-1}, \quad n = 1, 2, \ldots$	$0 < p < 1$	$\dfrac{1}{p}$	$\dfrac{1-p}{p^2}$
Poisson	$f_P(x\|\lambda,t)$ $= \dfrac{(\lambda t)^x e^{-\lambda t}}{x!}, \quad x = 0, 1, \ldots$	$\lambda > 0$ $t > 0$	λt	λt

4. Summary of Common Continuous Probability Models

Name	*Probability Density Function*	*Parameters*	*Mean*	*Variance*
General Normal	$f_n(x\|\mu,\sigma) = \frac{1}{\sigma\sqrt{2\pi}} e^{-\frac{1}{2}\left(\frac{x-\mu}{\sigma}\right)^2}, \quad -\infty < x < \infty$	$-\infty < \mu < \infty$ $\sigma > 0$	μ	σ^2
Standard Normal	$f_N(z) = \frac{1}{\sqrt{2\pi}} e^{-\frac{1}{2}z^2}, \quad -\infty < z < \infty$		0	1
Gamma	$f_\gamma(t\|x,\lambda) = \frac{\lambda e^{-\lambda t}(\lambda t)^{x-1}}{(x-1)!}, \quad t \geq 0$	$x > 0$ $\lambda > 0$	$\frac{x}{\lambda}$	$\frac{x}{\lambda^2}$
Erlang	$f_E(t\|x,\lambda) = \frac{\lambda e^{-\lambda t}(\lambda t)^{x-1}}{(x-1)!}, \quad t \geq 0$	$x = 1, 2, \ldots$ $\lambda > 0$	$\frac{x}{\lambda}$	$\frac{x}{\lambda^2}$
Exponential	$f_e(t\|\lambda) = \lambda e^{-\lambda t}, \quad t \geq 0$	$\lambda > 0$	$\frac{1}{\lambda}$	$\frac{1}{\lambda^2}$
Rectangular (Uniform)	$f_r(x\|a,b) = \frac{1}{b-a}, \quad a \leq x \leq b$	$-\infty < a < \infty$ $-\infty < b < \infty$	$\frac{a+b}{2}$	$\frac{(b-a)^2}{12}$

5. Basic Rules for Derivatives

In the following formulas, x is a variable, whereas c and n are fixed real numbers. In addition, u, v, and w represent functions of x.

1. Derivative of a constant is zero: $\dfrac{d}{dx}(c) = 0$

2. Derivative of an expression may be taken term by term:

$$\frac{d}{dx}(u + v - w) = \frac{du}{dx} + \frac{dv}{dx} - \frac{dw}{dx}$$

3. Constant may be factored out: $\dfrac{d}{dx}(cu) = c\dfrac{du}{dx}$

4. Power rule: $\dfrac{d}{dx}(x^n) = nx^{n-1}$

5. Product rule: $\dfrac{d}{dx}(uv) = u\dfrac{dv}{dx} + v\dfrac{du}{dx}$

6. Quotient rule: $\dfrac{d}{dx}\left(\dfrac{u}{v}\right) = \dfrac{v\dfrac{du}{dx} - u\dfrac{dv}{dx}}{v^2}$

7. Chain rule: $\dfrac{d}{dx}\Big(f(u)\Big) = \dfrac{d}{du}\Big(f(u)\Big)\dfrac{du}{dx}$; e.g., $\dfrac{d}{dx}(u^n) = nu^{n-1}\dfrac{du}{dx}$

8. Natural logarithmic function: $\dfrac{d}{dx}(\ln u) = \dfrac{1}{u}\dfrac{du}{dx}$

9. Exponential function: $\dfrac{d}{dx}(e^u) = e^u\dfrac{du}{dx}$

6. Basic Rules for Integration

In the following formulas, x is a variable, whereas a, c, and n are fixed real numbers. In addition, u, v, and w are functions of x.

1. Integration of an expression may be taken term by term:

$$\int (u + v - w)\,dx = \int u\,dx + \int v\,dx - \int w\,dx$$

2. Constant may be factored out: $\int au\,dx = a \int u\,dx$

3. Integral of x^n for $n \neq -1$: $\int x^n\,dx = \dfrac{x^{n+1}}{n+1} + c$

4. Integral of $\dfrac{1}{x}$: $\int \dfrac{dx}{x} = \ln x + c$

5. Integral of $\ln x$: $\int \ln x\,dx = x \ln x - x + c$

6. Integration by parts: $\int u\,dv = uv - \int v\,du$

7. Exponential Functions:

$$\int e^{ax}\,dx = \frac{e^{ax}}{a} + c$$

$$\int xe^{ax}\,dx = \frac{e^{ax}}{a^2}(ax - 1) + c$$

$$\int x^2e^{ax}\,dx = \frac{x^2\,e^{ax}}{a} - \frac{2e^{ax}}{a^3}(ax - 1) + c$$

Appendix II
Tables

A. FACTORIAL VALUES

B. BINOMIAL COEFFICIENTS

C. EXPONENTIAL FUNCTIONS

D. BINOMIAL DISTRIBUTION — INDIVIDUAL TERMS

E. BINOMIAL DISTRIBUTION — CUMULATIVE TERMS

F. POISSON DISTRIBUTION — INDIVIDUAL TERMS

G. POISSON DISTRIBUTION — CUMULATIVE TERMS

H. STANDARD NORMAL INTEGRAL

I. SQUARES AND SQUARE ROOTS

J. FOUR PLACE-MANTISSAS FOR COMMON LOGARITHMS

K. NATURAL (NAPIERIAN) LOGARITHMS

Table A

Factorial Values

$n! = c(10^x)$: where c is tabulated in the second column and x is tabulated in the third column. In other words, $n!$ is obtained by selecting the corresponding c and moving the decimal point x places to the right.

Example: $9! = 3.6288(10^5) = 362{,}880$

n	c	x	n	c	x	n	c	x	n	c	x
1	1.0000	0	26	4.0329	26	51	1.5511	66	76	1.8855	111
2	2.0000	0	27	1.0889	28	52	8.0658	67	77	1.4518	113
3	6.0000	0	28	3.0489	29	53	4.2749	69	78	1.1324	115
4	2.4000	1	29	8.8418	30	54	2.3084	71	79	8.9462	116
5	1.2000	2	30	2.6525	32	55	1.2696	73	80	7.1569	118
6	7.2000	2	31	8.2228	33	56	7.1100	74	81	5.7971	120
7	5.0400	3	32	2.6313	35	57	4.0527	76	82	4.7536	122
8	4.0320	4	33	8.6833	36	58	2.3506	78	83	3.9455	124
9	3.6288	5	34	2.9523	38	59	1.3868	80	84	3.3142	126
10	3.6288	6	35	1.0333	40	60	8.3210	81	85	2.8171	128
11	3.9917	7	36	3.7199	41	61	5.0758	83	86	2.4227	130
12	4.7900	8	37	1.3764	43	62	3.1470	85	87	2.1078	132
13	6.2270	9	38	5.2302	44	63	1.9826	87	88	1.8548	134
14	8.7178	10	39	2.0398	46	64	1.2689	89	89	1.6508	136
15	1.3077	12	40	8.1592	47	65	8.2477	90	90	1.4857	138
16	2.0923	13	41	3.3453	49	66	5.4434	92	91	1.3520	140
17	3.5569	14	42	1.4050	51	67	3.6471	94	92	1.2438	142
18	6.4024	15	43	6.0415	52	68	2.4800	96	93	1.1568	144
19	1.2165	17	44	2.6583	54	69	1.7112	98	94	1.0874	146
20	2.4329	18	45	1.1962	56	70	1.1979	100	95	1.0330	148
21	5.1091	19	46	5.5026	57	71	8.5048	101	96	9.9168	149
22	1.1240	21	47	2.5862	59	72	6.1234	103	97	9.6193	151
23	2.5852	22	48	1.2414	61	73	4.4701	105	98	9.4269	153
24	6.2045	23	49	6.0828	62	74	3.3079	107	99	9.3326	155
25	1.5511	25	50	3.0414	64	75	2.4809	109	100	9.3326	157

Table B

Binomial Coefficients — Combinations of n Objects Taken r at a Time

$$\binom{n}{r} = \binom{n}{n-r} = \frac{n!}{r!(n-r)!}$$

Examples: $\binom{18}{6} = 18{,}564$ $\binom{18}{12} = \binom{18}{18-12} = 18{,}564$

n \ r	0	1	2	3	4	5	6	7	8	9	10
1	1	1									
2	1	2	1								
3	1	3	3	1							
4	1	4	6	4	1						
5	1	5	10	10	5	1					
6	1	6	15	20	15	6	1				
7	1	7	21	35	35	21	7	1			
8	1	8	28	56	70	56	28	8	1		
9	1	9	36	84	126	126	84	36	9	1	
10	1	10	45	120	210	252	210	120	45	10	1
11	1	11	55	165	330	462	462	330	165	55	11
12	1	12	66	220	495	792	924	792	495	220	66
13	1	13	78	286	715	1287	1716	1716	1287	715	286
14	1	14	91	364	1001	2002	3003	3432	3003	2002	1001
15	1	15	105	455	1365	3003	5005	6435	6435	5005	3003
16	1	16	120	560	1820	4368	8008	11440	12870	11440	8008
17	1	17	136	680	2380	6188	12376	19448	24310	24310	19448
18	1	18	153	816	3060	8568	18564	31824	43758	48620	43758
19	1	19	171	969	3876	11628	27132	50388	75582	92378	92378
20	1	20	190	1140	4845	15504	38760	77520	125970	167960	184756

Table C

Exponential Functions

x	e^x	e^{-x}	x	e^x	e^{-x}
0.0	1.0000	1.0000	3.5	33.1155	0.0302
0.1	1.1052	0.9048	3.6	36.5982	0.0273
0.2	1.2214	0.8187	3.7	40.4473	0.0247
0.3	1.3499	0.7408	3.8	44.7012	0.0224
0.4	1.4918	0.6703	3.9	49.4024	0.0202
0.5	1.6487	0.6065	4.0	54.5982	0.0183
0.6	1.8221	0.5488	4.1	60.3403	0.0166
0.7	2.0138	0.4966	4.2	66.6863	0.0150
0.8	2.2255	0.4493	4.3	73.6998	0.0136
0.9	2.4596	0.4066	4.4	81.4509	0.0123
1.0	2.7183	0.3679	4.5	90.0171	0.0111
1.1	3.0042	0.3329	4.6	99.4843	0.0101
1.2	3.3201	0.3012	4.7	109.9472	0.0091
1.3	3.6693	0.2725	4.8	121.5104	0.0082
1.4	4.0552	0.2466	4.9	134.2898	0.0074
1.5	4.4817	0.2231	5.0	148.4132	0.0067
1.6	4.9530	0.2019	5.1	164.0219	0.0061
1.7	5.4739	0.1827	5.2	181.2722	0.0055
1.8	6.0496	0.1653	5.3	200.3368	0.0050
1.9	6.6859	0.1496	5.4	221.4064	0.0045
2.0	7.3891	0.1353	5.5	244.6919	0.0041
2.1	8.1662	0.1225	5.6	270.4264	0.0037
2.2	9.0250	0.1108	5.7	298.8674	0.0033
2.3	9.9742	0.1003	5.8	330.2996	0.0030
2.4	11.0232	0.0907	5.9	365.0375	0.0027
2.5	12.1825	0.0821	6.0	403.4288	0.0025
2.6	13.4637	0.0743	6.1	445.8578	0.0022
2.7	14.8797	0.0672	6.2	492.7490	0.0020
2.8	16.4446	0.0608	6.3	544.5719	0.0018
2.9	18.1741	0.0550	6.4	601.8450	0.0017
3.0	20.0855	0.0498	6.5	665.1416	0.0015
3.1	22.1980	0.0450	6.6	735.0952	0.0014
3.2	24.5325	0.0408	6.7	812.4058	0.0012
3.3	27.1126	0.0369	6.8	897.8473	0.0011
3.4	29.9641	0.0334	6.9	992.2747	0.0010
			7.0	1096.6332	0.0009

Table D

Binomial Distribution — Individual Terms

$$f_b(x|n,p) = \binom{n}{x} p^x(1-p)^{n-x}$$

N	X	P=.001	.005	.01	.02	.03	.04	.05	.10	.15	.20	.25	.30	1/3	.40	.50	
2	0	.9980	.9900	.9801	.9604	.9409	.9216	.9025	.8100	.7225	.6400	.5625	.4900	.4444	.3600	.2500	2
	1	.0020	.0099	.0198	.0392	.0582	.0768	.0950	.1800	.2550	.3200	.3750	.4200	.4444	.4800	.5000	1
	2	.0000	.0000	.0001	.0004	.0009	.0016	.0025	.0100	.0225	.0400	.0625	.0900	.1111	.1600	.2500	0
3	0	.9970	.9851	.9703	.9412	.9127	.8847	.8574	.7290	.6141	.5120	.4219	.3430	.2963	.2160	.1250	3
	1	.0030	.0149	.0294	.0576	.0847	.1106	.1354	.2430	.3251	.3840	.4219	.4410	.4444	.4320	.3750	2
	2	.0000	.0001	.0003	.0012	.0026	.0046	.0071	.0270	.0574	.0960	.1406	.1890	.2222	.2880	.3750	1
	3	.0000	.0000	.0000	.0000	.0000	.0001	.0001	.0010	.0034	.0080	.0156	.0270	.0370	.0640	.1250	0
4	0	.9960	.9801	.9606	.9224	.8853	.8493	.8145	.6561	.5220	.4096	.3164	.2401	.1975	.1296	.0625	4
	1	.0040	.0197	.0388	.0753	.1095	.1416	.1715	.2916	.3685	.4096	.4219	.4116	.3951	.3456	.2500	3
	2	.0000	.0001	.0006	.0023	.0051	.0088	.0135	.0486	.0975	.1536	.2109	.2646	.2963	.3456	.3750	2
	3	.0000	.0000	.0000	.0000	.0001	.0002	.0005	.0036	.0115	.0256	.0469	.0756	.0988	.1536	.2500	1
	4	.0000	.0000	.0000	.0000	.0000	.0000	.0000	.0001	.0005	.0016	.0039	.0081	.0123	.0256	.0625	0
5	0	.9950	.9752	.9510	.9039	.8587	.8154	.7738	.5905	.4437	.3277	.2373	.1681	.1317	.0778	.0313	5
	1	.0050	.0245	.0480	.0922	.1328	.1699	.2036	.3280	.3915	.4096	.3955	.3601	.3292	.2592	.1563	4
	2	.0000	.0002	.0010	.0038	.0082	.0142	.0214	.0729	.1382	.2048	.2637	.3087	.3292	.3456	.3125	3
	3	.0000	.0000	.0000	.0001	.0003	.0006	.0011	.0081	.0244	.0512	.0879	.1323	.1646	.2304	.3125	2
	4	.0000	.0000	.0000	.0000	.0000	.0000	.0000	.0004	.0022	.0064	.0146	.0283	.0412	.0768	.1563	1
	5	.0000	.0000	.0000	.0000	.0000	.0000	.0000	.0000	.0001	.0003	.0010	.0024	.0041	.0102	.0313	0
6	0	.9940	.9704	.9415	.8858	.8330	.7828	.7351	.5314	.3771	.2621	.1780	.1176	.0878	.0467	.0156	6
	1	.0060	.0293	.0571	.1085	.1546	.1957	.2321	.3543	.3993	.3932	.3560	.3025	.2634	.1866	.0938	5
	2	.0000	.0004	.0014	.0055	.0120	.0204	.0305	.0984	.1762	.2458	.2966	.3241	.3292	.3110	.2344	4
	3	.0000	.0000	.0000	.0002	.0005	.0011	.0021	.0146	.0415	.0819	.1318	.1852	.2195	.2765	.3125	3
	4	.0000	.0000	.0000	.0000	.0000	.0000	.0001	.0012	.0055	.0154	.0330	.0595	.0823	.1382	.2344	2
	5	.0000	.0000	.0000	.0000	.0000	.0000	.0000	.0001	.0004	.0015	.0044	.0102	.0165	.0369	.0938	1
	6	.0000	.0000	.0000	.0000	.0000	.0000	.0000	.0000	.0000	.0001	.0002	.0007	.0014	.0041	.0156	0
7	0	.9930	.9655	.9321	.8681	.8080	.7514	.6983	.4783	.3206	.2097	.1335	.0824	.0585	.0280	.0078	7
	1	.0070	.0340	.0659	.1240	.1749	.2192	.2573	.3720	.3960	.3670	.3115	.2471	.2048	.1306	.0547	6
	2	.0000	.0005	.0020	.0076	.0162	.0274	.0406	.1240	.2097	.2753	.3115	.3177	.3073	.2613	.1641	5
	3	.0000	.0000	.0000	.0003	.0008	.0019	.0036	.0230	.0617	.1147	.1730	.2269	.2561	.2903	.2734	4
	4	.0000	.0000	.0000	.0000	.0000	.0001	.0002	.0026	.0109	.0287	.0577	.0972	.1280	.1935	.2734	3
	5	.0000	.0000	.0000	.0000	.0000	.0000	.0000	.0002	.0012	.0043	.0115	.0250	.0384	.0774	.1641	2
	6	.0000	.0000	.0000	.0000	.0000	.0000	.0000	.0000	.0001	.0004	.0013	.0036	.0064	.0172	.0547	1
	7	.0000	.0000	.0000	.0000	.0000	.0000	.0000	.0000	.0000	.0000	.0001	.0002	.0005	.0016	.0078	0
8	0	.9920	.9607	.9227	.8508	.7837	.7214	.6634	.4305	.2725	.1678	.1001	.0576	.0390	.0168	.0039	8
	1	.0079	.0386	.0746	.1389	.1939	.2405	.2793	.3826	.3847	.3355	.2670	.1977	.1561	.0896	.0313	7
	2	.0000	.0007	.0026	.0099	.0210	.0351	.0515	.1488	.2376	.2936	.3115	.2965	.2731	.2090	.1094	6
	3	.0000	.0000	.0001	.0004	.0013	.0029	.0054	.0331	.0839	.1468	.2076	.2541	.2731	.2787	.2188	5
	4	.0000	.0000	.0000	.0000	.0001	.0002	.0004	.0046	.0185	.0459	.0865	.1361	.1707	.2322	.2734	4
	5	.0000	.0000	.0000	.0000	.0000	.0000	.0000	.0004	.0026	.0092	.0231	.0467	.0683	.1239	.2188	3
	6	.0000	.0000	.0000	.0000	.0000	.0000	.0000	.0000	.0002	.0011	.0038	.0100	.0171	.0413	.1094	2
	7	.0000	.0000	.0000	.0000	.0000	.0000	.0000	.0000	.0000	.0001	.0004	.0012	.0024	.0079	.0313	1
	8	.0000	.0000	.0000	.0000	.0000	.0000	.0000	.0000	.0000	.0000	.0000	.0001	.0002	.0007	.0039	0
9	0	.9910	.9559	.9135	.8337	.7602	.6925	.6302	.3874	.2316	.1342	.0751	.0404	.0260	.0101	.0020	9
	1	.0089	.0432	.0830	.1531	.2116	.2597	.2985	.3874	.3679	.3020	.2253	.1556	.1171	.0605	.0176	8
	2	.0000	.0009	.0034	.0125	.0262	.0433	.0629	.1722	.2597	.3020	.3003	.2668	.2341	.1612	.0703	7
		P=.999	.995	.99	.98	.97	.96	.95	.90	.85	.80	.75	.70	2/3	.60	.50	X

Table D (continued)

Binomial Distribution — Individual Terms

$$f_b(x|n,p) = \binom{n}{x} p^x (1-p)^{n-x}$$

N	X	P=.001	.005	.01	.02	.03	.04	.05	.10	.15	.20	.25	.30	1/3	.40	.50	
9	3	.0000	.0000	.0001	.0006	.0019	.0042	.0077	.0446	.1069	.1762	.2336	.2668	.2731	.2508	.1641	6
	4	.0000	.0000	.0000	.0000	.0001	.0003	.0006	.0074	.0283	.0661	.1168	.1715	.2048	.2508	.2461	5
	5	.0000	.0000	.0000	.0000	.0000	.0000	.0000	.0008	.0050	.0165	.0389	.0735	.1024	.1672	.2461	4
	6	.0000	.0000	.0000	.0000	.0000	.0000	.0000	.0001	.0006	.0028	.0087	.0210	.0341	.0743	.1641	3
	7	.0000	.0000	.0000	.0000	.0000	.0000	.0000	.0000	.0000	.0003	.0012	.0039	.0073	.0212	.0703	2
	8	.0000	.0000	.0000	.0000	.0000	.0000	.0000	.0000	.0000	.0000	.0001	.0004	.0009	.0035	.0176	1
	9	.0000	.0000	.0000	.0000	.0000	.0000	.0000	.0000	.0000	.0000	.0000	.0000	.0001	.0003	.0020	0
10	0	.9900	.9511	.9044	.8171	.7374	.6648	.5987	.3487	.1969	.1074	.0563	.0282	.0173	.0060	.0010	10
	1	.0099	.0478	.0914	.1667	.2281	.2770	.3151	.3874	.3474	.2684	.1877	.1211	.0867	.0403	.0098	9
	2	.0000	.0011	.0042	.0153	.0317	.0519	.0746	.1937	.2759	.3020	.2816	.2335	.1951	.1209	.0439	8
	3	.0000	.0000	.0001	.0008	.0026	.0058	.0105	.0574	.1298	.2013	.2503	.2668	.2601	.2150	.1172	7
	4	.0000	.0000	.0000	.0000	.0001	.0004	.0010	.0112	.0401	.0881	.1460	.2001	.2276	.2508	.2051	6
	5	.0000	.0000	.0000	.0000	.0000	.0000	.0001	.0015	.0085	.0264	.0584	.1029	.1366	.2007	.2461	5
	6	.0000	.0000	.0000	.0000	.0000	.0000	.0000	.0001	.0012	.0055	.0162	.0368	.0569	.1115	.2051	4
	7	.0000	.0000	.0000	.0000	.0000	.0000	.0000	.0000	.0001	.0008	.0031	.0090	.0163	.0425	.1172	3
	8	.0000	.0000	.0000	.0000	.0000	.0000	.0000	.0000	.0000	.0001	.0004	.0014	.0030	.0106	.0439	2
	9	.0000	.0000	.0000	.0000	.0000	.0000	.0000	.0000	.0000	.0000	.0000	.0001	.0003	.0016	.0098	1
	10	.0000	.0000	.0000	.0000	.0000	.0000	.0000	.0000	.0000	.0000	.0000	.0000	.0000	.0001	.0010	0
11	0	.9891	.9464	.8953	.8007	.7153	.6382	.5688	.3138	.1673	.0859	.0422	.0198	.0116	.0036	.0005	11
	1	.0109	.0523	.0995	.1798	.2433	.2925	.3293	.3835	.3248	.2362	.1549	.0932	.0636	.0266	.0054	10
	2	.0001	.0013	.0050	.0183	.0376	.0609	.0867	.2131	.2866	.2953	.2581	.1998	.1590	.0887	.0269	9
	3	.0000	.0000	.0002	.0011	.0035	.0076	.0137	.0710	.1517	.2215	.2581	.2568	.2384	.1774	.0806	8
	4	.0000	.0000	.0000	.0000	.0002	.0006	.0014	.0158	.0536	.1107	.1721	.2201	.2384	.2365	.1611	7
	5	.0000	.0000	.0000	.0000	.0000	.0000	.0001	.0025	.0132	.0388	.0803	.1321	.1669	.2207	.2256	6
	6	.0000	.0000	.0000	.0000	.0000	.0000	.0000	.0003	.0023	.0097	.0268	.0566	.0835	.1471	.2256	5
	7	.0000	.0000	.0000	.0000	.0000	.0000	.0000	.0000	.0003	.0017	.0064	.0173	.0298	.0701	.1611	4
	8	.0000	.0000	.0000	.0000	.0000	.0000	.0000	.0000	.0000	.0002	.0011	.0037	.0075	.0234	.0806	3
	9	.0000	.0000	.0000	.0000	.0000	.0000	.0000	.0000	.0000	.0000	.0001	.0005	.0012	.0052	.0269	2
	10	.0000	.0000	.0000	.0000	.0000	.0000	.0000	.0000	.0000	.0000	.0000	.0000	.0001	.0007	.0054	1
	11	.0000	.0000	.0000	.0000	.0000	.0000	.0000	.0000	.0000	.0000	.0000	.0000	.0000	.0000	.0005	0
12	0	.9881	.9416	.8864	.7847	.6938	.6127	.5404	.2824	.1422	.0687	.0317	.0138	.0077	.0022	.0002	12
	1	.0119	.0568	.1074	.1922	.2575	.3064	.3413	.3766	.3012	.2062	.1267	.0712	.0462	.0174	.0029	11
	2	.0001	.0016	.0060	.0216	.0438	.0702	.0988	.2301	.2924	.2835	.2323	.1678	.1272	.0639	.0161	10
	3	.0000	.0000	.0002	.0015	.0045	.0098	.0173	.0852	.1720	.2362	.2581	.2397	.2120	.1419	.0537	9
	4	.0000	.0000	.0000	.0001	.0003	.0009	.0021	.0213	.0683	.1329	.1936	.2311	.2384	.2128	.1208	8
	5	.0000	.0000	.0000	.0000	.0000	.0001	.0002	.0038	.0193	.0532	.1032	.1585	.1908	.2270	.1934	7
	6	.0000	.0000	.0000	.0000	.0000	.0000	.0000	.0005	.0040	.0155	.0401	.0792	.1113	.1766	.2256	6
	7	.0000	.0000	.0000	.0000	.0000	.0000	.0000	.0000	.0006	.0033	.0115	.0291	.0477	.1009	.1934	5
	8	.0000	.0000	.0000	.0000	.0000	.0000	.0000	.0000	.0001	.0005	.0024	.0078	.0149	.0420	.1208	4
	9	.0000	.0000	.0000	.0000	.0000	.0000	.0000	.0000	.0000	.0001	.0004	.0015	.0033	.0125	.0537	3
	10	.0000	.0000	.0000	.0000	.0000	.0000	.0000	.0000	.0000	.0000	.0000	.0002	.0005	.0025	.0161	2
	11	.0000	.0000	.0000	.0000	.0000	.0000	.0000	.0000	.0000	.0000	.0000	.0000	.0000	.0003	.0029	1
	12	.0000	.0000	.0000	.0000	.0000	.0000	.0000	.0000	.0000	.0000	.0000	.0000	.0000	.0000	.0002	0
13	0	.9871	.9369	.8775	.7690	.6730	.5882	.5133	.2542	.1209	.0550	.0238	.0097	.0051	.0013	.0001	13
	1	.0128	.0612	.1152	.2040	.2706	.3186	.3512	.3672	.2774	.1787	.1029	.0540	.0334	.0113	.0016	12
		P=.999	.995	.99	.98	.97	.96	.95	.90	.85	.80	.75	.70	2/3	.60	.50	X

Table D (continued)

Binomial Distribution — Individual Terms

$$f_b(x|n,p) = \binom{n}{x} p^x (1-p)^{n-x}$$

N	X	P=.001	.005	.01	.02	.03	.04	.05	.10	.15	.20	.25	.30	1/3	.40	.50	
13	2	.0001	.0018	.0070	.0250	.0502	.0797	.1109	.2448	.2937	.2680	.2059	.1388	.1002	.0453	.0095	11
	3	.0000	.0000	.0003	.0019	.0057	.0122	.0214	.0997	.1900	.2457	.2517	.2181	.1837	.1107	.0349	10
	4	.0000	.0000	.0000	.0001	.0004	.0013	.0028	.0277	.0838	.1535	.2097	.2337	.2296	.1845	.0873	9
	5	.0000	.0000	.0000	.0000	.0000	.0001	.0003	.0055	.0266	.0691	.1258	.1803	.2067	.2214	.1571	8
	6	.0000	.0000	.0000	.0000	.0000	.0000	.0000	.0008	.0063	.0230	.0559	.1030	.1378	.1968	.2095	7
	7	.0000	.0000	.0000	.0000	.0000	.0000	.0000	.0001	.0011	.0058	.0186	.0442	.0689	.1312	.2095	6
	8	.0000	.0000	.0000	.0000	.0000	.0000	.0000	.0000	.0001	.0011	.0047	.0142	.0258	.0656	.1571	5
	9	.0000	.0000	.0000	.0000	.0000	.0000	.0000	.0000	.0000	.0001	.0009	.0034	.0072	.0243	.0873	4
	10	.0000	.0000	.0000	.0000	.0000	.0000	.0000	.0000	.0000	.0000	.0001	.0006	.0014	.0065	.0349	3
	11	.0000	.0000	.0000	.0000	.0000	.0000	.0000	.0000	.0000	.0000	.0000	.0001	.0002	.0012	.0095	2
	12	.0000	.0000	.0000	.0000	.0000	.0000	.0000	.0000	.0000	.0000	.0000	.0000	.0000	.0001	.0016	1
	13	.0000	.0000	.0000	.0000	.0000	.0000	.0000	.0000	.0000	.0000	.0000	.0000	.0000	.0000	.0001	0
14	0	.9861	.9322	.8687	.7536	.6528	.5647	.4877	.2288	.1028	.0440	.0178	.0068	.0034	.0008	.0001	14
	1	.0138	.0656	.1229	.2153	.2827	.3294	.3593	.3559	.2539	.1539	.0832	.0407	.0240	.0073	.0009	13
	2	.0001	.0021	.0081	.0286	.0568	.0892	.1229	.2570	.2912	.2501	.1802	.1134	.0779	.0317	.0056	12
	3	.0000	.0000	.0003	.0023	.0070	.0149	.0259	.1142	.2056	.2501	.2402	.1943	.1559	.0845	.0222	11
	4	.0000	.0000	.0000	.0001	.0006	.0017	.0037	.0349	.0998	.1720	.2202	.2290	.2143	.1549	.0611	10
	5	.0000	.0000	.0000	.0000	.0000	.0001	.0004	.0078	.0352	.0860	.1468	.1963	.2143	.2066	.1222	9
	6	.0000	.0000	.0000	.0000	.0000	.0000	.0000	.0013	.0093	.0322	.0734	.1262	.1607	.2066	.1833	8
	7	.0000	.0000	.0000	.0000	.0000	.0000	.0000	.0002	.0019	.0092	.0280	.0618	.0918	.1574	.2095	7
	8	.0000	.0000	.0000	.0000	.0000	.0000	.0000	.0000	.0003	.0020	.0082	.0232	.0402	.0918	.1833	6
	9	.0000	.0000	.0000	.0000	.0000	.0000	.0000	.0000	.0000	.0003	.0018	.0066	.0134	.0408	.1222	5
	10	.0000	.0000	.0000	.0000	.0000	.0000	.0000	.0000	.0000	.0000	.0003	.0014	.0033	.0136	.0611	4
	11	.0000	.0000	.0000	.0000	.0000	.0000	.0000	.0000	.0000	.0000	.0000	.0002	.0006	.0033	.0222	3
	12	.0000	.0000	.0000	.0000	.0000	.0000	.0000	.0000	.0000	.0000	.0000	.0000	.0001	.0005	.0056	2
	13	.0000	.0000	.0000	.0000	.0000	.0000	.0000	.0000	.0000	.0000	.0000	.0000	.0000	.0001	.0009	1
	14	.0000	.0000	.0000	.0000	.0000	.0000	.0000	.0000	.0000	.0000	.0000	.0000	.0000	.0000	.0001	0
15	0	.9851	.9276	.8601	.7386	.6333	.5421	.4633	.2059	.0874	.0352	.0134	.0047	.0023	.0005	.0000	15
	1	.0148	.0699	.1303	.2261	.2938	.3388	.3658	.3432	.2312	.1319	.0668	.0305	.0171	.0047	.0005	14
	2	.0001	.0025	.0092	.0323	.0636	.0988	.1348	.2669	.2856	.2309	.1559	.0916	.0599	.0219	.0032	13
	3	.0000	.0001	.0004	.0029	.0085	.0178	.0307	.1285	.2184	.2501	.2252	.1700	.1299	.0634	.0139	12
	4	.0000	.0000	.0000	.0002	.0008	.0022	.0049	.0428	.1156	.1876	.2252	.2186	.1948	.1268	.0417	11
	5	.0000	.0000	.0000	.0000	.0001	.0002	.0006	.0105	.0449	.1032	.1651	.2061	.2143	.1859	.0916	10
	6	.0000	.0000	.0000	.0000	.0000	.0000	.0000	.0019	.0132	.0430	.0917	.1472	.1786	.2066	.1527	9
	7	.0000	.0000	.0000	.0000	.0000	.0000	.0000	.0003	.0030	.0138	.0393	.0811	.1148	.1771	.1964	8
	8	.0000	.0000	.0000	.0000	.0000	.0000	.0000	.0000	.0005	.0035	.0131	.0348	.0574	.1181	.1964	7
	9	.0000	.0000	.0000	.0000	.0000	.0000	.0000	.0000	.0001	.0007	.0034	.0116	.0223	.0612	.1527	6
	10	.0000	.0000	.0000	.0000	.0000	.0000	.0000	.0000	.0000	.0001	.0007	.0030	.0067	.0245	.0916	5
	11	.0000	.0000	.0000	.0000	.0000	.0000	.0000	.0000	.0000	.0000	.0001	.0006	.0015	.0074	.0417	4
	12	.0000	.0000	.0000	.0000	.0000	.0000	.0000	.0000	.0000	.0000	.0000	.0001	.0003	.0016	.0139	3
	13	.0000	.0000	.0000	.0000	.0000	.0000	.0000	.0000	.0000	.0000	.0000	.0000	.0000	.0003	.0032	2
	14	.0000	.0000	.0000	.0000	.0000	.0000	.0000	.0000	.0000	.0000	.0000	.0000	.0000	.0000	.0005	1
16	0	.9841	.9229	.8515	.7238	.6143	.5204	.4401	.1853	.0743	.0281	.0100	.0033	.0015	.0003	.0000	16
	1	.0158	.0742	.1376	.2363	.3040	.3469	.3706	.3294	.2096	.1126	.0535	.0228	.0122	.0030	.0002	15
	2	.0001	.0028	.0104	.0362	.0705	.1084	.1463	.2745	.2775	.2111	.1336	.0732	.0457	.0150	.0018	14
	3	.0000	.0001	.0005	.0034	.0102	.0211	.0359	.1423	.2285	.2463	.2079	.1465	.1066	.0468	.0085	13
	4	.0000	.0000	.0000	.0002	.0010	.0029	.0061	.0514	.1311	.2001	.2252	.2040	.1732	.1014	.0278	12
		P=.999	.995	.99	.98	.97	.96	.95	.90	.85	.80	.75	.70	2/3	.60	.50	x

Table D (continued)

Binomial Distribution — Individual Terms

$$f_b(x|n,p) = \binom{n}{x} p^x(1-p)^{n-x}$$

N	X	P=.001	.005	.01	.02	.03	.04	.05	.10	.15	.20	.25	.30	1/3	.40	.50	
16	5	.0000	.0000	.0000	.0000	.0001	.0003	.0008	.0137	.0555	.1201	.1802	.2099	.2078	.1623	.0667	11
	6	.0000	.0000	.0000	.0000	.0000	.0000	.0001	.0028	.0180	.0550	.1101	.1649	.1905	.1983	.1222	10
	7	.0000	.0000	.0000	.0000	.0000	.0000	.0000	.0004	.0045	.0197	.0524	.1010	.1361	.1889	.1746	9
	8	.0000	.0000	.0000	.0000	.0000	.0000	.0000	.0001	.0009	.0055	.0197	.0487	.0765	.1417	.1964	8
	9	.0000	.0000	.0000	.0000	.0000	.0000	.0000	.0000	.0001	.0012	.0058	.0185	.0340	.0840	.1746	7
	10	.0000	.0000	.0000	.0000	.0000	.0000	.0000	.0000	.0000	.0002	.0014	.0056	.0119	.0392	.1222	6
	11	.0000	.0000	.0000	.0000	.0000	.0000	.0000	.0000	.0000	.0000	.0002	.0013	.0032	.0142	.0667	5
	12	.0000	.0000	.0000	.0000	.0000	.0000	.0000	.0000	.0000	.0000	.0000	.0002	.0007	.0040	.0278	4
	13	.0000	.0000	.0000	.0000	.0000	.0000	.0000	.0000	.0000	.0000	.0000	.0000	.0001	.0008	.0085	3
	14	.0000	.0000	.0000	.0000	.0000	.0000	.0000	.0000	.0000	.0000	.0000	.0000	.0000	.0001	.0018	2
	15	.0000	.0000	.0000	.0000	.0000	.0000	.0000	.0000	.0000	.0000	.0000	.0000	.0000	.0000	.0002	1
17	0	.9831	.9183	.8429	.7093	.5958	.4996	.4181	.1668	.0631	.0225	.0075	.0023	.0010	.0002	.0000	17
	1	.0167	.0784	.1447	.2461	.3133	.3539	.3741	.3150	.1893	.0957	.0426	.0169	.0086	.0019	.0001	16
	2	.0001	.0032	.0117	.0402	.0775	.1180	.1575	.2800	.2673	.1914	.1136	.0581	.0345	.0102	.0010	15
	3	.0000	.0001	.0006	.0041	.0120	.0246	.0415	.1556	.2359	.2393	.1893	.1245	.0863	.0341	.0052	14
	4	.0000	.0000	.0000	.0003	.0013	.0036	.0076	.0605	.1457	.2093	.2209	.1868	.1510	.0796	.0182	13
	5	.0000	.0000	.0000	.0000	.0001	.0004	.0010	.0175	.0668	.1361	.1914	.2081	.1963	.1379	.0472	12
	6	.0000	.0000	.0000	.0000	.0000	.0000	.0001	.0039	.0236	.0680	.1276	.1784	.1963	.1839	.0944	11
	7	.0000	.0000	.0000	.0000	.0000	.0000	.0000	.0007	.0065	.0267	.0668	.1201	.1542	.1927	.1484	10
	8	.0000	.0000	.0000	.0000	.0000	.0000	.0000	.0001	.0014	.0084	.0279	.0644	.0964	.1606	.1855	9
	9	.0000	.0000	.0000	.0000	.0000	.0000	.0000	.0000	.0003	.0021	.0093	.0276	.0482	.1070	.1855	8
	10	.0000	.0000	.0000	.0000	.0000	.0000	.0000	.0000	.0000	.0004	.0025	.0095	.0193	.0571	.1484	7
	11	.0000	.0000	.0000	.0000	.0000	.0000	.0000	.0000	.0000	.0001	.0005	.0026	.0061	.0242	.0944	6
	12	.0000	.0000	.0000	.0000	.0000	.0000	.0000	.0000	.0000	.0000	.0001	.0006	.0015	.0081	.0472	5
	13	.0000	.0000	.0000	.0000	.0000	.0000	.0000	.0000	.0000	.0000	.0000	.0001	.0003	.0021	.0182	4
	14	.0000	.0000	.0000	.0000	.0000	.0000	.0000	.0000	.0000	.0000	.0000	.0000	.0000	.0004	.0052	3
	15	.0000	.0000	.0000	.0000	.0000	.0000	.0000	.0000	.0000	.0000	.0000	.0000	.0000	.0001	.0010	2
	16	.0000	.0000	.0000	.0000	.0000	.0000	.0000	.0000	.0000	.0000	.0000	.0000	.0000	.0000	.0001	1
18	0	.9822	.9137	.8345	.6951	.5780	.4796	.3972	.1501	.0536	.0180	.0056	.0016	.0007	.0001	.0000	18
	1	.0177	.0826	.1517	.2554	.3217	.3597	.3763	.3002	.1704	.0811	.0338	.0126	.0061	.0012	.0001	17
	2	.0002	.0035	.0130	.0443	.0846	.1274	.1683	.2835	.2556	.1723	.0958	.0458	.0259	.0069	.0006	16
	3	.0000	.0001	.0007	.0048	.0140	.0283	.0473	.1680	.2406	.2297	.1704	.1046	.0690	.0246	.0031	15
	4	.0000	.0000	.0000	.0004	.0016	.0044	.0093	.0700	.1592	.2153	.2130	.1681	.1294	.0614	.0117	14
	5	.0000	.0000	.0000	.0000	.0001	.0005	.0014	.0218	.0787	.1507	.1988	.2017	.1812	.1146	.0327	13
	6	.0000	.0000	.0000	.0000	.0000	.0000	.0002	.0052	.0301	.0816	.1436	.1873	.1963	.1655	.0708	12
	7	.0000	.0000	.0000	.0000	.0000	.0000	.0000	.0010	.0091	.0350	.0820	.1376	.1682	.1892	.1214	11
	8	.0000	.0000	.0000	.0000	.0000	.0000	.0000	.0002	.0022	.0120	.0376	.0811	.1157	.1734	.1669	10
	9	.0000	.0000	.0000	.0000	.0000	.0000	.0000	.0000	.0004	.0033	.0139	.0386	.0643	.1284	.1855	9
	10	.0000	.0000	.0000	.0000	.0000	.0000	.0000	.0000	.0001	.0008	.0042	.0149	.0289	.0771	.1669	8
	11	.0000	.0000	.0000	.0000	.0000	.0000	.0000	.0000	.0000	.0001	.0010	.0046	.0105	.0374	.1214	7
	12	.0000	.0000	.0000	.0000	.0000	.0000	.0000	.0000	.0000	.0000	.0002	.0012	.0031	.0145	.0708	6
	13	.0000	.0000	.0000	.0000	.0000	.0000	.0000	.0000	.0000	.0000	.0000	.0002	.0007	.0045	.0327	5
	14	.0000	.0000	.0000	.0000	.0000	.0000	.0000	.0000	.0000	.0000	.0000	.0000	.0001	.0011	.0117	4
	15	.0000	.0000	.0000	.0000	.0000	.0000	.0000	.0000	.0000	.0000	.0000	.0000	.0000	.0002	.0031	3
	16	.0000	.0000	.0000	.0000	.0000	.0000	.0000	.0000	.0000	.0000	.0000	.0000	.0000	.0000	.0006	2
	17	.0000	.0000	.0000	.0000	.0000	.0000	.0000	.0000	.0000	.0000	.0000	.0000	.0000	.0000	.0001	1
		P=.999	.995	.99	.98	.97	.96	.95	.90	.85	.80	.75	.70	2/3	.60	.50	x

Table D (csntinued)

Binomial Distribution — Individual Terms

$$f_b(x|n,p) = \binom{n}{x} p^x (1-p)^{n-x}$$

N	X	P=.001	.005	.01	.02	.03	.04	.05	.10	.15	.20	.25	.30	1/3	.40	.50	
19	0	.9812	.9092	.8262	.6812	.5606	.4604	.3774	.1351	.0456	.0144	.0042	.0011	.0005	.0001	.0000	19
	1	.0187	.0868	.1586	.2642	.3294	.3645	.3774	.2852	.1529	.0685	.0268	.0093	.0043	.0008	.0000	18
	2	.0002	.0039	.0144	.0485	.0917	.1367	.1787	.2852	.2428	.1540	.0803	.0358	.0193	.0046	.0003	17
	3	.0000	.0001	.0008	.0056	.0161	.0323	.0533	.1796	.2428	.2182	.1517	.0869	.0546	.0175	.0018	16
	4	.0000	.0000	.0000	.0005	.0020	.0054	.0112	.0798	.1714	.2182	.2023	.1491	.1093	.0467	.0074	15
	5	.0000	.0000	.0000	.0000	.0002	.0007	.0018	.0266	.0907	.1636	.2023	.1916	.1639	.0933	.0222	14
	6	.0000	.0000	.0000	.0000	.0000	.0001	.0002	.0069	.0374	.0955	.1574	.1916	.1912	.1451	.0518	13
	7	.0000	.0000	.0000	.0000	.0000	.0000	.0000	.0014	.0122	.0443	.0974	.1525	.1776	.1797	.0961	12
	8	.0000	.0000	.0000	.0000	.0000	.0000	.0000	.0002	.0032	.0166	.0487	.0981	.1332	.1797	.1442	11
	9	.0000	.0000	.0000	.0000	.0000	.0000	.0000	.0000	.0007	.0051	.0198	.0514	.0814	.1464	.1762	10
	10	.0000	.0000	.0000	.0000	.0000	.0000	.0000	.0000	.0001	.0013	.0066	.0220	.0407	.0976	.1762	9
	11	.0000	.0000	.0000	.0000	.0000	.0000	.0000	.0000	.0000	.0003	.0018	.0077	.0166	.0532	.1442	8
	12	.0000	.0000	.0000	.0000	.0000	.0000	.0000	.0000	.0000	.0000	.0004	.0022	.0055	.0237	.0961	7
	13	.0000	.0000	.0000	.0000	.0000	.0000	.0000	.0000	.0000	.0000	.0001	.0005	.0015	.0085	.0518	6
	14	.0000	.0000	.0000	.0000	.0000	.0000	.0000	.0000	.0000	.0000	.0000	.0001	.0003	.0024	.0222	5
	15	.0000	.0000	.0000	.0000	.0000	.0000	.0000	.0000	.0000	.0000	.0000	.0000	.0001	.0005	.0074	4
	16	.0000	.0000	.0000	.0000	.0000	.0000	.0000	.0000	.0000	.0000	.0000	.0000	.0000	.0001	.0018	3
	17	.0000	.0000	.0000	.0000	.0000	.0000	.0000	.0000	.0000	.0000	.0000	.0000	.0000	.0000	.0003	2
20	0	.9802	.9046	.8179	.6676	.5438	.4420	.3585	.1216	.0388	.0115	.0032	.0008	.0003	.0000	.0000	20
	1	.0196	.0909	.1652	.2725	.3364	.3683	.3774	.2702	.1368	.0576	.0211	.0068	.0030	.0005	.0000	19
	2	.0002	.0043	.0159	.0528	.0988	.1458	.1887	.2852	.2293	.1369	.0669	.0278	.0143	.0031	.0002	18
	3	.0000	.0001	.0010	.0065	.0183	.0364	.0596	.1901	.2428	.2054	.1339	.0716	.0429	.0123	.0011	17
	4	.0000	.0000	.0000	.0006	.0024	.0065	.0133	.0898	.1821	.2182	.1897	.1304	.0911	.0350	.0046	16
	5	.0000	.0000	.0000	.0000	.0002	.0009	.0022	.0319	.1028	.1746	.2023	.1789	.1457	.0746	.0148	15
	6	.0000	.0000	.0000	.0000	.0000	.0001	.0003	.0089	.0454	.1091	.1686	.1916	.1821	.1244	.0370	14
	7	.0000	.0000	.0000	.0000	.0000	.0000	.0000	.0020	.0160	.0545	.1124	.1643	.1821	.1659	.0739	13
	8	.0000	.0000	.0000	.0000	.0000	.0000	.0000	.0004	.0046	.0222	.0609	.1144	.1480	.1797	.1201	12
	9	.0000	.0000	.0000	.0000	.0000	.0000	.0000	.0001	.0011	.0074	.0271	.0654	.0987	.1597	.1602	11
	10	.0000	.0000	.0000	.0000	.0000	.0000	.0000	.0000	.0002	.0020	.0099	.0308	.0543	.1171	.1762	10
	11	.0000	.0000	.0000	.0000	.0000	.0000	.0000	.0000	.0000	.0005	.0030	.0120	.0247	.0710	.1602	9
	12	.0000	.0000	.0000	.0000	.0000	.0000	.0000	.0000	.0000	.0001	.0008	.0039	.0092	.0355	.1201	8
	13	.0000	.0000	.0000	.0000	.0000	.0000	.0000	.0000	.0000	.0000	.0002	.0010	.0028	.0146	.0739	7
	14	.0000	.0000	.0000	.0000	.0000	.0000	.0000	.0000	.0000	.0000	.0000	.0002	.0007	.0049	.0370	6
	15	.0000	.0000	.0000	.0000	.0000	.0000	.0000	.0000	.0000	.0000	.0000	.0000	.0001	.0013	.0148	5
	16	.0000	.0000	.0000	.0000	.0000	.0000	.0000	.0000	.0000	.0000	.0000	.0000	.0000	.0003	.0046	4
	17	.0000	.0000	.0000	.0000	.0000	.0000	.0000	.0000	.0000	.0000	.0000	.0000	.0000	.0000	.0011	3
	18	.0000	.0000	.0000	.0000	.0000	.0000	.0000	.0000	.0000	.0000	.0000	.0000	.0000	.0000	.0002	2
21	0	.9792	.9001	.8097	.6543	.5275	.4243	.3406	.1094	.0329	.0092	.0024	.0006	.0002	.0000	.0000	21
	1	.0206	.0950	.1718	.2804	.3426	.3713	.3764	.2553	.1221	.0484	.0166	.0050	.0021	.0003	.0000	20
	2	.0002	.0048	.0173	.0572	.1060	.1547	.1981	.2837	.2155	.1211	.0555	.0215	.0105	.0020	.0001	19
	3	.0000	.0002	.0011	.0074	.0208	.0408	.0660	.1996	.2408	.1917	.1172	.0585	.0333	.0086	.0006	18
	4	.0000	.0000	.0001	.0007	.0029	.0077	.0156	.0998	.1912	.2156	.1757	.1128	.0750	.0259	.0029	17
	5	.0000	.0000	.0000	.0000	.0003	.0011	.0028	.0377	.1147	.1833	.1992	.1643	.1275	.0588	.0097	16
	6	.0000	.0000	.0000	.0000	.0000	.0001	.0004	.0112	.0540	.1222	.1770	.1878	.1700	.1045	.0259	15
	7	.0000	.0000	.0000	.0000	.0000	.0000	.0000	.0027	.0204	.0655	.1265	.1725	.1821	.1493	.0554	14
	8	.0000	.0000	.0000	.0000	.0000	.0000	.0000	.0005	.0063	.0286	.0738	.1294	.1594	.1742	.0970	13
	9	.0000	.0000	.0000	.0000	.0000	.0000	.0000	.0001	.0016	.0103	.0355	.0801	.1151	.1677	.1402	12
		P=.999	.995	.99	.98	.97	.96	.95	.90	.85	.80	.75	.70	2/3	.60	.50	X

Table D (continued)

Binomial Distribution — Individual Terms

$$f_b(x|n,p) = \binom{n}{x} p^x(1-p)^{n-x}$$

N	X	P=.001	.005	.01	.02	.03	.04	.05	.10	.15	.20	.25	.30	1/3	.40	.50	
21	10	.0000	.0000	.0000	.0000	.0000	.0000	.0000	.0000	.0003	.0031	.0142	.0412	.0691	.1342	.1682	11
	11	.0000	.0000	.0000	.0000	.0000	.0000	.0000	.0000	.0001	.0008	.0047	.0176	.0345	.0895	.1682	10
	12	.0000	.0000	.0000	.0000	.0000	.0000	.0000	.0000	.0000	.0002	.0013	.0063	.0144	.0497	.1402	9
	13	.0000	.0000	.0000	.0000	.0000	.0000	.0000	.0000	.0000	.0000	.0003	.0019	.0050	.0229	.0970	8
	14	.0000	.0000	.0000	.0000	.0000	.0000	.0000	.0000	.0000	.0000	.0001	.0005	.0014	.0087	.0554	7
	15	.0000	.0000	.0000	.0000	.0000	.0000	.0000	.0000	.0000	.0000	.0000	.0001	.0003	.0027	.0259	6
	16	.0000	.0000	.0000	.0000	.0000	.0000	.0000	.0000	.0000	.0000	.0000	.0000	.0001	.0007	.0097	5
	17	.0000	.0000	.0000	.0000	.0000	.0000	.0000	.0000	.0000	.0000	.0000	.0000	.0000	.0001	.0029	4
	18	.0000	.0000	.0000	.0000	.0000	.0000	.0000	.0000	.0000	.0000	.0000	.0000	.0000	.0000	.0006	3
	19	.0000	.0000	.0000	.0000	.0000	.0000	.0000	.0000	.0000	.0000	.0000	.0000	.0000	.0000	.0001	2
22	0	.9782	.8956	.8016	.6412	.5117	.4073	.3235	.0985	.0280	.0074	.0018	.0004	.0001	.0000	.0000	22
	1	.0215	.0990	.1781	.2879	.3481	.3734	.3746	.2407	.1087	.0406	.0131	.0037	.0015	.0002	.0000	21
	2	.0002	.0052	.0189	.0617	.1131	.1634	.2070	.2808	.2015	.1065	.0458	.0166	.0077	.0014	.0001	20
	3	.0000	.0002	.0013	.0084	.0233	.0454	.0726	.2080	.2370	.1775	.1017	.0474	.0257	.0060	.0004	19
	4	.0000	.0000	.0001	.0008	.0034	.0090	.0182	.1098	.1987	.2108	.1611	.0965	.0611	.0190	.0017	18
	5	.0000	.0000	.0000	.0001	.0004	.0013	.0034	.0439	.1262	.1898	.1933	.1489	.1100	.0456	.0063	17
	6	.0000	.0000	.0000	.0000	.0000	.0002	.0005	.0138	.0631	.1344	.1826	.1808	.1558	.0862	.0178	16
	7	.0000	.0000	.0000	.0000	.0000	.0000	.0001	.0035	.0255	.0768	.1391	.1771	.1781	.1314	.0407	15
	8	.0000	.0000	.0000	.0000	.0000	.0000	.0000	.0007	.0084	.0360	.0869	.1423	.1670	.1642	.0762	14
	9	.0000	.0000	.0000	.0000	.0000	.0000	.0000	.0001	.0023	.0140	.0451	.0949	.1299	.1703	.1186	13
	10	.0000	.0000	.0000	.0000	.0000	.0000	.0000	.0000	.0005	.0046	.0195	.0529	.0844	.1476	.1542	12
	11	.0000	.0000	.0000	.0000	.0000	.0000	.0000	.0000	.0001	.0012	.0071	.0247	.0460	.1073	.1682	11
	12	.0000	.0000	.0000	.0000	.0000	.0000	.0000	.0000	.0000	.0003	.0022	.0097	.0211	.0656	.1542	10
	13	.0000	.0000	.0000	.0000	.0000	.0000	.0000	.0000	.0000	.0001	.0006	.0032	.0081	.0336	.1186	9
	14	.0000	.0000	.0000	.0000	.0000	.0000	.0000	.0000	.0000	.0000	.0001	.0009	.0026	.0144	.0762	8
	15	.0000	.0000	.0000	.0000	.0000	.0000	.0000	.0000	.0000	.0000	.0000	.0002	.0007	.0051	.0407	7
	16	.0000	.0000	.0000	.0000	.0000	.0000	.0000	.0000	.0000	.0000	.0000	.0000	.0002	.0015	.0178	6
	17	.0000	.0000	.0000	.0000	.0000	.0000	.0000	.0000	.0000	.0000	.0000	.0000	.0000	.0004	.0063	5
	18	.0000	.0000	.0000	.0000	.0000	.0000	.0000	.0000	.0000	.0000	.0000	.0000	.0000	.0001	.0017	4
	19	.0000	.0000	.0000	.0000	.0000	.0000	.0000	.0000	.0000	.0000	.0000	.0000	.0000	.0000	.0004	3
	20	.0000	.0000	.0000	.0000	.0000	.0000	.0000	.0000	.0000	.0000	.0000	.0000	.0000	.0000	.0001	2
23	0	.9773	.8911	.7936	.6283	.4963	.3911	.3074	.0886	.0238	.0059	.0013	.0003	.0001	.0000	.0000	23
	1	.0225	.1030	.1844	.2949	.3530	.3748	.3721	.2265	.0966	.0339	.0103	.0027	.0010	.0001	.0000	22
	2	.0002	.0057	.0205	.0662	.1201	.1718	.2154	.2768	.1875	.0933	.0376	.0127	.0056	.0009	.0000	21
	3	.0000	.0002	.0014	.0095	.0260	.0501	.0794	.2153	.2317	.1633	.0878	.0382	.0197	.0041	.0002	20
	4	.0000	.0000	.0001	.0010	.0040	.0104	.0209	.1196	.2044	.2042	.1463	.0818	.0493	.0138	.0011	19
	5	.0000	.0000	.0000	.0001	.0005	.0017	.0042	.0505	.1371	.1940	.1853	.1332	.0937	.0350	.0040	18
	6	.0000	.0000	.0000	.0000	.0000	.0002	.0007	.0168	.0726	.1455	.1853	.1712	.1405	.0700	.0120	17
	7	.0000	.0000	.0000	.0000	.0000	.0000	.0001	.0045	.0311	.0883	.1500	.1782	.1707	.1133	.0292	16
	8	.0000	.0000	.0000	.0000	.0000	.0000	.0000	.0010	.0110	.0442	.1000	.1527	.1707	.1511	.0584	15
	9	.0000	.0000	.0000	.0000	.0000	.0000	.0000	.0002	.0032	.0184	.0555	.1091	.1422	.1679	.0974	14
	10	.0000	.0000	.0000	.0000	.0000	.0000	.0000	.0000	.0008	.0064	.0259	.0655	.0996	.1567	.1364	13
	11	.0000	.0000	.0000	.0000	.0000	.0000	.0000	.0000	.0002	.0019	.0102	.0332	.0588	.1234	.1612	12
	12	.0000	.0000	.0000	.0000	.0000	.0000	.0000	.0000	.0000	.0005	.0034	.0142	.0294	.0823	.1612	11
	13	.0000	.0000	.0000	.0000	.0000	.0000	.0000	.0000	.0000	.0001	.0010	.0052	.0124	.0464	.1364	10
	14	.0000	.0000	.0000	.0000	.0000	.0000	.0000	.0000	.0000	.0000	.0002	.0016	.0044	.0221	.0974	9
		P=.999	.995	.99	.98	.97	.96	.95	.90	.85	.80	.75	.70	2/3	.60	.50	X

Table D (continued)

Binomial Distribution — Individual Terms

$$f_b(x|n,p) = \binom{n}{x} p^x (1-p)^{n-x}$$

N	X	P=.001	.005	.01	.02	.03	.04	.05	.10	.15	.20	.25	.30	1/3	.40	.50	
23	15	.0000	.0000	.0000	.0000	.0000	.0000	.0000	.0000	.0000	.0000	.0000	.0004	.0013	.0088	.0584	8
	16	.0000	.0000	.0000	.0000	.0000	.0000	.0000	.0000	.0000	.0000	.0000	.0001	.0003	.0029	.0292	7
	17	.0000	.0000	.0000	.0000	.0000	.0000	.0000	.0000	.0000	.0000	.0000	.0000	.0001	.0008	.0120	6
	18	.0000	.0000	.0000	.0000	.0000	.0000	.0000	.0000	.0000	.0000	.0000	.0000	.0000	.0002	.0040	5
	19	.0000	.0000	.0000	.0000	.0000	.0000	.0000	.0000	.0000	.0000	.0000	.0000	.0000	.0000	.0011	4
	20	.0000	.0000	.0000	.0000	.0000	.0000	.0000	.0000	.0000	.0000	.0000	.0000	.0000	.0000	.0002	3
24	0	.9763	.8867	.7857	.6158	.4814	.3754	.2920	.0798	.0202	.0047	.0010	.0002	.0001	.0000	.0000	24
	1	.0235	.1069	.1905	.3016	.3573	.3754	.3688	.2127	.0857	.0283	.0080	.0020	.0007	.0001	.0000	23
	2	.0003	.0062	.0221	.0708	.1271	.1799	.2232	.2718	.1739	.0815	.0308	.0097	.0041	.0006	.0000	22
	3	.0000	.0002	.0016	.0106	.0288	.0550	.0862	.2215	.2251	.1493	.0752	.0305	.0150	.0028	.0001	21
	4	.0000	.0000	.0001	.0011	.0047	.0120	.0238	.1292	.2085	.1960	.1316	.0687	.0395	.0099	.0006	20
	5	.0000	.0000	.0000	.0001	.0006	.0020	.0050	.0574	.1472	.1960	.1755	.1177	.0789	.0265	.0025	19
	6	.0000	.0000	.0000	.0000	.0001	.0003	.0008	.0202	.0822	.1552	.1853	.1598	.1249	.0560	.0080	18
	7	.0000	.0000	.0000	.0000	.0000	.0000	.0001	.0058	.0373	.0998	.1588	.1761	.1606	.0960	.0206	17
	8	.0000	.0000	.0000	.0000	.0000	.0000	.0000	.0014	.0140	.0530	.1125	.1604	.1707	.1360	.0438	16
	9	.0000	.0000	.0000	.0000	.0000	.0000	.0000	.0003	.0044	.0236	.0667	.1222	.1517	.1612	.0779	15
	10	.0000	.0000	.0000	.0000	.0000	.0000	.0000	.0000	.0012	.0088	.0333	.0785	.1138	.1612	.1169	14
	11	.0000	.0000	.0000	.0000	.0000	.0000	.0000	.0000	.0003	.0028	.0141	.0428	.0724	.1367	.1488	13
	12	.0000	.0000	.0000	.0000	.0000	.0000	.0000	.0000	.0000	.0008	.0051	.0199	.0392	.0988	.1612	12
	13	.0000	.0000	.0000	.0000	.0000	.0000	.0000	.0000	.0000	.0002	.0016	.0079	.0181	.0608	.1488	11
	14	.0000	.0000	.0000	.0000	.0000	.0000	.0000	.0000	.0000	.0000	.0004	.0026	.0071	.0318	.1169	10
	15	.0000	.0000	.0000	.0000	.0000	.0000	.0000	.0000	.0000	.0000	.0001	.0008	.0024	.0141	.0779	9
	16	.0000	.0000	.0000	.0000	.0000	.0000	.0000	.0000	.0000	.0000	.0000	.0002	.0007	.0053	.0438	8
	17	.0000	.0000	.0000	.0000	.0000	.0000	.0000	.0000	.0000	.0000	.0000	.0000	.0002	.0017	.0206	7
	18	.0000	.0000	.0000	.0000	.0000	.0000	.0000	.0000	.0000	.0000	.0000	.0000	.0000	.0004	.0080	6
	19	.0000	.0000	.0000	.0000	.0000	.0000	.0000	.0000	.0000	.0000	.0000	.0000	.0000	.0001	.0025	5
	20	.0000	.0000	.0000	.0000	.0000	.0000	.0000	.0000	.0000	.0000	.0000	.0000	.0000	.0000	.0006	4
	21	.0000	.0000	.0000	.0000	.0000	.0000	.0000	.0000	.0000	.0000	.0000	.0000	.0000	.0000	.0001	3
25	0	.9753	.8822	.7778	.6035	.4670	.3604	.2774	.0718	.0172	.0038	.0008	.0001	.0000	.0000	.0000	25
	1	.0244	.1108	.1964	.3079	.3611	.3754	.3650	.1994	.0759	.0236	.0063	.0014	.0005	.0000	.0000	24
	2	.0003	.0067	.0238	.0754	.1340	.1877	.2305	.2659	.1607	.0708	.0251	.0074	.0030	.0004	.0000	23
	3	.0000	.0003	.0018	.0118	.0318	.0600	.0930	.2265	.2174	.1358	.0641	.0243	.0114	.0019	.0001	22
	4	.0000	.0000	.0001	.0013	.0054	.0137	.0269	.1384	.2110	.1867	.1175	.0572	.0313	.0071	.0004	21
	5	.0000	.0000	.0000	.0001	.0007	.0024	.0060	.0646	.1564	.1960	.1645	.1030	.0658	.0199	.0016	20
	6	.0000	.0000	.0000	.0000	.0001	.0003	.0010	.0239	.0920	.1633	.1828	.1472	.1096	.0442	.0053	19
	7	.0000	.0000	.0000	.0000	.0000	.0000	.0001	.0072	.0441	.1108	.1654	.1712	.1487	.0800	.0143	18
	8	.0000	.0000	.0000	.0000	.0000	.0000	.0000	.0018	.0175	.0623	.1241	.1651	.1673	.1200	.0322	17
	9	.0000	.0000	.0000	.0000	.0000	.0000	.0000	.0004	.0058	.0294	.0781	.1336	.1580	.1511	.0609	16
	10	.0000	.0000	.0000	.0000	.0000	.0000	.0000	.0001	.0016	.0118	.0417	.0916	.1264	.1612	.0974	15
	11	.0000	.0000	.0000	.0000	.0000	.0000	.0000	.0000	.0004	.0040	.0189	.0536	.0862	.1465	.1328	14
	12	.0000	.0000	.0000	.0000	.0000	.0000	.0000	.0000	.0001	.0012	.0074	.0268	.0503	.1139	.1550	13
	13	.0000	.0000	.0000	.0000	.0000	.0000	.0000	.0000	.0000	.0003	.0025	.0115	.0251	.0760	.1550	12
	14	.0000	.0000	.0000	.0000	.0000	.0000	.0000	.0000	.0000	.0001	.0007	.0042	.0108	.0434	.1328	11
	15	.0000	.0000	.0000	.0000	.0000	.0000	.0000	.0000	.0000	.0000	.0002	.0013	.0040	.0212	.0974	10
	16	.0000	.0000	.0000	.0000	.0000	.0000	.0000	.0000	.0000	.0000	.0000	.0004	.0012	.0088	.0609	9
	17	.0000	.0000	.0000	.0000	.0000	.0000	.0000	.0000	.0000	.0000	.0000	.0001	.0003	.0031	.0322	8
		P=.999	.995	.99	.98	.97	.96	.95	.90	.85	.80	.75	.70	2/3	.60	.50	X

Table D (continued)

Binomial Distribution — Individual Terms

$$f_b(x|n,p) = \binom{n}{x} p^x (1-p)^{n-x}$$

N	X	P=.001	.005	.01	.02	.03	.04	.05	.10	.15	.20	.25	.30	1/3	.40	.50	
25	18	.0000	.0000	.0000	.0000	.0000	.0000	.0000	.0000	.0000	.0000	.0000	.0000	.0001	.0009	.0143	7
	19	.0000	.0000	.0000	.0000	.0000	.0000	.0000	.0000	.0000	.0000	.0000	.0000	.0000	.0002	.0053	6
	20	.0000	.0000	.0000	.0000	.0000	.0000	.0000	.0000	.0000	.0000	.0000	.0000	.0000	.0000	.0016	5
	21	.0000	.0000	.0000	.0000	.0000	.0000	.0000	.0000	.0000	.0000	.0000	.0000	.0000	.0000	.0004	4
	22	.0000	.0000	.0000	.0000	.0000	.0000	.0000	.0000	.0000	.0000	.0000	.0000	.0000	.0000	.0001	3
50	0	.9512	.7783	.6050	.3642	.2181	.1299	.0769	.0052	.0003	.0000	.0000	.0000	.0000	.0000	.0000	50
	1	.0476	.1956	.3056	.3716	.3372	.2706	.2025	.0286	.0026	.0002	.0000	.0000	.0000	.0000	.0000	49
	2	.0012	.0241	.0756	.1858	.2555	.2762	.2611	.0779	.0113	.0011	.0001	.0000	.0000	.0000	.0000	48
	3	.0000	.0019	.0122	.0607	.1264	.1842	.2199	.1386	.0319	.0044	.0004	.0000	.0000	.0000	.0000	47
	4	.0000	.0001	.0015	.0145	.0459	.0902	.1360	.1809	.0661	.0128	.0016	.0001	.0000	.0000	.0000	46
	5	.0000	.0000	.0001	.0027	.0131	.0346	.0658	.1849	.1072	.0295	.0049	.0006	.0001	.0000	.0000	45
	6	.0000	.0000	.0000	.0004	.0030	.0108	.0260	.1541	.1419	.0554	.0123	.0018	.0004	.0000	.0000	44
	7	.0000	.0000	.0000	.0001	.0006	.0028	.0086	.1076	.1575	.0870	.0259	.0048	.0012	.0000	.0000	43
	8	.0000	.0000	.0000	.0000	.0001	.0006	.0024	.0643	.1493	.1169	.0463	.0110	.0033	.0002	.0000	42
	9	.0000	.0000	.0000	.0000	.0000	.0001	.0006	.0333	.1230	.1364	.0721	.0220	.0077	.0005	.0000	41
	10	.0000	.0000	.0000	.0000	.0000	.0000	.0001	.0152	.0890	.1398	.0985	.0386	.0157	.0014	.0000	40
	11	.0000	.0000	.0000	.0000	.0000	.0000	.0000	.0061	.0571	.1271	.1194	.0602	.0286	.0035	.0000	39
	12	.0000	.0000	.0000	.0000	.0000	.0000	.0000	.0022	.0328	.1033	.1294	.0838	.0465	.0076	.0001	38
	13	.0000	.0000	.0000	.0000	.0000	.0000	.0000	.0007	.0169	.0755	.1261	.1050	.0679	.0147	.0003	37
	14	.0000	.0000	.0000	.0000	.0000	.0000	.0000	.0002	.0079	.0499	.1110	.1189	.0898	.0260	.0008	36
	15	.0000	.0000	.0000	.0000	.0000	.0000	.0000	.0001	.0033	.0299	.0888	.1223	.1077	.0415	.0020	35
	16	.0000	.0000	.0000	.0000	.0000	.0000	.0000	.0000	.0013	.0164	.0648	.1147	.1178	.0606	.0044	34
	17	.0000	.0000	.0000	.0000	.0000	.0000	.0000	.0000	.0005	.0082	.0432	.0983	.1178	.0808	.0087	33
	18	.0000	.0000	.0000	.0000	.0000	.0000	.0000	.0000	.0001	.0037	.0264	.0772	.1080	.0987	.0160	32
	19	.0000	.0000	.0000	.0000	.0000	.0000	.0000	.0000	.0000	.0016	.0148	.0558	.0910	.1109	.0270	31
	20	.0000	.0000	.0000	.0000	.0000	.0000	.0000	.0000	.0000	.0006	.0077	.0370	.0705	.1146	.0419	30
	21	.0000	.0000	.0000	.0000	.0000	.0000	.0000	.0000	.0000	.0002	.0036	.0227	.0503	.1091	.0598	29
	22	.0000	.0000	.0000	.0000	.0000	.0000	.0000	.0000	.0000	.0001	.0016	.0128	.0332	.0959	.0788	28
	23	.0000	.0000	.0000	.0000	.0000	.0000	.0000	.0000	.0000	.0000	.0006	.0067	.0202	.0778	.0960	27
	24	.0000	.0000	.0000	.0000	.0000	.0000	.0000	.0000	.0000	.0000	.0002	.0032	.0114	.0584	.1080	26
	25	.0000	.0000	.0000	.0000	.0000	.0000	.0000	.0000	.0000	.0000	.0001	.0014	.0059	.0405	.1123	25
	26	.0000	.0000	.0000	.0000	.0000	.0000	.0000	.0000	.0000	.0000	.0000	.0006	.0028	.0259	.1080	24
	27	.0000	.0000	.0000	.0000	.0000	.0000	.0000	.0000	.0000	.0000	.0000	.0002	.0013	.0154	.0960	23
	28	.0000	.0000	.0000	.0000	.0000	.0000	.0000	.0000	.0000	.0000	.0000	.0001	.0005	.0084	.0788	22
	29	.0000	.0000	.0000	.0000	.0000	.0000	.0000	.0000	.0000	.0000	.0000	.0000	.0002	.0043	.0598	21
	30	.0000	.0000	.0000	.0000	.0000	.0000	.0000	.0000	.0000	.0000	.0000	.0000	.0001	.0020	.0419	20
	31	.0000	.0000	.0000	.0000	.0000	.0000	.0000	.0000	.0000	.0000	.0000	.0000	.0000	.0009	.0270	19
	32	.0000	.0000	.0000	.0000	.0000	.0000	.0000	.0000	.0000	.0000	.0000	.0000	.0000	.0003	.0160	18
	33	.0000	.0000	.0000	.0000	.0000	.0000	.0000	.0000	.0000	.0000	.0000	.0000	.0000	.0001	.0087	17
	34	.0000	.0000	.0000	.0000	.0000	.0000	.0000	.0000	.0000	.0000	.0000	.0000	.0000	.0000	.0044	16
	35	.0000	.0000	.0000	.0000	.0000	.0000	.0000	.0000	.0000	.0000	.0000	.0000	.0000	.0000	.0020	15
	36	.0000	.0000	.0000	.0000	.0000	.0000	.0000	.0000	.0000	.0000	.0000	.0000	.0000	.0000	.0008	14
	37	.0000	.0000	.0000	.0000	.0000	.0000	.0000	.0000	.0000	.0000	.0000	.0000	.0000	.0000	.0003	13
	38	.0000	.0000	.0000	.0000	.0000	.0000	.0000	.0000	.0000	.0000	.0000	.0000	.0000	.0000	.0001	12
75	0	.9277	.6866	.4706	.2198	.1018	.0468	.0213	.0004	.0000	.0000	.0000	.0000	.0000	.0000	.0000	75
	1	.0696	.2588	.3565	.3364	.2362	.1463	.0843	.0031	.0001	.0000	.0000	.0000	.0000	.0000	.0000	74
		P=.999	.995	.99	.98	.97	.96	.95	.90	.85	.80	.75	.70	2/3	.60	.50	X

Table D (continued)

Binomial Distribution — Individual Terms

$$f_b(x|n,p) = \binom{n}{x} p^x (1-p)^{n-x}$$

N	X	P=.001	.005	.01	.02	.03	.04	.05	.10	.15	.20	.25	.30	1/3	.40	.50	
75	2	.0026	.0481	.1332	.2540	.2703	.2255	.1641	.0127	.0004	.0000	.0000	.0000	.0000	.0000	.0000	73
	3	.0001	.0059	.0327	.1261	.2034	.2287	.2101	.0343	.0019	.0001	.0000	.0000	.0000	.0000	.0000	72
	4	.0000	.0005	.0060	.0463	.1132	.1715	.1991	.0685	.0060	.0003	.0000	.0000	.0000	.0000	.0000	71
	5	.0000	.0000	.0009	.0134	.0497	.1015	.1488	.1081	.0150	.0009	.0000	.0000	.0000	.0000	.0000	70
	6	.0000	.0000	.0001	.0032	.0179	.0493	.0914	.1402	.0309	.0027	.0001	.0000	.0000	.0000	.0000	69
	7	.0000	.0000	.0000	.0006	.0055	.0203	.0474	.1535	.0538	.0065	.0004	.0000	.0000	.0000	.0000	68
	8	.0000	.0000	.0000	.0001	.0014	.0072	.0212	.1450	.0807	.0139	.0011	.0000	.0000	.0000	.0000	67
	9	.0000	.0000	.0000	.0000	.0003	.0022	.0083	.1199	.1060	.0258	.0027	.0001	.0000	.0000	.0000	66
	10	.0000	.0000	.0000	.0000	.0001	.0006	.0029	.0880	.1235	.0426	.0060	.0004	.0001	.0000	.0000	65
	11	.0000	.0000	.0000	.0000	.0000	.0002	.0009	.0578	.1288	.0630	.0118	.0011	.0001	.0000	.0000	64
	12	.0000	.0000	.0000	.0000	.0000	.0000	.0003	.0342	.1212	.0840	.0209	.0024	.0004	.0000	.0000	63
	13	.0000	.0000	.0000	.0000	.0000	.0000	.0001	.0184	.1037	.1017	.0338	.0050	.0010	.0000	.0000	62
	14	.0000	.0000	.0000	.0000	.0000	.0000	.0000	.0091	.0810	.1126	.0500	.0095	.0021	.0000	.0000	61
	15	.0000	.0000	.0000	.0000	.0000	.0000	.0000	.0041	.0581	.1145	.0677	.0166	.0043	.0001	.0000	60
	16	.0000	.0000	.0000	.0000	.0000	.0000	.0000	.0017	.0385	.1073	.0846	.0267	.0081	.0003	.0000	59
	17	.0000	.0000	.0000	.0000	.0000	.0000	.0000	.0007	.0236	.0931	.0979	.0397	.0141	.0007	.0000	58
	18	.0000	.0000	.0000	.0000	.0000	.0000	.0000	.0002	.0134	.0750	.1052	.0549	.0227	.0015	.0000	57
	19	.0000	.0000	.0000	.0000	.0000	.0000	.0000	.0001	.0071	.0563	.1052	.0705	.0340	.0030	.0000	56
	20	.0000	.0000	.0000	.0000	.0000	.0000	.0000	.0000	.0035	.0394	.0982	.0846	.0476	.0056	.0000	55
	21	.0000	.0000	.0000	.0000	.0000	.0000	.0000	.0000	.0016	.0258	.0857	.0950	.0623	.0097	.0001	54
	22	.0000	.0000	.0000	.0000	.0000	.0000	.0000	.0000	.0007	.0158	.0701	.0999	.0765	.0159	.0001	53
	23	.0000	.0000	.0000	.0000	.0000	.0000	.0000	.0000	.0003	.0091	.0539	.0987	.0881	.0244	.0003	52
	24	.0000	.0000	.0000	.0000	.0000	.0000	.0000	.0000	.0001	.0049	.0389	.0917	.0954	.0352	.0007	51
	25	.0000	.0000	.0000	.0000	.0000	.0000	.0000	.0000	.0000	.0025	.0265	.0801	.0973	.0479	.0014	50
	26	.0000	.0000	.0000	.0000	.0000	.0000	.0000	.0000	.0000	.0012	.0170	.0660	.0936	.0614	.0027	49
	27	.0000	.0000	.0000	.0000	.0000	.0000	.0000	.0000	.0000	.0005	.0103	.0514	.0849	.0742	.0049	48
	28	.0000	.0000	.0000	.0000	.0000	.0000	.0000	.0000	.0000	.0002	.0059	.0377	.0728	.0848	.0083	47
	29	.0000	.0000	.0000	.0000	.0000	.0000	.0000	.0000	.0000	.0001	.0032	.0262	.0590	.0917	.0135	46
	30	.0000	.0000	.0000	.0000	.0000	.0000	.0000	.0000	.0000	.0000	.0016	.0172	.0452	.0937	.0207	45
	31	.0000	.0000	.0000	.0000	.0000	.0000	.0000	.0000	.0000	.0000	.0008	.0107	.0328	.0907	.0300	44
	32	.0000	.0000	.0000	.0000	.0000	.0000	.0000	.0000	.0000	.0000	.0004	.0063	.0226	.0831	.0413	43
	33	.0000	.0000	.0000	.0000	.0000	.0000	.0000	.0000	.0000	.0000	.0002	.0035	.0147	.0722	.0538	42
	34	.0000	.0000	.0000	.0000	.0000	.0000	.0000	.0000	.0000	.0000	.0001	.0019	.0091	.0595	.0665	41
	35	.0000	.0000	.0000	.0000	.0000	.0000	.0000	.0000	.0000	.0000	.0000	.0009	.0053	.0464	.0779	40
	36	.0000	.0000	.0000	.0000	.0000	.0000	.0000	.0000	.0000	.0000	.0000	.0004	.0030	.0344	.0865	39
	37	.0000	.0000	.0000	.0000	.0000	.0000	.0000	.0000	.0000	.0000	.0000	.0002	.0016	.0242	.0912	38
	38	.0000	.0000	.0000	.0000	.0000	.0000	.0000	.0000	.0000	.0000	.0000	.0001	.0008	.0161	.0912	37
	39	.0000	.0000	.0000	.0000	.0000	.0000	.0000	.0000	.0000	.0000	.0000	.0000	.0004	.0102	.0865	36
	40	.0000	.0000	.0000	.0000	.0000	.0000	.0000	.0000	.0000	.0000	.0000	.0000	.0002	.0061	.0779	35
	41	.0000	.0000	.0000	.0000	.0000	.0000	.0000	.0000	.0000	.0000	.0000	.0000	.0001	.0035	.0665	34
	42	.0000	.0000	.0000	.0000	.0000	.0000	.0000	.0000	.0000	.0000	.0000	.0000	.0000	.0019	.0538	33
	43	.0000	.0000	.0000	.0000	.0000	.0000	.0000	.0000	.0000	.0000	.0000	.0000	.0000	.0010	.0413	32
	44	.0000	.0000	.0000	.0000	.0000	.0000	.0000	.0000	.0000	.0000	.0000	.0000	.0000	.0005	.0300	31
	45	.0000	.0000	.0000	.0000	.0000	.0000	.0000	.0000	.0000	.0000	.0000	.0000	.0000	.0002	.0207	30
	46	.0000	.0000	.0000	.0000	.0000	.0000	.0000	.0000	.0000	.0000	.0000	.0000	.0000	.0001	.0135	29
	47	.0000	.0000	.0000	.0000	.0000	.0000	.0000	.0000	.0000	.0000	.0000	.0000	.0000	.0000	.0083	28
	48	.0000	.0000	.0000	.0000	.0000	.0000	.0000	.0000	.0000	.0000	.0000	.0000	.0000	.0000	.0049	27
		P=.999	.995	.99	.98	.97	.96	.95	.90	.85	.80	.75	.70	2/3	.60	.50	X

Table D (continued)

Binomial Distribution — Individual Terms

$$f_b(x|n,p) = \binom{n}{x} p^x(1-p)^{n-x}$$

N	X	P=.001	.005	.01	.02	.03	.04	.05	.10	.15	.20	.25	.30	1/3	.40	.50	
75	49	.0000	.0000	.0000	.0000	.0000	.0000	.0000	.0000	.0000	.0000	.0000	.0000	.0000	.0000	.0027	26
	50	.0000	.0000	.0000	.0000	.0000	.0000	.0000	.0000	.0000	.0000	.0000	.0000	.0000	.0000	.0014	25
	51	.0000	.0000	.0000	.0000	.0000	.0000	.0000	.0000	.0000	.0000	.0000	.0000	.0000	.0000	.0007	24
	52	.0000	.0000	.0000	.0000	.0000	.0000	.0000	.0000	.0000	.0000	.0000	.0000	.0000	.0000	.0003	23
	53	.0000	.0000	.0000	.0000	.0000	.0000	.0000	.0000	.0000	.0000	.0000	.0000	.0000	.0000	.0001	22
	54	.0000	.0000	.0000	.0000	.0000	.0000	.0000	.0000	.0000	.0000	.0000	.0000	.0000	.0000	.0001	21
100	0	.9048	.6058	.3660	.1326	.0476	.0169	.0059	.0000	.0000	.0000	.0000	.0000	.0000	.0000	.0000	100
	1	.0906	.3044	.3697	.2707	.1471	.0703	.0312	.0003	.0000	.0000	.0000	.0000	.0000	.0000	.0000	99
	2	.0045	.0757	.1849	.2734	.2252	.1450	.0812	.0016	.0000	.0000	.0000	.0000	.0000	.0000	.0000	98
	3	.0001	.0124	.0610	.1823	.2275	.1973	.1396	.0059	.0001	.0000	.0000	.0000	.0000	.0000	.0000	97
	4	.0000	.0015	.0149	.0902	.1706	.1994	.1781	.0159	.0003	.0000	.0000	.0000	.0000	.0000	.0000	96
	5	.0000	.0001	.0029	.0353	.1013	.1595	.1800	.0339	.0011	.0000	.0000	.0000	.0000	.0000	.0000	95
	6	.0000	.0000	.0005	.0114	.0496	.1052	.1500	.0596	.0031	.0001	.0000	.0000	.0000	.0000	.0000	94
	7	.0000	.0000	.0001	.0031	.0206	.0589	.1060	.0889	.0075	.0002	.0000	.0000	.0000	.0000	.0000	93
	8	.0000	.0000	.0000	.0007	.0074	.0285	.0649	.1148	.0153	.0006	.0000	.0000	.0000	.0000	.0000	92
	9	.0000	.0000	.0000	.0002	.0023	.0121	.0349	.1304	.0276	.0015	.0000	.0000	.0000	.0000	.0000	91
	10	.0000	.0000	.0000	.0000	.0007	.0046	.0167	.1319	.0444	.0034	.0001	.0000	.0000	.0000	.0000	90
	11	.0000	.0000	.0000	.0000	.0002	.0016	.0072	.1199	.0640	.0069	.0003	.0000	.0000	.0000	.0000	89
	12	.0000	.0000	.0000	.0000	.0000	.0005	.0028	.0988	.0838	.0128	.0006	.0000	.0000	.0000	.0000	88
	13	.0000	.0000	.0000	.0000	.0000	.0001	.0010	.0743	.1001	.0216	.0014	.0000	.0000	.0000	.0000	87
	14	.0000	.0000	.0000	.0000	.0000	.0000	.0003	.0513	.1098	.0335	.0030	.0001	.0000	.0000	.0000	86
	15	.0000	.0000	.0000	.0000	.0000	.0000	.0001	.0327	.1111	.0481	.0057	.0002	.0000	.0000	.0000	85
	16	.0000	.0000	.0000	.0000	.0000	.0000	.0000	.0193	.1041	.0638	.0100	.0006	.0001	.0000	.0000	84
	17	.0000	.0000	.0000	.0000	.0000	.0000	.0000	.0106	.0908	.0789	.0165	.0012	.0001	.0000	.0000	83
	18	.0000	.0000	.0000	.0000	.0000	.0000	.0000	.0054	.0739	.0909	.0254	.0024	.0003	.0000	.0000	82
	19	.0000	.0000	.0000	.0000	.0000	.0000	.0000	.0026	.0563	.0981	.0365	.0044	.0006	.0000	.0000	81
	20	.0000	.0000	.0000	.0000	.0000	.0000	.0000	.0012	.0402	.0993	.0493	.0076	.0013	.0000	.0000	80
	21	.0000	.0000	.0000	.0000	.0000	.0000	.0000	.0005	.0270	.0946	.0626	.0124	.0024	.0000	.0000	79
	22	.0000	.0000	.0000	.0000	.0000	.0000	.0000	.0002	.0171	.0849	.0749	.0190	.0043	.0001	.0000	78
	23	.0000	.0000	.0000	.0000	.0000	.0000	.0000	.0001	.0103	.0720	.0847	.0277	.0073	.0001	.0000	77
	24	.0000	.0000	.0000	.0000	.0000	.0000	.0000	.0000	.0058	.0577	.0906	.0380	.0117	.0003	.0000	76
	25	.0000	.0000	.0000	.0000	.0000	.0000	.0000	.0000	.0031	.0439	.0918	.0496	.0178	.0006	.0000	75
	26	.0000	.0000	.0000	.0000	.0000	.0000	.0000	.0000	.0016	.0316	.0883	.0613	.0256	.0012	.0000	74
	27	.0000	.0000	.0000	.0000	.0000	.0000	.0000	.0000	.0008	.0217	.0806	.0720	.0351	.0022	.0000	73
	28	.0000	.0000	.0000	.0000	.0000	.0000	.0000	.0000	.0004	.0141	.0701	.0804	.0458	.0038	.0000	72
	29	.0000	.0000	.0000	.0000	.0000	.0000	.0000	.0000	.0002	.0088	.0580	.0856	.0569	.0063	.0000	71
	30	.0000	.0000	.0000	.0000	.0000	.0000	.0000	.0000	.0001	.0052	.0458	.0868	.0673	.0100	.0000	70
	31	.0000	.0000	.0000	.0000	.0000	.0000	.0000	.0000	.0000	.0029	.0344	.0840	.0760	.0151	.0001	69
	32	.0000	.0000	.0000	.0000	.0000	.0000	.0000	.0000	.0000	.0016	.0248	.0776	.0819	.0217	.0001	68
	33	.0000	.0000	.0000	.0000	.0000	.0000	.0000	.0000	.0000	.0008	.0170	.0685	.0844	.0297	.0002	67
	34	.0000	.0000	.0000	.0000	.0000	.0000	.0000	.0000	.0000	.0004	.0112	.0579	.0831	.0391	.0005	66
	35	.0000	.0000	.0000	.0000	.0000	.0000	.0000	.0000	.0000	.0002	.0070	.0468	.0784	.0491	.0009	65
	36	.0000	.0000	.0000	.0000	.0000	.0000	.0000	.0000	.0000	.0001	.0042	.0362	.0708	.0591	.0016	64
	37	.0000	.0000	.0000	.0000	.0000	.0000	.0000	.0000	.0000	.0000	.0024	.0268	.0612	.0682	.0027	63
	38	.0000	.0000	.0000	.0000	.0000	.0000	.0000	.0000	.0000	.0000	.0013	.0191	.0507	.0754	.0045	62
	39	.0000	.0000	.0000	.0000	.0000	.0000	.0000	.0000	.0000	.0000	.0007	.0130	.0403	.0799	.0071	61
		P=.999	.995	.99	.98	.97	.96	.95	.90	.85	.80	.75	.70	2/3	.60	.50	X

Table D (continued)

Binomial Distribution — Individual Terms

$$f_b(x|n,p) = \binom{n}{x} p^x (1 - p)^{n-x}$$

N	X	P=.001	.005	.01	.02	.03	.04	.05	.10	.15	.20	.25	.30	1/3	.40	.50	
100	40	.0000	.0000	.0000	.0000	.0000	.0000	.0000	.0000	.0000	.0000	.0004	.0085	.0308	.0812	.0108	60
	41	.0000	.0000	.0000	.0000	.0000	.0000	.0000	.0000	.0000	.0000	.0002	.0053	.0225	.0792	.0159	59
	42	.0000	.0000	.0000	.0000	.0000	.0000	.0000	.0000	.0000	.0000	.0001	.0032	.0158	.0742	.0223	58
	43	.0000	.0000	.0000	.0000	.0000	.0000	.0000	.0000	.0000	.0000	.0000	.0019	.0107	.0667	.0301	57
	44	.0000	.0000	.0000	.0000	.0000	.0000	.0000	.0000	.0000	.0000	.0000	.0010	.0069	.0576	.0390	56
	45	.0000	.0000	.0000	.0000	.0000	.0000	.0000	.0000	.0000	.0000	.0000	.0005	.0043	.0478	.0485	55
	46	.0000	.0000	.0000	.0000	.0000	.0000	.0000	.0000	.0000	.0000	.0000	.0003	.0026	.0381	.0580	54
	47	.0000	.0000	.0000	.0000	.0000	.0000	.0000	.0000	.0000	.0000	.0000	.0001	.0015	.0292	.0666	53
	48	.0000	.0000	.0000	.0000	.0000	.0000	.0000	.0000	.0000	.0000	.0000	.0001	.0008	.0215	.0735	52
	49	.0000	.0000	.0000	.0000	.0000	.0000	.0000	.0000	.0000	.0000	.0000	.0000	.0004	.0152	.0780	51
	50	.0000	.0000	.0000	.0000	.0000	.0000	.0000	.0000	.0000	.0000	.0000	.0000	.0002	.0103	.0796	50
	51	.0000	.0000	.0000	.0000	.0000	.0000	.0000	.0000	.0000	.0000	.0000	.0000	.0001	.0068	.0780	49
	52	.0000	.0000	.0000	.0000	.0000	.0000	.0000	.0000	.0000	.0000	.0000	.0000	.0001	.0042	.0735	48
	53	.0000	.0000	.0000	.0000	.0000	.0000	.0000	.0000	.0000	.0000	.0000	.0000	.0000	.0026	.0666	47
	54	.0000	.0000	.0000	.0000	.0000	.0000	.0000	.0000	.0000	.0000	.0000	.0000	.0000	.0015	.0580	46
	55	.0000	.0000	.0000	.0000	.0000	.0000	.0000	.0000	.0000	.0000	.0000	.0000	.0000	.0008	.0485	45
	56	.0000	.0000	.0000	.0000	.0000	.0000	.0000	.0000	.0000	.0000	.0000	.0000	.0000	.0004	.0390	44
	57	.0000	.0000	.0000	.0000	.0000	.0000	.0000	.0000	.0000	.0000	.0000	.0000	.0000	.0002	.0301	43
	58	.0000	.0000	.0000	.0000	.0000	.0000	.0000	.0000	.0000	.0000	.0000	.0000	.0000	.0001	.0223	42
	59	.0000	.0000	.0000	.0000	.0000	.0000	.0000	.0000	.0000	.0000	.0000	.0000	.0000	.0001	.0159	41
	60	.0000	.0000	.0000	.0000	.0000	.0000	.0000	.0000	.0000	.0000	.0000	.0000	.0000	.0000	.0108	40
	61	.0000	.0000	.0000	.0000	.0000	.0000	.0000	.0000	.0000	.0000	.0000	.0000	.0000	.0000	.0071	39
	62	.0000	.0000	.0000	.0000	.0000	.0000	.0000	.0000	.0000	.0000	.0000	.0000	.0000	.0000	.0045	38
	63	.0000	.0000	.0000	.0000	.0000	.0000	.0000	.0000	.0000	.0000	.0000	.0000	.0000	.0000	.0027	37
	64	.0000	.0000	.0000	.0000	.0000	.0000	.0000	.0000	.0000	.0000	.0000	.0000	.0000	.0000	.0016	36
	65	.0000	.0000	.0000	.0000	.0000	.0000	.0000	.0000	.0000	.0000	.0000	.0000	.0000	.0000	.0009	35
	66	.0000	.0000	.0000	.0000	.0000	.0000	.0000	.0000	.0000	.0000	.0000	.0000	.0000	.0000	.0005	34
	67	.0000	.0000	.0000	.0000	.0000	.0000	.0000	.0000	.0000	.0000	.0000	.0000	.0000	.0000	.0002	33
	68	.0000	.0000	.0000	.0000	.0000	.0000	.0000	.0000	.0000	.0000	.0000	.0000	.0000	.0000	.0001	32
	69	.0000	.0000	.0000	.0000	.0000	.0000	.0000	.0000	.0000	.0000	.0000	.0000	.0000	.0000	.0001	31

Table E

Binomial Distribution — Cumulative Terms

$$F_b(x|n,p) = \sum_{t=0}^{x} \binom{n}{t} p^t(1-p)^{n-t}$$

N	X	P=.001	.01	.03	.05	.10	.20	.25	.50	.60	.65	.70	.75	.80	.90	.95	X
2	0	.9980	.9801	.9409	.9025	.8100	.6400	.5625	.2500	.1600	.1225	.0900	.0625	.0400	.0100	.0025	0
	1	1.0000	.9999	.9991	.9975	.9900	.9600	.9375	.7500	.6400	.5775	.5100	.4375	.3600	.1900	.0975	1
3	0	.9970	.9703	.9127	.8574	.7290	.5120	.4219	.1250	.0640	.0429	.0270	.0156	.0080	.0010	.0001	0
	1	1.0000	.9997	.9974	.9927	.9720	.8960	.8438	.5000	.3520	.2817	.2160	.1563	.1040	.0280	.0072	1
	2	1.0000	1.0000	1.0000	.9999	.9990	.9920	.9844	.8750	.7840	.7254	.6570	.5781	.4880	.2710	.1426	2
4	0	.9960	.9606	.8853	.8145	.6561	.4096	.3164	.0625	.0256	.0150	.0081	.0039	.0016	.0001	.0000	0
	1	1.0000	.9994	.9948	.9860	.9477	.8192	.7383	.3125	.1792	.1265	.0837	.0508	.0272	.0037	.0005	1
	2	1.0000	1.0000	.9999	.9995	.9963	.9728	.9492	.6875	.5248	.4370	.3483	.2617	.1808	.0523	.0140	2
	3	1.0000	1.0000	1.0000	1.0000	.9999	.9984	.9961	.9375	.8704	.8215	.7599	.6836	.5904	.3439	.1855	3
5	0	.9950	.9510	.8587	.7738	.5905	.3277	.2373	.0313	.0102	.0053	.0024	.0010	.0003	.0000	.0000	0
	1	1.0000	.9990	.9915	.9774	.9185	.7373	.6328	.1875	.0870	.0540	.0308	.0156	.0067	.0005	.0000	1
	2	1.0000	1.0000	.9997	.9988	.9914	.9421	.8965	.5000	.3174	.2352	.1631	.1035	.0579	.0086	.0012	2
	3	1.0000	1.0000	1.0000	1.0000	.9995	.9933	.9844	.8125	.6630	.5716	.4718	.3672	.2627	.0815	.0226	3
	4	1.0000	1.0000	1.0000	1.0000	1.0000	.9997	.9990	.9688	.9222	.8840	.8319	.7627	.6723	.4095	.2262	4
6	0	.9940	.9415	.8330	.7351	.5314	.2621	.1780	.0156	.0041	.0018	.0007	.0002	.0001	.0000	.0000	0
	1	1.0000	.9985	.9875	.9672	.8857	.6554	.5339	.1094	.0410	.0223	.0109	.0046	.0016	.0001	.0000	1
	2	1.0000	1.0000	.9995	.9978	.9841	.9011	.8306	.3438	.1792	.1174	.0705	.0376	.0170	.0013	.0001	2
	3	1.0000	1.0000	1.0000	.9999	.9987	.9830	.9624	.6563	.4557	.3529	.2557	.1694	.0989	.0158	.0022	3
	4	1.0000	1.0000	1.0000	1.0000	.9999	.9984	.9954	.8906	.7667	.6809	.5798	.4661	.3446	.1143	.0328	4
	5	1.0000	1.0000	1.0000	1.0000	1.0000	.9999	.9998	.9844	.9533	.9246	.8824	.8220	.7379	.4686	.2649	5
7	0	.9930	.9321	.8080	.6983	.4783	.2097	.1335	.0078	.0016	.0006	.0002	.0001	.0000	.0000	.0000	0
	1	1.0000	.9980	.9829	.9556	.8503	.5767	.4449	.0625	.0188	.0090	.0038	.0013	.0004	.0000	.0000	1
	2	1.0000	1.0000	.9991	.9962	.9743	.8520	.7564	.2266	.0963	.0556	.0288	.0129	.0047	.0002	.0000	2
	3	1.0000	1.0000	1.0000	.9998	.9973	.9667	.9294	.5000	.2898	.1998	.1260	.0706	.0333	.0027	.0002	3
	4	1.0000	1.0000	1.0000	1.0000	.9998	.9953	.9871	.7734	.5801	.4677	.3529	.2436	.1480	.0257	.0038	4
	5	1.0000	1.0000	1.0000	1.0000	1.0000	.9996	.9987	.9375	.8414	.7662	.6706	.5551	.4233	.1497	.0444	5
	6	1.0000	1.0000	1.0000	1.0000	1.0000	1.0000	.9999	.9922	.9720	.9510	.9176	.8665	.7903	.5217	.3017	6
8	0	.9920	.9227	.7837	.6634	.4305	.1678	.1001	.0039	.0007	.0002	.0001	.0000	.0000	.0000	.0000	0
	1	1.0000	.9973	.9777	.9428	.8131	.5033	.3671	.0352	.0085	.0036	.0013	.0004	.0001	.0000	.0000	1
	2	1.0000	.9999	.9986	.9942	.9619	.7969	.6785	.1445	.0498	.0253	.0113	.0042	.0012	.0000	.0000	2
	3	1.0000	1.0000	.9999	.9996	.9950	.9437	.8862	.3633	.1737	.1061	.0580	.0273	.0104	.0004	.0000	3
	4	1.0000	1.0000	1.0000	1.0000	.9996	.9896	.9727	.6367	.4059	.2936	.1941	.1138	.0563	.0050	.0004	4
	5	1.0000	1.0000	1.0000	1.0000	1.0000	.9988	.9958	.8555	.6846	.5722	.4482	.3215	.2031	.0381	.0058	5
	6	1.0000	1.0000	1.0000	1.0000	1.0000	.9999	.9996	.9648	.8936	.8309	.7447	.6329	.4967	.1869	.0572	6
	7	1.0000	1.0000	1.0000	1.0000	1.0000	1.0000	1.0000	.9961	.9832	.9681	.9424	.8999	.8322	.5695	.3366	7
9	0	.9910	.9135	.7602	.6302	.3874	.1342	.0751	.0020	.0003	.0001	.0000	.0000	.0000	.0000	.0000	0
	1	1.0000	.9966	.9718	.9288	.7748	.4362	.3003	.0195	.0038	.0014	.0004	.0001	.0000	.0000	.0000	1
	2	1.0000	.9999	.9980	.9916	.9470	.7382	.6007	.0898	.0250	.0112	.0043	.0013	.0003	.0000	.0000	2
	3	1.0000	1.0000	.9999	.9994	.9917	.9144	.8343	.2539	.0994	.0536	.0253	.0100	.0031	.0001	.0000	3
	4	1.0000	1.0000	1.0000	1.0000	.9991	.9804	.9511	.5000	.2666	.1717	.0988	.0489	.0196	.0009	.0000	4
	5	1.0000	1.0000	1.0000	1.0000	.9999	.9969	.9900	.7461	.5174	.3911	.2703	.1657	.0856	.0083	.0006	5
	6	1.0000	1.0000	1.0000	1.0000	1.0000	.9997	.9987	.9102	.7682	.6627	.5372	.3993	.2618	.0530	.0084	6
	7	1.0000	1.0000	1.0000	1.0000	1.0000	1.0000	.9999	.9805	.9295	.8789	.8040	.6997	.5638	.2252	.0712	7
	8	1.0000	1.0000	1.0000	1.0000	1.0000	1.0000	1.0000	.9980	.9899	.9793	.9596	.9249	.8658	.6126	.3698	8
		P=.001	.01	.03	.05	.10	.20	.25	.50	.60	.65	.70	.75	.80	.90	.95	X

Table E (continued)

Binomial Distribution — Cumulative Terms

$$F_b(x|n,p) = \sum_{t=0}^{x} \binom{n}{t} p^t (1-p)^{n-t}$$

N	x	P=.001	.01	.03	.05	.10	.20	.25	.50	.60	.65	.70	.75	.80	.90	.95	x
10	0	.9900	.9044	.7374	.5987	.3487	.1074	.0563	.0010	.0001	.0000	.0000	.0000	.0000	.0000	.0000	0
	1	1.0000	.9957	.9655	.9139	.7361	.3758	.2440	.0107	.0017	.0005	.0001	.0000	.0000	.0000	.0000	1
	2	1.0000	.9999	.9972	.9885	.9298	.6778	.5256	.0547	.0123	.0048	.0016	.0004	.0001	.0000	.0000	2
	3	1.0000	1.0000	.9999	.9990	.9872	.8791	.7759	.1719	.0548	.0260	.0106	.0035	.0009	.0000	.0000	3
	4	1.0000	1.0000	1.0000	.9999	.9984	.9672	.9219	.3770	.1662	.0949	.0473	.0197	.0064	.0001	.0000	4
	5	1.0000	1.0000	1.0000	1.0000	.9999	.9936	.9803	.6230	.3669	.2485	.1503	.0781	.0328	.0016	.0001	5
	6	1.0000	1.0000	1.0000	1.0000	1.0000	.9991	.9965	.8281	.6177	.4862	.3504	.2241	.1209	.0128	.0010	6
	7	1.0000	1.0000	1.0000	1.0000	1.0000	.9999	.9996	.9453	.8327	.7384	.6172	.4744	.3222	.0702	.0115	7
	8	1.0000	1.0000	1.0000	1.0000	1.0000	1.0000	1.0000	.9893	.9536	.9140	.8507	.7560	.6242	.2639	.0861	8
	9	1.0000	1.0000	1.0000	1.0000	1.0000	1.0000	1.0000	.9990	.9940	.9865	.9718	.9437	.8926	.6513	.4013	9
11	0	.9891	.8953	.7153	.5688	.3138	.0859	.0422	.0005	.0000	.0000	.0000	.0000	.0000	.0000	.0000	0
	1	.9999	.9948	.9587	.8981	.6974	.3221	.1971	.0059	.0007	.0002	.0000	.0000	.0000	.0000	.0000	1
	2	1.0000	.9998	.9963	.9848	.9104	.6174	.4552	.0327	.0059	.0020	.0006	.0001	.0000	.0000	.0000	2
	3	1.0000	1.0000	.9998	.9984	.9815	.8389	.7133	.1133	.0293	.0122	.0043	.0012	.0002	.0000	.0000	3
	4	1.0000	1.0000	1.0000	.9999	.9972	.9496	.8854	.2744	.0994	.0501	.0216	.0076	.0020	.0000	.0000	4
	5	1.0000	1.0000	1.0000	1.0000	.9997	.9883	.9657	.5000	.2465	.1487	.0782	.0343	.0117	.0003	.0000	5
	6	1.0000	1.0000	1.0000	1.0000	1.0000	.9980	.9924	.7256	.4672	.3317	.2103	.1146	.0504	.0028	.0001	6
	7	1.0000	1.0000	1.0000	1.0000	1.0000	.9998	.9988	.8867	.7037	.5744	.4304	.2867	.1611	.0185	.0016	7
	8	1.0000	1.0000	1.0000	1.0000	1.0000	1.0000	.9999	.9673	.8811	.7999	.6873	.5448	.3826	.0896	.0152	8
	9	1.0000	1.0000	1.0000	1.0000	1.0000	1.0000	1.0000	.9941	.9698	.9394	.8870	.8029	.6779	.3026	.1019	9
	10	1.0000	1.0000	1.0000	1.0000	1.0000	1.0000	1.0000	.9995	.9964	.9912	.9802	.9578	.9141	.6862	.4312	10
12	0	.9881	.8864	.6938	.5404	.2824	.0687	.0317	.0002	.0000	.0000	.0000	.0000	.0000	.0000	.0000	0
	1	.9999	.9938	.9514	.8816	.6590	.2749	.1584	.0032	.0003	.0001	.0000	.0000	.0000	.0000	.0000	1
	2	1.0000	.9998	.9952	.9804	.8891	.5583	.3907	.0193	.0028	.0008	.0002	.0000	.0000	.0000	.0000	2
	3	1.0000	1.0000	.9997	.9978	.9744	.7946	.6488	.0730	.0153	.0056	.0017	.0004	.0001	.0000	.0000	3
	4	1.0000	1.0000	1.0000	.9998	.9957	.9274	.8424	.1938	.0573	.0255	.0095	.0028	.0006	.0000	.0000	4
	5	1.0000	1.0000	1.0000	1.0000	.9995	.9806	.9456	.3872	.1582	.0846	.0386	.0143	.0039	.0001	.0000	5
	6	1.0000	1.0000	1.0000	1.0000	.9999	.9961	.9857	.6128	.3348	.2127	.1178	.0544	.0194	.0005	.0000	6
	7	1.0000	1.0000	1.0000	1.0000	1.0000	.9994	.9972	.8062	.5618	.4167	.2763	.1576	.0726	.0043	.0002	7
	8	1.0000	1.0000	1.0000	1.0000	1.0000	.9999	.9996	.9270	.7747	.6533	.5075	.3512	.2054	.0256	.0022	8
	9	1.0000	1.0000	1.0000	1.0000	1.0000	1.0000	1.0000	.9807	.9166	.8487	.7472	.6093	.4417	.1109	.0196	9
	10	1.0000	1.0000	1.0000	1.0000	1.0000	1.0000	1.0000	.9968	.9804	.9576	.9150	.8416	.7251	.3410	.1184	10
	11	1.0000	1.0000	1.0000	1.0000	1.0000	1.0000	1.0000	.9998	.9978	.9943	.9862	.9683	.9313	.7176	.4596	11
13	0	.9871	.8775	.6730	.5133	.2542	.0550	.0238	.0001	.0000	.0000	.0000	.0000	.0000	.0000	.0000	0
	1	.9999	.9928	.9436	.8646	.6213	.2336	.1267	.0017	.0001	.0000	.0000	.0000	.0000	.0000	.0000	1
	2	1.0000	.9997	.9938	.9755	.8661	.5017	.3326	.0112	.0013	.0003	.0001	.0000	.0000	.0000	.0000	2
	3	1.0000	1.0000	.9995	.9969	.9658	.7473	.5843	.0461	.0078	.0025	.0007	.0001	.0000	.0000	.0000	3
	4	1.0000	1.0000	1.0000	.9997	.9935	.9009	.7940	.1334	.0321	.0126	.0040	.0010	.0002	.0000	.0000	4
	5	1.0000	1.0000	1.0000	1.0000	.9991	.9700	.9198	.2905	.0977	.0462	.0182	.0056	.0012	.0000	.0000	5
	6	1.0000	1.0000	1.0000	1.0000	.9999	.9930	.9757	.5000	.2288	.1295	.0624	.0243	.0070	.0001	.0000	6
	7	1.0000	1.0000	1.0000	1.0000	1.0000	.9988	.9944	.7095	.4256	.2841	.1654	.0802	.0300	.0009	.0000	7
	8	1.0000	1.0000	1.0000	1.0000	1.0000	.9998	.9990	.8666	.6470	.4995	.3457	.2060	.0991	.0065	.0003	8
	9	1.0000	1.0000	1.0000	1.0000	1.0000	1.0000	.9999	.9539	.8314	.7217	.5794	.4157	.2527	.0342	.0031	9
	10	1.0000	1.0000	1.0000	1.0000	1.0000	1.0000	1.0000	.9888	.9421	.8868	.7975	.6674	.4983	.1339	.0245	10
	11	1.0000	1.0000	1.0000	1.0000	1.0000	1.0000	1.0000	.9983	.9874	.9704	.9363	.8733	.7664	.3787	.1354	11
	12	1.0000	1.0000	1.0000	1.0000	1.0000	1.0000	1.0000	.9999	.9987	.9963	.9903	.9762	.9450	.7458	.4867	12
		P=.001	.01	.03	.05	.10	.20	.25	.50	.60	.65	.70	.75	.80	.90	.95	x

Table E (continued)

Binomial Distribution — Cumulative Terms

$$F_b(x|n,p) = \sum_{t=0}^{x} \binom{n}{t} p^t(1-p)^{n-t}$$

N	X	P=.001	.01	.03	.05	.10	.20	.25	.50	.60	.65	.70	.75	.80	.90	.95	X
14	0	.9861	.8687	.6528	.4877	.2288	.0440	.0178	.0001	.0000	.0000	.0000	.0000	.0000	.0000	.0000	0
	1	.9999	.9916	.9355	.8470	.5846	.1979	.1010	.0009	.0001	.0000	.0000	.0000	.0000	.0000	.0000	1
	2	1.0000	.9997	.9923	.9699	.8416	.4481	.2811	.0065	.0006	.0001	.0000	.0000	.0000	.0000	.0000	2
	3	1.0000	1.0000	.9994	.9958	.9559	.6982	.5213	.0287	.0039	.0011	.0002	.0000	.0000	.0000	.0000	3
	4	1.0000	1.0000	1.0000	.9996	.9908	.8702	.7415	.0898	.0175	.0060	.0017	.0003	.0000	.0000	.0000	4
	5	1.0000	1.0000	1.0000	1.0000	.9985	.9561	.8883	.2120	.0583	.0243	.0083	.0022	.0004	.0000	.0000	5
	6	1.0000	1.0000	1.0000	1.0000	.9998	.9884	.9617	.3953	.1501	.0753	.0315	.0103	.0024	.0000	.0000	6
	7	1.0000	1.0000	1.0000	1.0000	1.0000	.9976	.9897	.6047	.3075	.1836	.0933	.0383	.0116	.0002	.0000	7
	8	1.0000	1.0000	1.0000	1.0000	1.0000	.9996	.9978	.7880	.5141	.3595	.2195	.1117	.0439	.0015	.0000	8
	9	1.0000	1.0000	1.0000	1.0000	1.0000	1.0000	.9997	.9102	.7207	.5773	.4158	.2585	.1298	.0092	.0004	9
	10	1.0000	1.0000	1.0000	1.0000	1.0000	1.0000	1.0000	.9713	.8757	.7795	.6448	.4787	.3018	.0441	.0042	10
	11	1.0000	1.0000	1.0000	1.0000	1.0000	1.0000	1.0000	.9935	.9602	.9161	.8392	.7189	.5519	.1584	.0301	11
	12	1.0000	1.0000	1.0000	1.0000	1.0000	1.0000	1.0000	.9991	.9919	.9795	.9525	.8990	.8021	.4154	.1530	12
	13	1.0000	1.0000	1.0000	1.0000	1.0000	1.0000	1.0000	.9999	.9992	.9976	.9932	.9822	.9560	.7712	.5123	13
15	0	.9851	.8601	.6333	.4633	.2059	.0352	.0134	.0000	.0000	.0000	.0000	.0000	.0000	.0000	.0000	0
	1	.9999	.9904	.9270	.8290	.5490	.1671	.0802	.0005	.0000	.0000	.0000	.0000	.0000	.0000	.0000	1
	2	1.0000	.9996	.9906	.9638	.8159	.3980	.2361	.0037	.0003	.0001	.0000	.0000	.0000	.0000	.0000	2
	3	1.0000	1.0000	.9992	.9945	.9444	.6482	.4613	.0176	.0019	.0005	.0001	.0000	.0000	.0000	.0000	3
	4	1.0000	1.0000	.9999	.9994	.9873	.8358	.6865	.0592	.0093	.0028	.0007	.0001	.0000	.0000	.0000	4
	5	1.0000	1.0000	1.0000	.9999	.9977	.9389	.8516	.1509	.0338	.0124	.0037	.0008	.0001	.0000	.0000	5
	6	1.0000	1.0000	1.0000	1.0000	.9997	.9819	.9434	.3036	.0950	.0422	.0152	.0042	.0008	.0000	.0000	6
	7	1.0000	1.0000	1.0000	1.0000	1.0000	.9958	.9827	.5000	.2131	.1132	.0500	.0173	.0042	.0000	.0000	7
	8	1.0000	1.0000	1.0000	1.0000	1.0000	.9992	.9958	.6964	.3902	.2452	.1311	.0566	.0181	.0003	.0000	8
	9	1.0000	1.0000	1.0000	1.0000	1.0000	.9999	.9992	.8491	.5968	.4357	.2784	.1484	.0611	.0022	.0001	9
	10	1.0000	1.0000	1.0000	1.0000	1.0000	1.0000	.9999	.9408	.7827	.6481	.4845	.3135	.1642	.0127	.0006	10
	11	1.0000	1.0000	1.0000	1.0000	1.0000	1.0000	1.0000	.9824	.9095	.8273	.7031	.5387	.3518	.0556	.0055	11
	12	1.0000	1.0000	1.0000	1.0000	1.0000	1.0000	1.0000	.9963	.9729	.9383	.8732	.7639	.6020	.1841	.0362	12
	13	1.0000	1.0000	1.0000	1.0000	1.0000	1.0000	1.0000	.9995	.9948	.9858	.9647	.9198	.8329	.4510	.1710	13
	14	1.0000	1.0000	1.0000	1.0000	1.0000	1.0000	1.0000	1.0000	.9995	.9984	.9953	.9866	.9648	.7941	.5367	14
16	0	.9841	.8515	.6143	.4401	.1853	.0281	.0100	.0000	.0000	.0000	.0000	.0000	.0000	.0000	.0000	0
	1	.9999	.9891	.9182	.8108	.5147	.1407	.0635	.0003	.0000	.0000	.0000	.0000	.0000	.0000	.0000	1
	2	1.0000	.9995	.9887	.9571	.7892	.3518	.1971	.0021	.0001	.0000	.0000	.0000	.0000	.0000	.0000	2
	3	1.0000	1.0000	.9989	.9930	.9316	.5981	.4050	.0106	.0009	.0002	.0000	.0000	.0000	.0000	.0000	3
	4	1.0000	1.0000	.9999	.9991	.9830	.7982	.6302	.0384	.0049	.0013	.0003	.0000	.0000	.0000	.0000	4
	5	1.0000	1.0000	1.0000	.9999	.9967	.9183	.8103	.1051	.0191	.0062	.0016	.0003	.0000	.0000	.0000	5
	6	1.0000	1.0000	1.0000	1.0000	.9995	.9733	.9204	.2272	.0583	.0229	.0071	.0016	.0002	.0000	.0000	6
	7	1.0000	1.0000	1.0000	1.0000	.9999	.9930	.9729	.4018	.1423	.0671	.0257	.0075	.0015	.0000	.0000	7
	8	1.0000	1.0000	1.0000	1.0000	1.0000	.9985	.9925	.5982	.2839	.1594	.0744	.0271	.0070	.0001	.0000	8
	9	1.0000	1.0000	1.0000	1.0000	1.0000	.9998	.9984	.7728	.4728	.3119	.1753	.0796	.0267	.0005	.0000	9
	10	1.0000	1.0000	1.0000	1.0000	1.0000	1.0000	.9997	.8949	.6712	.5100	.3402	.1897	.0817	.0033	.0001	10
	11	1.0000	1.0000	1.0000	1.0000	1.0000	1.0000	1.0000	.9616	.8334	.7108	.5501	.3698	.2018	.0170	.0009	11
	12	1.0000	1.0000	1.0000	1.0000	1.0000	1.0000	1.0000	.9894	.9349	.8661	.7541	.5950	.4019	.0684	.0070	12
	13	1.0000	1.0000	1.0000	1.0000	1.0000	1.0000	1.0000	.9979	.9817	.9549	.9006	.8029	.6482	.2108	.0429	13
	14	1.0000	1.0000	1.0000	1.0000	1.0000	1.0000	1.0000	.9997	.9967	.9902	.9739	.9365	.8593	.4853	.1892	14
	15	1.0000	1.0000	1.0000	1.0000	1.0000	1.0000	1.0000	1.0000	.9997	.9990	.9967	.9900	.9719	.8147	.5599	15
17	0	.9831	.8429	.5958	.4181	.1668	.0225	.0075	.0000	.0000	.0000	.0000	.0000	.0000	.0000	.0000	0
		P=.001	.01	.03	.05	.10	.20	.25	.50	.60	.65	.70	.75	.80	.90	.95	X

Table E (continued)

Binomial Distribution — Cumulative Terms

$$F_b(x|n,p) = \sum_{t=0}^{x} \binom{n}{t} p^t (1-p)^{n-t}$$

N	X	P=.001	.01	.03	.05	.10	.20	.25	.50	.60	.65	.70	.75	.80	.90	.95	X
17	1	.9999	.9877	.9091	.7922	.4818	.1182	.0501	.0001	.0000	.0000	.0000	.0000	.0000	.0000	.0000	1
	2	1.0000	.9994	.9866	.9497	.7618	.3096	.1637	.0012	.0001	.0000	.0000	.0000	.0000	.0000	.0000	2
	3	1.0000	1.0000	.9986	.9912	.9174	.5489	.3530	.0064	.0005	.0001	.0000	.0000	.0000	.0000	.0000	3
	4	1.0000	1.0000	.9999	.9988	.9779	.7582	.5739	.0245	.0025	.0006	.0001	.0000	.0000	.0000	.0000	4
	5	1.0000	1.0000	1.0000	.9999	.9953	.8943	.7653	.0717	.0106	.0030	.0007	.0001	.0000	.0000	.0000	5
	6	1.0000	1.0000	1.0000	1.0000	.9992	.9623	.8929	.1662	.0348	.0120	.0032	.0006	.0001	.0000	.0000	6
	7	1.0000	1.0000	1.0000	1.0000	.9999	.9891	.9598	.3145	.0919	.0383	.0127	.0031	.0005	.0000	.0000	7
	8	1.0000	1.0000	1.0000	1.0000	1.0000	.9974	.9876	.5000	.1989	.0994	.0403	.0124	.0026	.0000	.0000	8
	9	1.0000	1.0000	1.0000	1.0000	1.0000	.9995	.9969	.6855	.3595	.2128	.1046	.0402	.0109	.0001	.0000	9
	10	1.0000	1.0000	1.0000	1.0000	1.0000	.9999	.9994	.8338	.5522	.3812	.2248	.1071	.0377	.0006	.0000	10
	11	1.0000	1.0000	1.0000	1.0000	1.0000	1.0000	.9999	.9283	.7361	.5803	.4032	.2347	.1057	.0047	.0001	11
	12	1.0000	1.0000	1.0000	1.0000	1.0000	1.0000	1.0000	.9755	.8740	.7652	.6113	.4261	.2418	.0221	.0012	12
	13	1.0000	1.0000	1.0000	1.0000	1.0000	1.0000	1.0000	.9936	.9536	.8972	.7981	.6470	.4511	.0826	.0088	13
	14	1.0000	1.0000	1.0000	1.0000	1.0000	1.0000	1.0000	.9988	.9877	.9673	.9226	.8363	.6904	.2382	.0503	14
	15	1.0000	1.0000	1.0000	1.0000	1.0000	1.0000	1.0000	.9999	.9979	.9933	.9807	.9499	.8818	.5182	.2078	15
	16	1.0000	1.0000	1.0000	1.0000	1.0000	1.0000	1.0000	1.0000	.9998	.9993	.9977	.9925	.9775	.8332	.5819	16
18	0	.9822	.8345	.5780	.3972	.1501	.0180	.0056	.0000	.0000	.0000	.0000	.0000	.0000	.0000	.0000	0
	1	.9998	.9862	.8997	.7735	.4503	.0991	.0395	.0001	.0000	.0000	.0000	.0000	.0000	.0000	.0000	1
	2	1.0000	.9993	.9843	.9419	.7338	.2713	.1353	.0007	.0000	.0000	.0000	.0000	.0000	.0000	.0000	2
	3	1.0000	1.0000	.9982	.9891	.9018	.5010	.3057	.0038	.0002	.0000	.0000	.0000	.0000	.0000	.0000	3
	4	1.0000	1.0000	.9998	.9985	.9718	.7164	.5187	.0154	.0013	.0003	.0000	.0000	.0000	.0000	.0000	4
	5	1.0000	1.0000	1.0000	.9998	.9936	.8671	.7175	.0481	.0058	.0014	.0003	.0000	.0000	.0000	.0000	5
	6	1.0000	1.0000	1.0000	1.0000	.9988	.9487	.8610	.1189	.0203	.0062	.0014	.0002	.0000	.0000	.0000	6
	7	1.0000	1.0000	1.0000	1.0000	.9998	.9837	.9431	.2403	.0576	.0212	.0061	.0012	.0002	.0000	.0000	7
	8	1.0000	1.0000	1.0000	1.0000	1.0000	.9957	.9807	.4073	.1347	.0597	.0210	.0054	.0009	.0000	.0000	8
	9	1.0000	1.0000	1.0000	1.0000	1.0000	.9991	.9946	.5927	.2632	.1391	.0596	.0193	.0043	.0000	.0000	9
	10	1.0000	1.0000	1.0000	1.0000	1.0000	.9998	.9988	.7597	.4366	.2717	.1407	.0569	.0163	.0002	.0000	10
	11	1.0000	1.0000	1.0000	1.0000	1.0000	1.0000	.9998	.8811	.6257	.4509	.2783	.1390	.0513	.0012	.0000	11
	12	1.0000	1.0000	1.0000	1.0000	1.0000	1.0000	1.0000	.9519	.7912	.6450	.4656	.2825	.1329	.0064	.0002	12
	13	1.0000	1.0000	1.0000	1.0000	1.0000	1.0000	1.0000	.9846	.9058	.8114	.6673	.4813	.2836	.0282	.0015	13
	14	1.0000	1.0000	1.0000	1.0000	1.0000	1.0000	1.0000	.9962	.9672	.9217	.8354	.6943	.4990	.0982	.0109	14
	15	1.0000	1.0000	1.0000	1.0000	1.0000	1.0000	1.0000	.9993	.9918	.9764	.9400	.8647	.7287	.2662	.0581	15
	16	1.0000	1.0000	1.0000	1.0000	1.0000	1.0000	1.0000	.9999	.9987	.9954	.9858	.9605	.9009	.5497	.2265	16
	17	1.0000	1.0000	1.0000	1.0000	1.0000	1.0000	1.0000	1.0000	.9999	.9996	.9984	.9944	.9820	.8499	.6028	17
19	0	.9812	.8262	.5606	.3774	.1351	.0144	.0042	.0000	.0000	.0000	.0000	.0000	.0000	.0000	.0000	0
	1	.9998	.9847	.8900	.7547	.4203	.0829	.0310	.0000	.0000	.0000	.0000	.0000	.0000	.0000	.0000	1
	2	1.0000	.9991	.9817	.9335	.7054	.2369	.1113	.0004	.0000	.0000	.0000	.0000	.0000	.0000	.0000	2
	3	1.0000	1.0000	.9978	.9868	.8850	.4551	.2631	.0022	.0001	.0000	.0000	.0000	.0000	.0000	.0000	3
	4	1.0000	1.0000	.9998	.9980	.9648	.6733	.4654	.0096	.0006	.0001	.0000	.0000	.0000	.0000	.0000	4
	5	1.0000	1.0000	1.0000	.9998	.9914	.8369	.6678	.0318	.0031	.0007	.0001	.0000	.0000	.0000	.0000	5
	6	1.0000	1.0000	1.0000	1.0000	.9983	.9324	.8251	.0835	.0116	.0031	.0006	.0001	.0000	.0000	.0000	6
	7	1.0000	1.0000	1.0000	1.0000	.9997	.9767	.9225	.1796	.0352	.0114	.0028	.0005	.0000	.0000	.0000	7
	8	1.0000	1.0000	1.0000	1.0000	1.0000	.9933	.9713	.3238	.0885	.0347	.0105	.0023	.0003	.0000	.0000	8
	9	1.0000	1.0000	1.0000	1.0000	1.0000	.9984	.9911	.5000	.1861	.0875	.0326	.0089	.0016	.0000	.0000	9
	10	1.0000	1.0000	1.0000	1.0000	1.0000	.9997	.9977	.6762	.3325	.1855	.0839	.0287	.0067	.0000	.0000	10
	11	1.0000	1.0000	1.0000	1.0000	1.0000	.9999	.9995	.8204	.5122	.3344	.1820	.0775	.0233	.0003	.0000	11
		P=.001	.01	.03	.05	.10	.20	.25	.50	.60	.65	.70	.75	.80	.90	.95	X

Table E (continued)

Binomial Distribution — Cumulative Terms

$$F_b(x|n,p) = \sum_{t=0}^{x} \binom{n}{t} p^t(1-p)^{n-t}$$

N	X	P=.001	.01	.03	.05	.10	.20	.25	.50	.60	.65	.70	.75	.80	.90	.95	X
19	12	1.0000	1.0000	1.0000	1.0000	1.0000	1.0000	.9999	.9165	.6919	.5188	.3345	.1749	.0676	.0017	.0000	12
	13	1.0000	1.0000	1.0000	1.0000	1.0000	1.0000	1.0000	.9682	.8371	.7032	.5261	.3322	.1631	.0086	.0002	13
	14	1.0000	1.0000	1.0000	1.0000	1.0000	1.0000	1.0000	.9904	.9304	.8500	.7178	.5346	.3267	.0352	.0020	14
	15	1.0000	1.0000	1.0000	1.0000	1.0000	1.0000	1.0000	.9978	.9770	.9409	.8668	.7369	.5449	.1150	.0132	15
	16	1.0000	1.0000	1.0000	1.0000	1.0000	1.0000	1.0000	.9996	.9945	.9830	.9538	.8887	.7631	.2946	.0665	16
	17	1.0000	1.0000	1.0000	1.0000	1.0000	1.0000	1.0000	1.0000	.9992	.9969	.9896	.9690	.9171	.5797	.2453	17
	18	1.0000	1.0000	1.0000	1.0000	1.0000	1.0000	1.0000	1.0000	.9999	.9997	.9989	.9958	.9856	.8649	.6226	18
20	0	.9802	.8179	.5438	.3585	.1216	.0115	.0032	.0000	.0000	.0000	.0000	.0000	.0000	.0000	.0000	0
	1	.9998	.9831	.8802	.7358	.3917	.0692	.0243	.0000	.0000	.0000	.0000	.0000	.0000	.0000	.0000	1
	2	1.0000	.9990	.9790	.9245	.6769	.2061	.0913	.0002	.0000	.0000	.0000	.0000	.0000	.0000	.0000	2
	3	1.0000	1.0000	.9973	.9841	.8670	.4114	.2252	.0013	.0000	.0000	.0000	.0000	.0000	.0000	.0000	3
	4	1.0000	1.0000	.9997	.9974	.9568	.6296	.4148	.0059	.0003	.0000	.0000	.0000	.0000	.0000	.0000	4
	5	1.0000	1.0000	1.0000	.9997	.9887	.8042	.6172	.0207	.0016	.0003	.0000	.0000	.0000	.0000	.0000	5
	6	1.0000	1.0000	1.0000	1.0000	.9976	.9133	.7858	.0577	.0065	.0015	.0003	.0000	.0000	.0000	.0000	6
	7	1.0000	1.0000	1.0000	1.0000	.9996	.9679	.8982	.1316	.0210	.0060	.0013	.0002	.0000	.0000	.0000	7
	8	1.0000	1.0000	1.0000	1.0000	.9999	.9900	.9591	.2517	.0565	.0196	.0051	.0009	.0001	.0000	.0000	8
	9	1.0000	1.0000	1.0000	1.0000	1.0000	.9974	.9861	.4119	.1275	.0532	.0171	.0039	.0006	.0000	.0000	9
	10	1.0000	1.0000	1.0000	1.0000	1.0000	.9994	.9961	.5881	.2447	.1218	.0480	.0139	.0026	.0000	.0000	10
	11	1.0000	1.0000	1.0000	1.0000	1.0000	.9999	.9991	.7483	.4044	.2376	.1133	.0409	.0100	.0001	.0000	11
	12	1.0000	1.0000	1.0000	1.0000	1.0000	1.0000	.9998	.8684	.5841	.3990	.2277	.1018	.0321	.0004	.0000	12
	13	1.0000	1.0000	1.0000	1.0000	1.0000	1.0000	1.0000	.9423	.7500	.5834	.3920	.2142	.0867	.0024	.0000	13
	14	1.0000	1.0000	1.0000	1.0000	1.0000	1.0000	1.0000	.9793	.8744	.7546	.5836	.3828	.1958	.0113	.0003	14
	15	1.0000	1.0000	1.0000	1.0000	1.0000	1.0000	1.0000	.9941	.9490	.8818	.7625	.5852	.3704	.0432	.0026	15
	16	1.0000	1.0000	1.0000	1.0000	1.0000	1.0000	1.0000	.9987	.9840	.9556	.8929	.7748	.5886	.1330	.0159	16
	17	1.0000	1.0000	1.0000	1.0000	1.0000	1.0000	1.0000	.9998	.9964	.9879	.9645	.9087	.7939	.3231	.0755	17
	18	1.0000	1.0000	1.0000	1.0000	1.0000	1.0000	1.0000	1.0000	.9995	.9979	.9924	.9757	.9308	.6083	.2642	18
	19	1.0000	1.0000	1.0000	1.0000	1.0000	1.0000	1.0000	1.0000	1.0000	.9998	.9992	.9968	.9885	.8784	.6415	19
21	0	.9792	.8097	.5275	.3406	.1094	.0092	.0024	.0000	.0000	.0000	.0000	.0000	.0000	.0000	.0000	0
	1	.9998	.9815	.8701	.7170	.3647	.0576	.0190	.0000	.0000	.0000	.0000	.0000	.0000	.0000	.0000	1
	2	1.0000	.9988	.9760	.9151	.6484	.1787	.0745	.0001	.0000	.0000	.0000	.0000	.0000	.0000	.0000	2
	3	1.0000	.9999	.9968	.9811	.8480	.3704	.1917	.0007	.0000	.0000	.0000	.0000	.0000	.0000	.0000	3
	4	1.0000	1.0000	.9997	.9968	.9478	.5860	.3674	.0036	.0002	.0000	.0000	.0000	.0000	.0000	.0000	4
	5	1.0000	1.0000	1.0000	.9996	.9856	.7693	.5666	.0133	.0008	.0001	.0000	.0000	.0000	.0000	.0000	5
	6	1.0000	1.0000	1.0000	1.0000	.9967	.8915	.7436	.0392	.0036	.0007	.0001	.0000	.0000	.0000	.0000	6
	7	1.0000	1.0000	1.0000	1.0000	.9994	.9569	.8701	.0946	.0123	.0031	.0006	.0001	.0000	.0000	.0000	7
	8	1.0000	1.0000	1.0000	1.0000	.9999	.9856	.9439	.1917	.0352	.0108	.0024	.0004	.0000	.0000	.0000	8
	9	1.0000	1.0000	1.0000	1.0000	1.0000	.9959	.9794	.3318	.0849	.0313	.0087	.0017	.0002	.0000	.0000	9
	10	1.0000	1.0000	1.0000	1.0000	1.0000	.9990	.9936	.5000	.1744	.0772	.0264	.0064	.0010	.0000	.0000	10
	11	1.0000	1.0000	1.0000	1.0000	1.0000	.9998	.9983	.6682	.3086	.1623	.0676	.0206	.0041	.0000	.0000	11
	12	1.0000	1.0000	1.0000	1.0000	1.0000	1.0000	.9996	.8083	.4763	.2941	.1477	.0561	.0144	.0001	.0000	12
	13	1.0000	1.0000	1.0000	1.0000	1.0000	1.0000	.9999	.9054	.6505	.4635	.2770	.1299	.0431	.0006	.0000	13
	14	1.0000	1.0000	1.0000	1.0000	1.0000	1.0000	1.0000	.9608	.7998	.6433	.4495	.2564	.1085	.0033	.0000	14
	15	1.0000	1.0000	1.0000	1.0000	1.0000	1.0000	1.0000	.9867	.9043	.7991	.6373	.4334	.2307	.0144	.0004	15
	16	1.0000	1.0000	1.0000	1.0000	1.0000	1.0000	1.0000	.9964	.9630	.9076	.8016	.6326	.4140	.0522	.0032	16
	17	1.0000	1.0000	1.0000	1.0000	1.0000	1.0000	1.0000	.9993	.9890	.9669	.9144	.8083	.6296	.1520	.0189	17
	18	1.0000	1.0000	1.0000	1.0000	1.0000	1.0000	1.0000	.9999	.9976	.9914	.9729	.9255	.8213	.3516	.0849	18
	19	1.0000	1.0000	1.0000	1.0000	1.0000	1.0000	1.0000	1.0000	.9997	.9985	.9944	.9810	.9424	.6353	.2830	19
		P=.001	.01	.03	.05	.10	.20	.25	.50	.60	.65	.70	.75	.80	.90	.95	X

Table E (continued)

Binomial Distribution — Cumulative Terms

$$F_b(x|n,p) = \sum_{t=0}^{x} \binom{n}{t} p^t (1-p)^{n-t}$$

N	x	P=.001	.01	.03	.05	.10	.20	.25	.50	.60	.65	.70	.75	.80	.90	.95	x
21	20	1.0000	1.0000	1.0000	1.0000	1.0000	1.0000	1.0000	1.0000	1.0000	.9999	.9994	.9976	.9908	.8906	.6594	20
22	0	.9782	.8016	.5117	.3235	.0985	.0074	.0018	.0000	.0000	.0000	.0000	.0000	.0000	.0000	.0000	0
	1	.9998	.9798	.8598	.6982	.3392	.0480	.0149	.0000	.0000	.0000	.0000	.0000	.0000	.0000	.0000	1
	2	1.0000	.9987	.9728	.9052	.6200	.1545	.0606	.0001	.0000	.0000	.0000	.0000	.0000	.0000	.0000	2
	3	1.0000	.9999	.9962	.9778	.8281	.3320	.1624	.0004	.0000	.0000	.0000	.0000	.0000	.0000	.0000	3
	4	1.0000	1.0000	.9996	.9960	.9379	.5429	.3235	.0022	.0001	.0000	.0000	.0000	.0000	.0000	.0000	4
	5	1.0000	1.0000	1.0000	.9994	.9818	.7326	.5168	.0085	.0004	.0001	.0000	.0000	.0000	.0000	.0000	5
	6	1.0000	1.0000	1.0000	.9999	.9956	.8670	.6994	.0262	.0019	.0003	.0000	.0000	.0000	.0000	.0000	6
	7	1.0000	1.0000	1.0000	1.0000	.9991	.9439	.8385	.0669	.0070	.0016	.0002	.0000	.0000	.0000	.0000	7
	8	1.0000	1.0000	1.0000	1.0000	.9999	.9799	.9254	.1431	.0215	.0058	.0011	.0001	.0000	.0000	.0000	8
	9	1.0000	1.0000	1.0000	1.0000	1.0000	.9939	.9705	.2617	.0551	.0180	.0043	.0007	.0001	.0000	.0000	9
	10	1.0000	1.0000	1.0000	1.0000	1.0000	.9984	.9900	.4159	.1207	.0474	.0140	.0029	.0003	.0000	.0000	10
	11	1.0000	1.0000	1.0000	1.0000	1.0000	.9997	.9971	.5841	.2280	.1070	.0387	.0100	.0016	.0000	.0000	11
	12	1.0000	1.0000	1.0000	1.0000	1.0000	.9999	.9993	.7383	.3756	.2084	.0916	.0295	.0061	.0000	.0000	12
	13	1.0000	1.0000	1.0000	1.0000	1.0000	1.0000	.9999	.8569	.5460	.3534	.1865	.0746	.0201	.0001	.0000	13
	14	1.0000	1.0000	1.0000	1.0000	1.0000	1.0000	1.0000	.9331	.7102	.5264	.3287	.1615	.0561	.0009	.0000	14
	15	1.0000	1.0000	1.0000	1.0000	1.0000	1.0000	1.0000	.9738	.8416	.6978	.5058	.3006	.1330	.0044	.0001	15
	16	1.0000	1.0000	1.0000	1.0000	1.0000	1.0000	1.0000	.9915	.9278	.8371	.6866	.4832	.2674	.0182	.0006	16
	17	1.0000	1.0000	1.0000	1.0000	1.0000	1.0000	1.0000	.9978	.9734	.9284	.8355	.6765	.4571	.0621	.0040	17
	18	1.0000	1.0000	1.0000	1.0000	1.0000	1.0000	1.0000	.9996	.9924	.9755	.9319	.8376	.6680	.1719	.0222	18
	19	1.0000	1.0000	1.0000	1.0000	1.0000	1.0000	1.0000	.9999	.9984	.9939	.9793	.9394	.8455	.3800	.0948	19
	20	1.0000	1.0000	1.0000	1.0000	1.0000	1.0000	1.0000	1.0000	.9998	.9990	.9959	.9851	.9520	.6608	.3018	20
	21	1.0000	1.0000	1.0000	1.0000	1.0000	1.0000	1.0000	1.0000	1.0000	.9999	.9996	.9982	.9926	.9015	.6765	21
23	0	.9773	.7936	.4963	.3074	.0886	.0059	.0013	.0000	.0000	.0000	.0000	.0000	.0000	.0000	.0000	0
	1	.9998	.9780	.8493	.6794	.3151	.0398	.0116	.0000	.0000	.0000	.0000	.0000	.0000	.0000	.0000	1
	2	1.0000	.9985	.9695	.8948	.5920	.1332	.0492	.0000	.0000	.0000	.0000	.0000	.0000	.0000	.0000	2
	3	1.0000	.9999	.9955	.9742	.8073	.2965	.1370	.0002	.0000	.0000	.0000	.0000	.0000	.0000	.0000	3
	4	1.0000	1.0000	.9995	.9951	.9269	.5007	.2832	.0013	.0000	.0000	.0000	.0000	.0000	.0000	.0000	4
	5	1.0000	1.0000	1.0000	.9992	.9774	.6947	.4685	.0053	.0002	.0000	.0000	.0000	.0000	.0000	.0000	5
	6	1.0000	1.0000	1.0000	.9999	.9942	.8402	.6537	.0173	.0010	.0002	.0000	.0000	.0000	.0000	.0000	6
	7	1.0000	1.0000	1.0000	1.0000	.9988	.9285	.8037	.0466	.0040	.0008	.0001	.0000	.0000	.0000	.0000	7
	8	1.0000	1.0000	1.0000	1.0000	.9998	.9727	.9037	.1050	.0128	.0030	.0005	.0001	.0000	.0000	.0000	8
	9	1.0000	1.0000	1.0000	1.0000	1.0000	.9911	.9592	.2024	.0349	.0100	.0021	.0003	.0000	.0000	.0000	9
	10	1.0000	1.0000	1.0000	1.0000	1.0000	.9975	.9851	.3388	.0813	.0283	.0072	.0012	.0001	.0000	.0000	10
	11	1.0000	1.0000	1.0000	1.0000	1.0000	.9994	.9954	.5000	.1636	.0682	.0214	.0046	.0006	.0000	.0000	11
	12	1.0000	1.0000	1.0000	1.0000	1.0000	.9999	.9988	.6612	.2871	.1425	.0546	.0149	.0025	.0000	.0000	12
	13	1.0000	1.0000	1.0000	1.0000	1.0000	1.0000	.9997	.7976	.4438	.2592	.1201	.0408	.0089	.0000	.0000	13
	14	1.0000	1.0000	1.0000	1.0000	1.0000	1.0000	.9999	.8950	.6116	.4140	.2291	.0963	.0273	.0002	.0000	14
	15	1.0000	1.0000	1.0000	1.0000	1.0000	1.0000	1.0000	.9534	.7627	.5864	.3819	.1963	.0715	.0012	.0000	15
	16	1.0000	1.0000	1.0000	1.0000	1.0000	1.0000	1.0000	.9827	.8760	.7466	.5601	.3463	.1598	.0058	.0001	16
	17	1.0000	1.0000	1.0000	1.0000	1.0000	1.0000	1.0000	.9947	.9460	.8691	.7312	.5315	.3053	.0226	.0008	17
	18	1.0000	1.0000	1.0000	1.0000	1.0000	1.0000	1.0000	.9987	.9810	.9449	.8644	.7168	.4993	.0731	.0049	18
	19	1.0000	1.0000	1.0000	1.0000	1.0000	1.0000	1.0000	.9998	.9948	.9819	.9462	.8630	.7035	.1927	.0258	19
	20	1.0000	1.0000	1.0000	1.0000	1.0000	1.0000	1.0000	1.0000	.9990	.9957	.9843	.9508	.8668	.4080	.1052	20
	21	1.0000	1.0000	1.0000	1.0000	1.0000	1.0000	1.0000	1.0000	.9999	.9993	.9970	.9884	.9602	.6849	.3206	21
	22	1.0000	1.0000	1.0000	1.0000	1.0000	1.0000	1.0000	1.0000	1.0000	.9999	.9997	.9987	.9941	.9114	.6926	22
		P=.001	.01	.03	.05	.10	.20	.25	.50	.60	.65	.70	.75	.80	.90	.95	x

Table E (continued)

Binomial Distribution — Cumulative Terms

$$F_b(x|n,p) = \sum_{t=0}^{x} \binom{n}{t} p^t(1-p)^{n-t}$$

N	X	P=.001	.01	.03	.05	.10	.20	.25	.50	.60	.65	.70	.75	.80	.90	.95	X
24	0	.9763	.7857	.4814	.2920	.0798	.0047	.0010	.0000	.0000	.0000	.0000	.0000	.0000	.0000	.0000	0
	1	.9997	.9761	.8388	.6608	.2925	.0331	.0090	.0000	.0000	.0000	.0000	.0000	.0000	.0000	.0000	1
	2	1.0000	.9983	.9659	.8841	.5643	.1145	.0398	.0000	.0000	.0000	.0000	.0000	.0000	.0000	.0000	2
	3	1.0000	.9999	.9947	.9702	.7857	.2639	.1150	.0001	.0000	.0000	.0000	.0000	.0000	.0000	.0000	3
	4	1.0000	1.0000	.9994	.9940	.9149	.4599	.2466	.0008	.0000	.0000	.0000	.0000	.0000	.0000	.0000	4
	5	1.0000	1.0000	.9999	.9990	.9723	.6559	.4222	.0033	.0001	.0000	.0000	.0000	.0000	.0000	.0000	5
	6	1.0000	1.0000	1.0000	.9999	.9925	.8111	.6074	.0113	.0005	.0001	.0000	.0000	.0000	.0000	.0000	6
	7	1.0000	1.0000	1.0000	1.0000	.9983	.9108	.7662	.0320	.0022	.0004	.0000	.0000	.0000	.0000	.0000	7
	8	1.0000	1.0000	1.0000	1.0000	.9997	.9638	.8787	.0758	.0075	.0016	.0002	.0000	.0000	.0000	.0000	8
	9	1.0000	1.0000	1.0000	1.0000	.9999	.9874	.9453	.1537	.0217	.0055	.0010	.0001	.0000	.0000	.0000	9
	10	1.0000	1.0000	1.0000	1.0000	1.0000	.9962	.9787	.2706	.0535	.0164	.0036	.0005	.0000	.0000	.0000	10
	11	1.0000	1.0000	1.0000	1.0000	1.0000	.9990	.9928	.4194	.1143	.0423	.0115	.0021	.0002	.0000	.0000	11
	12	1.0000	1.0000	1.0000	1.0000	1.0000	.9998	.9979	.5806	.2130	.0942	.0314	.0072	.0010	.0000	.0000	12
	13	1.0000	1.0000	1.0000	1.0000	1.0000	1.0000	.9995	.7294	.3498	.1833	.0742	.0213	.0038	.0000	.0000	13
	14	1.0000	1.0000	1.0000	1.0000	1.0000	1.0000	.9999	.8463	.5109	.3133	.1528	.0547	.0126	.0001	.0000	14
	15	1.0000	1.0000	1.0000	1.0000	1.0000	1.0000	1.0000	.9242	.6721	.4743	.2750	.1213	.0362	.0003	.0000	15
	16	1.0000	1.0000	1.0000	1.0000	1.0000	1.0000	1.0000	.9680	.8081	.6425	.4353	.2338	.0892	.0017	.0000	16
	17	1.0000	1.0000	1.0000	1.0000	1.0000	1.0000	1.0000	.9887	.9040	.7894	.6114	.3926	.1889	.0075	.0001	17
	18	1.0000	1.0000	1.0000	1.0000	1.0000	1.0000	1.0000	.9967	.9600	.8956	.7712	.5778	.3441	.0277	.0010	18
	19	1.0000	1.0000	1.0000	1.0000	1.0000	1.0000	1.0000	.9992	.9866	.9578	.8889	.7534	.5401	.0851	.0060	19
	20	1.0000	1.0000	1.0000	1.0000	1.0000	1.0000	1.0000	.9999	.9965	.9867	.9576	.8850	.7361	.2143	.0298	20
	21	1.0000	1.0000	1.0000	1.0000	1.0000	1.0000	1.0000	1.0000	.9993	.9970	.9881	.9602	.8855	.4357	.1159	21
	22	1.0000	1.0000	1.0000	1.0000	1.0000	1.0000	1.0000	1.0000	.9999	.9995	.9978	.9910	.9669	.7075	.3392	22
	23	1.0000	1.0000	1.0000	1.0000	1.0000	1.0000	1.0000	1.0000	1.0000	1.0000	.9998	.9990	.9953	.9202	.7080	23
25	0	.9753	.7778	.4670	.2774	.0718	.0038	.0008	.0000	.0000	.0000	.0000	.0000	.0000	.0000	.0000	0
	1	.9997	.9742	.8280	.6424	.2712	.0274	.0070	.0000	.0000	.0000	.0000	.0000	.0000	.0000	.0000	1
	2	1.0000	.9980	.9620	.8729	.5371	.0982	.0321	.0000	.0000	.0000	.0000	.0000	.0000	.0000	.0000	2
	3	1.0000	.9999	.9938	.9659	.7636	.2340	.0962	.0001	.0000	.0000	.0000	.0000	.0000	.0000	.0000	3
	4	1.0000	1.0000	.9992	.9928	.9020	.4207	.2137	.0005	.0000	.0000	.0000	.0000	.0000	.0000	.0000	4
	5	1.0000	1.0000	.9999	.9988	.9666	.6167	.3783	.0020	.0001	.0000	.0000	.0000	.0000	.0000	.0000	5
	6	1.0000	1.0000	1.0000	.9998	.9905	.7800	.5611	.0073	.0003	.0000	.0000	.0000	.0000	.0000	.0000	6
	7	1.0000	1.0000	1.0000	1.0000	.9977	.8909	.7265	.0216	.0012	.0002	.0000	.0000	.0000	.0000	.0000	7
	8	1.0000	1.0000	1.0000	1.0000	.9995	.9532	.8506	.0539	.0043	.0008	.0001	.0000	.0000	.0000	.0000	8
	9	1.0000	1.0000	1.0000	1.0000	.9999	.9827	.9287	.1148	.0132	.0029	.0005	.0000	.0000	.0000	.0000	9
	10	1.0000	1.0000	1.0000	1.0000	1.0000	.9944	.9703	.2122	.0344	.0093	.0018	.0002	.0000	.0000	.0000	10
	11	1.0000	1.0000	1.0000	1.0000	1.0000	.9985	.9893	.3450	.0778	.0255	.0060	.0009	.0001	.0000	.0000	11
	12	1.0000	1.0000	1.0000	1.0000	1.0000	.9996	.9966	.5000	.1538	.0604	.0175	.0034	.0004	.0000	.0000	12
	13	1.0000	1.0000	1.0000	1.0000	1.0000	.9999	.9991	.6550	.2677	.1254	.0442	.0107	.0015	.0000	.0000	13
	14	1.0000	1.0000	1.0000	1.0000	1.0000	1.0000	.9998	.7878	.4142	.2288	.0978	.0297	.0056	.0000	.0000	14
	15	1.0000	1.0000	1.0000	1.0000	1.0000	1.0000	1.0000	.8852	.5754	.3697	.1894	.0713	.0173	.0001	.0000	15
	16	1.0000	1.0000	1.0000	1.0000	1.0000	1.0000	1.0000	.9461	.7265	.5332	.3231	.1494	.0468	.0005	.0000	16
	17	1.0000	1.0000	1.0000	1.0000	1.0000	1.0000	1.0000	.9784	.8464	.6939	.4882	.2735	.1091	.0023	.0000	17
	18	1.0000	1.0000	1.0000	1.0000	1.0000	1.0000	1.0000	.9927	.9264	.8266	.6593	.4389	.2200	.0095	.0002	18
	19	1.0000	1.0000	1.0000	1.0000	1.0000	1.0000	1.0000	.9980	.9706	.9174	.8065	.6217	.3833	.0334	.0012	19
	20	1.0000	1.0000	1.0000	1.0000	1.0000	1.0000	1.0000	.9995	.9905	.9680	.9095	.7863	.5793	.0980	.0072	20
	21	1.0000	1.0000	1.0000	1.0000	1.0000	1.0000	1.0000	.9999	.9976	.9903	.9668	.9038	.7660	.2364	.0341	21
	22	1.0000	1.0000	1.0000	1.0000	1.0000	1.0000	1.0000	1.0000	.9996	.9979	.9910	.9679	.9018	.4629	.1271	22
		P=.001	.01	.03	.05	.10	.20	.25	.50	.60	.65	.70	.75	.80	.90	.95	X

Table E (continued)

Binomial Distribution — Cumulative Terms

$$F_b(x|n,p) = \sum_{t=0}^{x} \binom{n}{t} p^t (1-p)^{n-t}$$

N	X	P=.001	.01	.03	.05	.10	.20	.25	.50	.60	.65	.70	.75	.80	.90	.95	X
25	23	1.0000	1.0000	1.0000	1.0000	1.0000	1.0000	1.0000	1.0000	.9999	.9997	.9984	.9930	.9726	.7288	.3576	23
	24	1.0000	1.0000	1.0000	1.0000	1.0000	1.0000	1.0000	1.0000	1.0000	1.0000	.9999	.9992	.9962	.9282	.7226	24
50	0	.9512	.6050	.2181	.0769	.0052	.0000	.0000	.0000	.0000	.0000	.0000	.0000	.0000	.0000	.0000	0
	1	.9988	.9106	.5553	.2794	.0338	.0002	.0000	.0000	.0000	.0000	.0000	.0000	.0000	.0000	.0000	1
	2	1.0000	.9862	.8108	.5405	.1117	.0013	.0001	.0000	.0000	.0000	.0000	.0000	.0000	.0000	.0000	2
	3	1.0000	.9984	.9372	.7604	.2503	.0057	.0005	.0000	.0000	.0000	.0000	.0000	.0000	.0000	.0000	3
	4	1.0000	.9999	.9832	.8964	.4312	.0185	.0021	.0000	.0000	.0000	.0000	.0000	.0000	.0000	.0000	4
	5	1.0000	1.0000	.9963	.9622	.6161	.0480	.0070	.0000	.0000	.0000	.0000	.0000	.0000	.0000	.0000	5
	6	1.0000	1.0000	.9993	.9882	.7702	.1034	.0194	.0000	.0000	.0000	.0000	.0000	.0000	.0000	.0000	6
	7	1.0000	1.0000	.9999	.9968	.8779	.1904	.0453	.0000	.0000	.0000	.0000	.0000	.0000	.0000	.0000	7
	8	1.0000	1.0000	1.0000	.9992	.9421	.3073	.0916	.0000	.0000	.0000	.0000	.0000	.0000	.0000	.0000	8
	9	1.0000	1.0000	1.0000	.9998	.9755	.4437	.1637	.0000	.0000	.0000	.0000	.0000	.0000	.0000	.0000	9
	10	1.0000	1.0000	1.0000	1.0000	.9906	.5836	.2622	.0000	.0000	.0000	.0000	.0000	.0000	.0000	.0000	10
	11	1.0000	1.0000	1.0000	1.0000	.9968	.7107	.3816	.0000	.0000	.0000	.0000	.0000	.0000	.0000	.0000	11
	12	1.0000	1.0000	1.0000	1.0000	.9990	.8139	.5110	.0002	.0000	.0000	.0000	.0000	.0000	.0000	.0000	12
	13	1.0000	1.0000	1.0000	1.0000	.9997	.8894	.6370	.0005	.0000	.0000	.0000	.0000	.0000	.0000	.0000	13
	14	1.0000	1.0000	1.0000	1.0000	.9999	.9393	.7481	.0013	.0000	.0000	.0000	.0000	.0000	.0000	.0000	14
	15	1.0000	1.0000	1.0000	1.0000	1.0000	.9692	.8369	.0033	.0000	.0000	.0000	.0000	.0000	.0000	.0000	15
	16	1.0000	1.0000	1.0000	1.0000	1.0000	.9856	.9017	.0077	.0001	.0000	.0000	.0000	.0000	.0000	.0000	16
	17	1.0000	1.0000	1.0000	1.0000	1.0000	.9937	.9449	.0164	.0002	.0000	.0000	.0000	.0000	.0000	.0000	17
	18	1.0000	1.0000	1.0000	1.0000	1.0000	.9975	.9713	.0325	.0005	.0000	.0000	.0000	.0000	.0000	.0000	18
	19	1.0000	1.0000	1.0000	1.0000	1.0000	.9991	.9861	.0595	.0014	.0001	.0000	.0000	.0000	.0000	.0000	19
	20	1.0000	1.0000	1.0000	1.0000	1.0000	.9997	.9937	.1013	.0034	.0003	.0000	.0000	.0000	.0000	.0000	20
	21	1.0000	1.0000	1.0000	1.0000	1.0000	.9999	.9974	.1611	.0076	.0007	.0000	.0000	.0000	.0000	.0000	21
	22	1.0000	1.0000	1.0000	1.0000	1.0000	1.0000	.9990	.2399	.0160	.0019	.0001	.0000	.0000	.0000	.0000	22
	23	1.0000	1.0000	1.0000	1.0000	1.0000	1.0000	.9996	.3359	.0314	.0045	.0003	.0000	.0000	.0000	.0000	23
	24	1.0000	1.0000	1.0000	1.0000	1.0000	1.0000	.9999	.4439	.0573	.0100	.0009	.0000	.0000	.0000	.0000	24
	25	1.0000	1.0000	1.0000	1.0000	1.0000	1.0000	1.0000	.5561	.0978	.0207	.0024	.0001	.0000	.0000	.0000	25
	26	1.0000	1.0000	1.0000	1.0000	1.0000	1.0000	1.0000	.6641	.1562	.0396	.0056	.0004	.0000	.0000	.0000	26
	27	1.0000	1.0000	1.0000	1.0000	1.0000	1.0000	1.0000	.7601	.2340	.0710	.0123	.0010	.0000	.0000	.0000	27
	28	1.0000	1.0000	1.0000	1.0000	1.0000	1.0000	1.0000	.8389	.3299	.1187	.0251	.0026	.0001	.0000	.0000	28
	29	1.0000	1.0000	1.0000	1.0000	1.0000	1.0000	1.0000	.8987	.4390	.1861	.0478	.0063	.0003	.0000	.0000	29
	30	1.0000	1.0000	1.0000	1.0000	1.0000	1.0000	1.0000	.9405	.5535	.2736	.0848	.0139	.0009	.0000	.0000	30
	31	1.0000	1.0000	1.0000	1.0000	1.0000	1.0000	1.0000	.9675	.6644	.3784	.1406	.0287	.0025	.0000	.0000	31
	32	1.0000	1.0000	1.0000	1.0000	1.0000	1.0000	1.0000	.9836	.7631	.4940	.2178	.0551	.0063	.0000	.0000	32
	33	1.0000	1.0000	1.0000	1.0000	1.0000	1.0000	1.0000	.9923	.8439	.6111	.3161	.0983	.0144	.0000	.0000	33
	34	1.0000	1.0000	1.0000	1.0000	1.0000	1.0000	1.0000	.9967	.9045	.7199	.4308	.1631	.0308	.0000	.0000	34
	35	1.0000	1.0000	1.0000	1.0000	1.0000	1.0000	1.0000	.9987	.9460	.8122	.5532	.2519	.0607	.0001	.0000	35
	36	1.0000	1.0000	1.0000	1.0000	1.0000	1.0000	1.0000	.9995	.9720	.8837	.6721	.3630	.1106	.0003	.0000	36
	37	1.0000	1.0000	1.0000	1.0000	1.0000	1.0000	1.0000	.9998	.9867	.9339	.7771	.4890	.1861	.0010	.0000	37
	38	1.0000	1.0000	1.0000	1.0000	1.0000	1.0000	1.0000	1.0000	.9943	.9658	.8610	.6184	.2893	.0032	.0000	38
	39	1.0000	1.0000	1.0000	1.0000	1.0000	1.0000	1.0000	1.0000	.9978	.9840	.9211	.7378	.4164	.0094	.0000	39
	40	1.0000	1.0000	1.0000	1.0000	1.0000	1.0000	1.0000	1.0000	.9992	.9933	.9598	.8363	.5563	.0245	.0002	40
	41	1.0000	1.0000	1.0000	1.0000	1.0000	1.0000	1.0000	1.0000	.9998	.9975	.9817	.9084	.6927	.0579	.0008	41
	42	1.0000	1.0000	1.0000	1.0000	1.0000	1.0000	1.0000	1.0000	.9999	.9992	.9927	.9547	.8096	.1221	.0032	42
	43	1.0000	1.0000	1.0000	1.0000	1.0000	1.0000	1.0000	1.0000	1.0000	.9998	.9975	.9806	.8966	.2298	.0118	43
	44	1.0000	1.0000	1.0000	1.0000	1.0000	1.0000	1.0000	1.0000	1.0000	.9999	.9993	.9930	.9520	.3839	.0378	44
		P=.001	.01	.03	.05	.10	.20	.25	.50	.60	.65	.70	.75	.80	.90	.95	X

Table E (continued)

Binomial Distribution — Cumulative Terms

$$F_b(x|n,p) = \sum_{t=0}^{x} \binom{n}{t} p^t (1-p)^{n-t}$$

N	X	P=.001	.01	.03	.05	.10	.20	.25	.50	.60	.65	.70	.75	.80	.90	.95	X
50	45	1.0000	1.0000	1.0000	1.0000	1.0000	1.0000	1.0000	1.0000	1.0000	1.0000	.9998	.9979	.9815	.5688	.1036	45
	46	1.0000	1.0000	1.0000	1.0000	1.0000	1.0000	1.0000	1.0000	1.0000	1.0000	1.0000	.9995	.9943	.7497	.2396	46
	47	1.0000	1.0000	1.0000	1.0000	1.0000	1.0000	1.0000	1.0000	1.0000	1.0000	1.0000	.9999	.9987	.8883	.4595	47
	48	1.0000	1.0000	1.0000	1.0000	1.0000	1.0000	1.0000	1.0000	1.0000	1.0000	1.0000	1.0000	.9998	.9662	.7206	48
	49	1.0000	1.0000	1.0000	1.0000	1.0000	1.0000	1.0000	1.0000	1.0000	1.0000	1.0000	1.0000	1.0000	.9948	.9231	49
75	0	.9277	.4706	.1018	.0213	.0004	.0000	.0000	.0000	.0000	.0000	.0000	.0000	.0000	.0000	.0000	0
	1	.9974	.8271	.3380	.1056	.0035	.0000	.0000	.0000	.0000	.0000	.0000	.0000	.0000	.0000	.0000	1
	2	.9999	.9603	.6083	.2697	.0161	.0000	.0000	.0000	.0000	.0000	.0000	.0000	.0000	.0000	.0000	2
	3	1.0000	.9931	.8118	.4798	.0504	.0001	.0000	.0000	.0000	.0000	.0000	.0000	.0000	.0000	.0000	3
	4	1.0000	.9990	.9250	.6789	.1189	.0003	.0000	.0000	.0000	.0000	.0000	.0000	.0000	.0000	.0000	4
	5	1.0000	.9999	.9747	.8276	.2271	.0012	.0000	.0000	.0000	.0000	.0000	.0000	.0000	.0000	.0000	5
	6	1.0000	1.0000	.9927	.9190	.3673	.0039	.0002	.0000	.0000	.0000	.0000	.0000	.0000	.0000	.0000	6
	7	1.0000	1.0000	.9981	.9664	.5208	.0104	.0005	.0000	.0000	.0000	.0000	.0000	.0000	.0000	.0000	7
	8	1.0000	1.0000	.9996	.9876	.6658	.0243	.0016	.0000	.0000	.0000	.0000	.0000	.0000	.0000	.0000	8
	9	1.0000	1.0000	.9999	.9959	.7858	.0501	.0044	.0000	.0000	.0000	.0000	.0000	.0000	.0000	.0000	9
	10	1.0000	1.0000	1.0000	.9988	.8737	.0928	.0103	.0000	.0000	.0000	.0000	.0000	.0000	.0000	.0000	10
	11	1.0000	1.0000	1.0000	.9997	.9315	.1557	.0221	.0000	.0000	.0000	.0000	.0000	.0000	.0000	.0000	11
	12	1.0000	1.0000	1.0000	.9999	.9657	.2397	.0431	.0000	.0000	.0000	.0000	.0000	.0000	.0000	.0000	12
	13	1.0000	1.0000	1.0000	1.0000	.9841	.3414	.0769	.0000	.0000	.0000	.0000	.0000	.0000	.0000	.0000	13
	14	1.0000	1.0000	1.0000	1.0000	.9932	.4540	.1269	.0000	.0000	.0000	.0000	.0000	.0000	.0000	.0000	14
	15	1.0000	1.0000	1.0000	1.0000	.9973	.5685	.1946	.0000	.0000	.0000	.0000	.0000	.0000	.0000	.0000	15
	16	1.0000	1.0000	1.0000	1.0000	.9990	.6759	.2792	.0000	.0000	.0000	.0000	.0000	.0000	.0000	.0000	16
	17	1.0000	1.0000	1.0000	1.0000	.9996	.7690	.3772	.0000	.0000	.0000	.0000	.0000	.0000	.0000	.0000	17
	18	1.0000	1.0000	1.0000	1.0000	.9999	.8440	.4823	.0000	.0000	.0000	.0000	.0000	.0000	.0000	.0000	18
	19	1.0000	1.0000	1.0000	1.0000	1.0000	.9003	.5875	.0000	.0000	.0000	.0000	.0000	.0000	.0000	.0000	19
	20	1.0000	1.0000	1.0000	1.0000	1.0000	.9397	.6857	.0000	.0000	.0000	.0000	.0000	.0000	.0000	.0000	20
	21	1.0000	1.0000	1.0000	1.0000	1.0000	.9654	.7714	.0001	.0000	.0000	.0000	.0000	.0000	.0000	.0000	21
	22	1.0000	1.0000	1.0000	1.0000	1.0000	.9813	.8415	.0002	.0000	.0000	.0000	.0000	.0000	.0000	.0000	22
	23	1.0000	1.0000	1.0000	1.0000	1.0000	.9904	.8954	.0005	.0000	.0000	.0000	.0000	.0000	.0000	.0000	23
	24	1.0000	1.0000	1.0000	1.0000	1.0000	.9953	.9343	.0012	.0000	.0000	.0000	.0000	.0000	.0000	.0000	24
	25	1.0000	1.0000	1.0000	1.0000	1.0000	.9978	.9607	.0026	.0000	.0000	.0000	.0000	.0000	.0000	.0000	25
	26	1.0000	1.0000	1.0000	1.0000	1.0000	.9991	.9777	.0053	.0000	.0000	.0000	.0000	.0000	.0000	.0000	26
	27	1.0000	1.0000	1.0000	1.0000	1.0000	.9996	.9879	.0101	.0000	.0000	.0000	.0000	.0000	.0000	.0000	27
	28	1.0000	1.0000	1.0000	1.0000	1.0000	.9998	.9938	.0185	.0001	.0000	.0000	.0000	.0000	.0000	.0000	28
	29	1.0000	1.0000	1.0000	1.0000	1.0000	.9999	.9970	.0320	.0002	.0000	.0000	.0000	.0000	.0000	.0000	29
	30	1.0000	1.0000	1.0000	1.0000	1.0000	1.0000	.9986	.0527	.0004	.0000	.0000	.0000	.0000	.0000	.0000	30
	31	1.0000	1.0000	1.0000	1.0000	1.0000	1.0000	.9994	.0827	.0008	.0000	.0000	.0000	.0000	.0000	.0000	31
	32	1.0000	1.0000	1.0000	1.0000	1.0000	1.0000	.9997	.1240	.0018	.0001	.0000	.0000	.0000	.0000	.0000	32
	33	1.0000	1.0000	1.0000	1.0000	1.0000	1.0000	.9999	.1778	.0037	.0002	.0000	.0000	.0000	.0000	.0000	33
	34	1.0000	1.0000	1.0000	1.0000	1.0000	1.0000	1.0000	.2443	.0072	.0004	.0000	.0000	.0000	.0000	.0000	34
	35	1.0000	1.0000	1.0000	1.0000	1.0000	1.0000	1.0000	.3222	.0133	.0009	.0000	.0000	.0000	.0000	.0000	35
	36	1.0000	1.0000	1.0000	1.0000	1.0000	1.0000	1.0000	.4088	.0235	.0019	.0001	.0000	.0000	.0000	.0000	36
	37	1.0000	1.0000	1.0000	1.0000	1.0000	1.0000	1.0000	.5000	.0396	.0038	.0001	.0000	.0000	.0000	.0000	37
	38	1.0000	1.0000	1.0000	1.0000	1.0000	1.0000	1.0000	.5912	.0637	.0074	.0003	.0000	.0000	.0000	.0000	38
	39	1.0000	1.0000	1.0000	1.0000	1.0000	1.0000	1.0000	.6778	.0981	.0138	.0008	.0000	.0000	.0000	.0000	39
	40	1.0000	1.0000	1.0000	1.0000	1.0000	1.0000	1.0000	.7557	.1446	.0245	.0017	.0000	.0000	.0000	.0000	40
	41	1.0000	1.0000	1.0000	1.0000	1.0000	1.0000	1.0000	.8221	.2041	.0414	.0036	.0001	.0000	.0000	.0000	41
		P=.001	.01	.03	.05	.10	.20	.25	.50	.60	.65	.70	.75	.80	.90	.95	X

Table E (continued)

Binomial Distribution — Cumulative Terms

$$F_b(x|n,p) = \sum_{t=0}^{x} \binom{n}{t} p^t (1-p)^{n-t}$$

N	X	P=.001	.01	.03	.05	.10	.20	.25	.50	.60	.65	.70	.75	.80	.90	.95	X
75	42	1.0000	1.0000	1.0000	1.0000	1.0000	1.0000	1.0000	.8760	.2763	.0668	.0071	.0003	.0000	.0000	.0000	42
	43	1.0000	1.0000	1.0000	1.0000	1.0000	1.0000	1.0000	.9173	.3594	.1030	.0134	.0006	.0000	.0000	.0000	43
	44	1.0000	1.0000	1.0000	1.0000	1.0000	1.0000	1.0000	.9473	.4501	.1519	.0242	.0014	.0000	.0000	.0000	44
	45	1.0000	1.0000	1.0000	1.0000	1.0000	1.0000	1.0000	.9680	.5438	.2144	.0414	.0030	.0001	.0000	.0000	45
	46	1.0000	1.0000	1.0000	1.0000	1.0000	1.0000	1.0000	.9815	.6354	.2902	.0676	.0062	.0002	.0000	.0000	46
	47	1.0000	1.0000	1.0000	1.0000	1.0000	1.0000	1.0000	.9898	.7203	.3770	.1053	.0120	.0004	.0000	.0000	47
	48	1.0000	1.0000	1.0000	1.0000	1.0000	1.0000	1.0000	.9947	.7945	.4711	.1567	.0223	.0009	.0000	.0000	48
	49	1.0000	1.0000	1.0000	1.0000	1.0000	1.0000	1.0000	.9974	.8558	.5673	.2227	.0393	.0021	.0000	.0000	49
	50	1.0000	1.0000	1.0000	1.0000	1.0000	1.0000	1.0000	.9988	.9037	.6603	.3029	.0657	.0047	.0000	.0000	50
	51	1.0000	1.0000	1.0000	1.0000	1.0000	1.0000	1.0000	.9995	.9389	.7449	.3945	.1046	.0096	.0000	.0000	51
	52	1.0000	1.0000	1.0000	1.0000	1.0000	1.0000	1.0000	.9998	.9633	.8174	.4932	.1585	.0187	.0000	.0000	52
	53	1.0000	1.0000	1.0000	1.0000	1.0000	1.0000	1.0000	.9999	.9791	.8759	.5932	.2286	.0345	.0000	.0000	53
	54	1.0000	1.0000	1.0000	1.0000	1.0000	1.0000	1.0000	1.0000	.9888	.9201	.6882	.3143	.0603	.0000	.0000	54
	55	1.0000	1.0000	1.0000	1.0000	1.0000	1.0000	1.0000	1.0000	.9944	.9515	.7729	.4125	.0997	.0000	.0000	55
	56	1.0000	1.0000	1.0000	1.0000	1.0000	1.0000	1.0000	1.0000	.9973	.9723	.8434	.5176	.1560	.0001	.0000	56
	57	1.0000	1.0000	1.0000	1.0000	1.0000	1.0000	1.0000	1.0000	.9988	.9851	.8983	.6228	.2310	.0003	.0000	57
	58	1.0000	1.0000	1.0000	1.0000	1.0000	1.0000	1.0000	1.0000	.9995	.9926	.9380	.7208	.3241	.0010	.0000	58
	59	1.0000	1.0000	1.0000	1.0000	1.0000	1.0000	1.0000	1.0000	.9998	.9965	.9647	.8054	.4315	.0027	.0000	59
	60	1.0000	1.0000	1.0000	1.0000	1.0000	1.0000	1.0000	1.0000	.9999	.9985	.9813	.8731	.5460	.0068	.0000	60
	61	1.0000	1.0000	1.0000	1.0000	1.0000	1.0000	1.0000	1.0000	1.0000	.9994	.9909	.9231	.6586	.0159	.0000	61
	62	1.0000	1.0000	1.0000	1.0000	1.0000	1.0000	1.0000	1.0000	1.0000	.9998	.9959	.9569	.7603	.0343	.0001	62
	63	1.0000	1.0000	1.0000	1.0000	1.0000	1.0000	1.0000	1.0000	1.0000	.9999	.9983	.9779	.8443	.0685	.0003	63
	64	1.0000	1.0000	1.0000	1.0000	1.0000	1.0000	1.0000	1.0000	1.0000	1.0000	.9994	.9897	.9072	.1263	.0012	64
	65	1.0000	1.0000	1.0000	1.0000	1.0000	1.0000	1.0000	1.0000	1.0000	1.0000	.9998	.9956	.9499	.2142	.0041	65
	66	1.0000	1.0000	1.0000	1.0000	1.0000	1.0000	1.0000	1.0000	1.0000	1.0000	.9999	.9984	.9757	.3342	.0124	66
	67	1.0000	1.0000	1.0000	1.0000	1.0000	1.0000	1.0000	1.0000	1.0000	1.0000	1.0000	.9995	.9896	.4792	.0336	67
	68	1.0000	1.0000	1.0000	1.0000	1.0000	1.0000	1.0000	1.0000	1.0000	1.0000	1.0000	.9998	.9961	.6327	.0810	68
	69	1.0000	1.0000	1.0000	1.0000	1.0000	1.0000	1.0000	1.0000	1.0000	1.0000	1.0000	1.0000	.9988	.7729	.1724	69
	70	1.0000	1.0000	1.0000	1.0000	1.0000	1.0000	1.0000	1.0000	1.0000	1.0000	1.0000	1.0000	.9997	.8811	.3211	70
	71	1.0000	1.0000	1.0000	1.0000	1.0000	1.0000	1.0000	1.0000	1.0000	1.0000	1.0000	1.0000	.9999	.9496	.5202	71
	72	1.0000	1.0000	1.0000	1.0000	1.0000	1.0000	1.0000	1.0000	1.0000	1.0000	1.0000	1.0000	1.0000	.9839	.7303	72
	73	1.0000	1.0000	1.0000	1.0000	1.0000	1.0000	1.0000	1.0000	1.0000	1.0000	1.0000	1.0000	1.0000	.9965	.8944	73
	74	1.0000	1.0000	1.0000	1.0000	1.0000	1.0000	1.0000	1.0000	1.0000	1.0000	1.0000	1.0000	1.0000	.9996	.9787	74
100	0	.9048	.3660	.0476	.0059	.0000	.0000	.0000	.0000	.0000	.0000	.0000	.0000	.0000	.0000	.0000	0
	1	.9954	.7358	.1946	.0371	.0003	.0000	.0000	.0000	.0000	.0000	.0000	.0000	.0000	.0000	.0000	1
	2	.9998	.9206	.4198	.1183	.0019	.0000	.0000	.0000	.0000	.0000	.0000	.0000	.0000	.0000	.0000	2
	3	1.0000	.9816	.6472	.2578	.0078	.0000	.0000	.0000	.0000	.0000	.0000	.0000	.0000	.0000	.0000	3
	4	1.0000	.9966	.8179	.4360	.0237	.0000	.0000	.0000	.0000	.0000	.0000	.0000	.0000	.0000	.0000	4
	5	1.0000	.9995	.9192	.6160	.0576	.0000	.0000	.0000	.0000	.0000	.0000	.0000	.0000	.0000	.0000	5
	6	1.0000	.9999	.9688	.7660	.1172	.0001	.0000	.0000	.0000	.0000	.0000	.0000	.0000	.0000	.0000	6
	7	1.0000	1.0000	.9894	.8720	.2060	.0003	.0000	.0000	.0000	.0000	.0000	.0000	.0000	.0000	.0000	7
	8	1.0000	1.0000	.9968	.9369	.3209	.0009	.0000	.0000	.0000	.0000	.0000	.0000	.0000	.0000	.0000	8
	9	1.0000	1.0000	.9991	.9718	.4513	.0023	.0000	.0000	.0000	.0000	.0000	.0000	.0000	.0000	.0000	9
	10	1.0000	1.0000	.9998	.9885	.5832	.0057	.0001	.0000	.0000	.0000	.0000	.0000	.0000	.0000	.0000	10
	11	1.0000	1.0000	1.0000	.9957	.7030	.0126	.0004	.0000	.0000	.0000	.0000	.0000	.0000	.0000	.0000	11
	12	1.0000	1.0000	1.0000	.9985	.8018	.0253	.0010	.0000	.0000	.0000	.0000	.0000	.0000	.0000	.0000	12
	13	1.0000	1.0000	1.0000	.9995	.8761	.0469	.0025	.0000	.0000	.0000	.0000	.0000	.0000	.0000	.0000	13
		P=.001	.01	.03	.05	.10	.20	.25	.50	.60	.65	.70	.75	.80	.90	.95	X

Table E (continued)

Binomial Distribution — Cumulative Terms

$$F_b(x|n,p) = \sum_{t=0}^{x} \binom{n}{t} p^t (1-p)^{n-t}$$

N	X	P=.001	.01	.03	.05	.10	.20	.25	.50	.60	.65	.70	.75	.80	.90	.95	X
100	14	1.0000	1.0000	1.0000	.9999	.9274	.0804	.0054	.0000	.0000	.0000	.0000	.0000	.0000	.0000	.0000	14
	15	1.0000	1.0000	1.0000	1.0000	.9601	.1285	.0111	.0000	.0000	.0000	.0000	.0000	.0000	.0000	.0000	15
	16	1.0000	1.0000	1.0000	1.0000	.9794	.1923	.0211	.0000	.0000	.0000	.0000	.0000	.0000	.0000	.0000	16
	17	1.0000	1.0000	1.0000	1.0000	.9900	.2712	.0376	.0000	.0000	.0000	.0000	.0000	.0000	.0000	.0000	17
	18	1.0000	1.0000	1.0000	1.0000	.9954	.3621	.0630	.0000	.0000	.0000	.0000	.0000	.0000	.0000	.0000	18
	19	1.0000	1.0000	1.0000	1.0000	.9980	.4602	.0995	.0000	.0000	.0000	.0000	.0000	.0000	.0000	.0000	19
	20	1.0000	1.0000	1.0000	1.0000	.9992	.5595	.1488	.0000	.0000	.0000	.0000	.0000	.0000	.0000	.0000	20
	21	1.0000	1.0000	1.0000	1.0000	.9997	.6540	.2114	.0000	.0000	.0000	.0000	.0000	.0000	.0000	.0000	21
	22	1.0000	1.0000	1.0000	1.0000	.9999	.7389	.2864	.0000	.0000	.0000	.0000	.0000	.0000	.0000	.0000	22
	23	1.0000	1.0000	1.0000	1.0000	1.0000	.8109	.3711	.0000	.0000	.0000	.0000	.0000	.0000	.0000	.0000	23
	24	1.0000	1.0000	1.0000	1.0000	1.0000	.8686	.4617	.0000	.0000	.0000	.0000	.0000	.0000	.0000	.0000	24
	25	1.0000	1.0000	1.0000	1.0000	1.0000	.9125	.5535	.0000	.0000	.0000	.0000	.0000	.0000	.0000	.0000	25
	26	1.0000	1.0000	1.0000	1.0000	1.0000	.9442	.6417	.0000	.0000	.0000	.0000	.0000	.0000	.0000	.0000	26
	27	1.0000	1.0000	1.0000	1.0000	1.0000	.9658	.7224	.0000	.0000	.0000	.0000	.0000	.0000	.0000	.0000	27
	28	1.0000	1.0000	1.0000	1.0000	1.0000	.9800	.7925	.0000	.0000	.0000	.0000	.0000	.0000	.0000	.0000	28
	29	1.0000	1.0000	1.0000	1.0000	1.0000	.9887	.8505	.0000	.0000	.0000	.0000	.0000	.0000	.0000	.0000	29
	30	1.0000	1.0000	1.0000	1.0000	1.0000	.9939	.8962	.0000	.0000	.0000	.0000	.0000	.0000	.0000	.0000	30
	31	1.0000	1.0000	1.0000	1.0000	1.0000	.9969	.9306	.0001	.0000	.0000	.0000	.0000	.0000	.0000	.0000	31
	32	1.0000	1.0000	1.0000	1.0000	1.0000	.9984	.9554	.0002	.0000	.0000	.0000	.0000	.0000	.0000	.0000	32
	33	1.0000	1.0000	1.0000	1.0000	1.0000	.9993	.9724	.0004	.0000	.0000	.0000	.0000	.0000	.0000	.0000	33
	34	1.0000	1.0000	1.0000	1.0000	1.0000	.9997	.9836	.0009	.0000	.0000	.0000	.0000	.0000	.0000	.0000	34
	35	1.0000	1.0000	1.0000	1.0000	1.0000	.9998	.9906	.0018	.0000	.0000	.0000	.0000	.0000	.0000	.0000	35
	36	1.0000	1.0000	1.0000	1.0000	1.0000	.9999	.9948	.0033	.0000	.0000	.0000	.0000	.0000	.0000	.0000	36
	37	1.0000	1.0000	1.0000	1.0000	1.0000	1.0000	.9972	.0060	.0000	.0000	.0000	.0000	.0000	.0000	.0000	37
	38	1.0000	1.0000	1.0000	1.0000	1.0000	1.0000	.9986	.0105	.0000	.0000	.0000	.0000	.0000	.0000	.0000	38
	39	1.0000	1.0000	1.0000	1.0000	1.0000	1.0000	.9993	.0176	.0000	.0000	.0000	.0000	.0000	.0000	.0000	39
	40	1.0000	1.0000	1.0000	1.0000	1.0000	1.0000	.9997	.0284	.0000	.0000	.0000	.0000	.0000	.0000	.0000	40
	41	1.0000	1.0000	1.0000	1.0000	1.0000	1.0000	.9998	.0443	.0001	.0000	.0000	.0000	.0000	.0000	.0000	41
	42	1.0000	1.0000	1.0000	1.0000	1.0000	1.0000	.9999	.0666	.0002	.0000	.0000	.0000	.0000	.0000	.0000	42
	43	1.0000	1.0000	1.0000	1.0000	1.0000	1.0000	1.0000	.0967	.0004	.0000	.0000	.0000	.0000	.0000	.0000	43
	44	1.0000	1.0000	1.0000	1.0000	1.0000	1.0000	1.0000	.1356	.0009	.0000	.0000	.0000	.0000	.0000	.0000	44
	45	1.0000	1.0000	1.0000	1.0000	1.0000	1.0000	1.0000	.1841	.0017	.0000	.0000	.0000	.0000	.0000	.0000	45
	46	1.0000	1.0000	1.0000	1.0000	1.0000	1.0000	1.0000	.2421	.0032	.0001	.0000	.0000	.0000	.0000	.0000	46
	47	1.0000	1.0000	1.0000	1.0000	1.0000	1.0000	1.0000	.3086	.0058	.0002	.0000	.0000	.0000	.0000	.0000	47
	48	1.0000	1.0000	1.0000	1.0000	1.0000	1.0000	1.0000	.3822	.0100	.0004	.0000	.0000	.0000	.0000	.0000	48
	49	1.0000	1.0000	1.0000	1.0000	1.0000	1.0000	1.0000	.4602	.0168	.0007	.0000	.0000	.0000	.0000	.0000	49
	50	1.0000	1.0000	1.0000	1.0000	1.0000	1.0000	1.0000	.5398	.0271	.0015	.0000	.0000	.0000	.0000	.0000	50
	51	1.0000	1.0000	1.0000	1.0000	1.0000	1.0000	1.0000	.6178	.0423	.0027	.0001	.0000	.0000	.0000	.0000	51
	52	1.0000	1.0000	1.0000	1.0000	1.0000	1.0000	1.0000	.6913	.0638	.0050	.0001	.0000	.0000	.0000	.0000	52
	53	1.0000	1.0000	1.0000	1.0000	1.0000	1.0000	1.0000	.7579	.0930	.0088	.0003	.0000	.0000	.0000	.0000	53
	54	1.0000	1.0000	1.0000	1.0000	1.0000	1.0000	1.0000	.8159	.1311	.0150	.0005	.0000	.0000	.0000	.0000	54
	55	1.0000	1.0000	1.0000	1.0000	1.0000	1.0000	1.0000	.8644	.1789	.0246	.0011	.0000	.0000	.0000	.0000	55
	56	1.0000	1.0000	1.0000	1.0000	1.0000	1.0000	1.0000	.9033	.2365	.0389	.0021	.0000	.0000	.0000	.0000	56
	57	1.0000	1.0000	1.0000	1.0000	1.0000	1.0000	1.0000	.9334	.3033	.0594	.0040	.0001	.0000	.0000	.0000	57
	58	1.0000	1.0000	1.0000	1.0000	1.0000	1.0000	1.0000	.9557	.3775	.0877	.0072	.0001	.0000	.0000	.0000	58
	59	1.0000	1.0000	1.0000	1.0000	1.0000	1.0000	1.0000	.9715	.4567	.1250	.0125	.0003	.0000	.0000	.0000	59
		P=.001	.01	.03	.05	.10	.20	.25	.50	.60	.65	.70	.75	.80	.90	.95	X

Table E (continued)

Binomial Distribution — Cumulative Terms

$$F(x|n,p) = \sum_{t=0}^{x} \binom{n}{t} p^t (1-p)^{n-t}$$

N	X	P=.001	.01	.03	.05	.10	.20	.25	.50	.60	.65	.70	.75	.80	.90	.95	X
100	60	1.0000	1.0000	1.0000	1.0000	1.0000	1.0000	1.0000	.9824	.5379	.1724	.0210	.0007	.0000	.0000	.0000	60
	61	1.0000	1.0000	1.0000	1.0000	1.0000	1.0000	1.0000	.9895	.6178	.2301	.0340	.0014	.0000	.0000	.0000	61
	62	1.0000	1.0000	1.0000	1.0000	1.0000	1.0000	1.0000	.9940	.6932	.2975	.0530	.0027	.0000	.0000	.0000	62
	63	1.0000	1.0000	1.0000	1.0000	1.0000	1.0000	1.0000	.9967	.7614	.3731	.0799	.0052	.0001	.0000	.0000	63
	64	1.0000	1.0000	1.0000	1.0000	1.0000	1.0000	1.0000	.9982	.8205	.4542	.1161	.0094	.0001	.0000	.0000	64
	65	1.0000	1.0000	1.0000	1.0000	1.0000	1.0000	1.0000	.9991	.8697	.5376	.1629	.0164	.0003	.0000	.0000	65
	66	1.0000	1.0000	1.0000	1.0000	1.0000	1.0000	1.0000	.9996	.9087	.6197	.2207	.0276	.0007	.0000	.0000	66
	67	1.0000	1.0000	1.0000	1.0000	1.0000	1.0000	1.0000	.9998	.9385	.6971	.2893	.0446	.0016	.0000	.0000	67
	68	1.0000	1.0000	1.0000	1.0000	1.0000	1.0000	1.0000	.9999	.9601	.7669	.3669	.0693	.0031	.0000	.0000	68
	69	1.0000	1.0000	1.0000	1.0000	1.0000	1.0000	1.0000	1.0000	.9752	.8270	.4509	.1038	.0061	.0000	.0000	69
	70	1.0000	1.0000	1.0000	1.0000	1.0000	1.0000	1.0000	1.0000	.9852	.8764	.5377	.1495	.0112	.0000	.0000	70
	71	1.0000	1.0000	1.0000	1.0000	1.0000	1.0000	1.0000	1.0000	.9916	.9152	.6232	.2075	.0200	.0000	.0000	71
	72	1.0000	1.0000	1.0000	1.0000	1.0000	1.0000	1.0000	1.0000	.9954	.9442	.7036	.2776	.0342	.0000	.0000	72
	73	1.0000	1.0000	1.0000	1.0000	1.0000	1.0000	1.0000	1.0000	.9976	.9648	.7756	.3583	.0558	.0000	.0000	73
	74	1.0000	1.0000	1.0000	1.0000	1.0000	1.0000	1.0000	1.0000	.9988	.9788	.8369	.4465	.0875	.0000	.0000	74
	75	1.0000	1.0000	1.0000	1.0000	1.0000	1.0000	1.0000	1.0000	.9994	.9879	.8864	.5383	.1314	.0000	.0000	75
	76	1.0000	1.0000	1.0000	1.0000	1.0000	1.0000	1.0000	1.0000	.9997	.9934	.9245	.6289	.1891	.0000	.0000	76
	77	1.0000	1.0000	1.0000	1.0000	1.0000	1.0000	1.0000	1.0000	.9999	.9966	.9521	.7136	.2611	.0001	.0000	77
	78	1.0000	1.0000	1.0000	1.0000	1.0000	1.0000	1.0000	1.0000	.9999	.9983	.9712	.7886	.3460	.0003	.0000	78
	79	1.0000	1.0000	1.0000	1.0000	1.0000	1.0000	1.0000	1.0000	1.0000	.9992	.9835	.8512	.4405	.0008	.0000	79
	80	1.0000	1.0000	1.0000	1.0000	1.0000	1.0000	1.0000	1.0000	1.0000	.9996	.9911	.9005	.5398	.0020	.0000	80
	81	1.0000	1.0000	1.0000	1.0000	1.0000	1.0000	1.0000	1.0000	1.0000	.9998	.9955	.9370	.6379	.0046	.0000	81
	82	1.0000	1.0000	1.0000	1.0000	1.0000	1.0000	1.0000	1.0000	1.0000	.9999	.9978	.9624	.7288	.0100	.0000	82
	83	1.0000	1.0000	1.0000	1.0000	1.0000	1.0000	1.0000	1.0000	1.0000	1.0000	.9990	.9789	.8077	.0206	.0000	83
	84	1.0000	1.0000	1.0000	1.0000	1.0000	1.0000	1.0000	1.0000	1.0000	1.0000	.9996	.9889	.8715	.0399	.0000	84
	85	1.0000	1.0000	1.0000	1.0000	1.0000	1.0000	1.0000	1.0000	1.0000	1.0000	.9998	.9946	.9195	.0726	.0001	85
	86	1.0000	1.0000	1.0000	1.0000	1.0000	1.0000	1.0000	1.0000	1.0000	1.0000	.9999	.9975	.9531	.1239	.0005	86
	87	1.0000	1.0000	1.0000	1.0000	1.0000	1.0000	1.0000	1.0000	1.0000	1.0000	1.0000	.9990	.9747	.1982	.0015	87
	88	1.0000	1.0000	1.0000	1.0000	1.0000	1.0000	1.0000	1.0000	1.0000	1.0000	1.0000	.9996	.9874	.2970	.0043	88
	89	1.0000	1.0000	1.0000	1.0000	1.0000	1.0000	1.0000	1.0000	1.0000	1.0000	1.0000	.9999	.9943	.4168	.0115	89
	90	1.0000	1.0000	1.0000	1.0000	1.0000	1.0000	1.0000	1.0000	1.0000	1.0000	1.0000	.9999	.9977	.5487	.0282	90
	91	1.0000	1.0000	1.0000	1.0000	1.0000	1.0000	1.0000	1.0000	1.0000	1.0000	1.0000	1.0000	.9991	.6791	.0631	91
	92	1.0000	1.0000	1.0000	1.0000	1.0000	1.0000	1.0000	1.0000	1.0000	1.0000	1.0000	1.0000	.9997	.7939	.1280	92
	93	1.0000	1.0000	1.0000	1.0000	1.0000	1.0000	1.0000	1.0000	1.0000	1.0000	1.0000	1.0000	.9999	.8828	.2340	93
	94	1.0000	1.0000	1.0000	1.0000	1.0000	1.0000	1.0000	1.0000	1.0000	1.0000	1.0000	1.0000	1.0000	.9424	.3840	94
	95	1.0000	1.0000	1.0000	1.0000	1.0000	1.0000	1.0000	1.0000	1.0000	1.0000	1.0000	1.0000	1.0000	.9763	.5640	95
	96	1.0000	1.0000	1.0000	1.0000	1.0000	1.0000	1.0000	1.0000	1.0000	1.0000	1.0000	1.0000	1.0000	.9922	.7422	96
	97	1.0000	1.0000	1.0000	1.0000	1.0000	1.0000	1.0000	1.0000	1.0000	1.0000	1.0000	1.0000	1.0000	.9980	.8817	97
	98	1.0000	1.0000	1.0000	1.0000	1.0000	1.0000	1.0000	1.0000	1.0000	1.0000	1.0000	1.0000	1.0000	.9997	.9629	98
	99	1.0000	1.0000	1.0000	1.0000	1.0000	1.0000	1.0000	1.0000	1.0000	1.0000	1.0000	1.0000	1.0000	1.0000	.9941	99

Table F

Poisson Distribution — Individual Terms

$$f_P(x|\lambda,t) = \frac{(\lambda t)^x e^{-\lambda t}}{x!}$$

x	.001	.002	.003	.004	.005	.006	.007	.008	.009	0.010	x
0	.9990	.9980	.9970	.9960	.9950	.9940	.9930	.9920	.9910	.9900	0
1	.0010	.0020	.0030	.0040	.0050	.0060	.0070	.0079	.0089	.0099	1
2	.0000	.0000	.0000	.0000	.0000	.0000	.0000	.0000	.0000	.0000	2

x	.010	.020	.030	.040	.050	.060	.070	.080	.090	0.100	x
0	.9900	.9802	.9704	.9608	.9512	.9418	.9324	.9231	.9139	.9048	0
1	.0099	.0196	.0291	.0384	.0476	.0565	.0653	.0738	.0823	.0905	1
2	.0000	.0002	.0004	.0008	.0012	.0017	.0023	.0030	.0037	.0045	2
3	.0000	.0000	.0000	.0000	.0000	.0000	.0001	.0001	.0001	.0002	3
4	.0000	.0000	.0000	.0000	.0000	.0000	.0000	.0000	.0000	.0000	4

x	0.10	0.20	0.30	0.40	0.50	0.60	0.70	0.80	0.90	1.00	x
0	.9048	.8187	.7408	.6703	.6065	.5488	.4966	.4493	.4066	.3679	0
1	.0905	.1637	.2222	.2681	.3033	.3293	.3476	.3595	.3659	.3679	1
2	.0045	.0164	.0333	.0536	.0758	.0988	.1217	.1438	.1647	.1839	2
3	.0002	.0011	.0033	.0072	.0126	.0198	.0284	.0383	.0494	.0613	3
4	.0000	.0001	.0003	.0007	.0016	.0030	.0050	.0077	.0111	.0153	4
5	.0000	.0000	.0000	.0001	.0002	.0004	.0007	.0012	.0020	.0031	5
6	.0000	.0000	.0000	.0000	.0000	.0000	.0001	.0002	.0003	.0005	6
7	.0000	.0000	.0000	.0000	.0000	.0000	.0000	.0000	.0000	.0001	7
8	.0000	.0000	.0000	.0000	.0000	.0000	.0000	.0000	.0000	.0000	8

x	1.10	1.20	1.30	1.40	1.50	1.60	1.70	1.80	1.90	2.00	x
0	.3329	.3012	.2725	.2466	.2231	.2019	.1827	.1653	.1496	.1353	0
1	.3662	.3614	.3543	.3452	.3347	.3230	.3106	.2975	.2842	.2707	1
2	.2014	.2169	.2303	.2417	.2510	.2584	.2640	.2678	.2700	.2707	2
3	.0738	.0867	.0998	.1128	.1255	.1378	.1496	.1607	.1710	.1804	3
4	.0203	.0260	.0324	.0395	.0471	.0551	.0636	.0723	.0812	.0902	4
5	.0045	.0062	.0084	.0111	.0141	.0176	.0216	.0260	.0309	.0361	5
6	.0008	.0012	.0018	.0026	.0035	.0047	.0061	.0078	.0098	.0120	6
7	.0001	.0002	.0003	.0005	.0008	.0011	.0015	.0020	.0027	.0034	7
8	.0000	.0000	.0001	.0001	.0001	.0002	.0003	.0005	.0006	.0009	8
9	.0000	.0000	.0000	.0000	.0000	.0000	.0001	.0001	.0001	.0002	9
10	.0000	.0000	.0000	.0000	.0000	.0000	.0000	.0000	.0000	.0000	10

x	2.10	2.20	2.30	2.40	2.50	2.60	2.70	2.80	2.90	3.00	x
0	.1225	.1108	.1003	.0907	.0821	.0743	.0672	.0608	.0550	.0498	0
1	.2572	.2438	.2306	.2177	.2052	.1931	.1815	.1703	.1596	.1494	1
2	.2700	.2681	.2652	.2613	.2565	.2510	.2450	.2384	.2314	.2240	2
3	.1890	.1966	.2033	.2090	.2138	.2176	.2205	.2225	.2237	.2240	3
4	.0992	.1082	.1169	.1254	.1336	.1414	.1488	.1557	.1622	.1680	4
5	.0417	.0476	.0538	.0602	.0668	.0735	.0804	.0872	.0940	.1008	5
6	.0146	.0174	.0206	.0241	.0278	.0319	.0362	.0407	.0455	.0504	6
7	.0044	.0055	.0068	.0083	.0099	.0118	.0139	.0163	.0188	.0216	7
8	.0011	.0015	.0019	.0025	.0031	.0038	.0047	.0057	.0068	.0081	8
9	.0003	.0004	.0005	.0007	.0009	.0011	.0014	.0018	.0022	.0027	9
10	.0001	.0001	.0001	.0002	.0002	.0003	.0004	.0005	.0006	.0008	10
11	.0000	.0000	.0000	.0000	.0000	.0001	.0001	.0001	.0002	.0002	11
12	.0000	.0000	.0000	.0000	.0000	.0000	.0000	.0000	.0000	.0001	12
13	.0000	.0000	.0000	.0000	.0000	.0000	.0000	.0000	.0000	.0000	13

x	3.10	3.20	3.30	3.40	3.50	3.60	3.70	3.80	3.90	4.00	x
0	.0450	.0408	.0369	.0334	.0302	.0273	.0247	.0224	.0202	.0183	0
1	.1397	.1304	.1217	.1135	.1057	.0984	.0915	.0850	.0789	.0733	1
2	.2165	.2087	.2008	.1929	.1850	.1771	.1692	.1615	.1539	.1465	2
3	.2237	.2226	.2209	.2186	.2158	.2125	.2087	.2046	.2001	.1954	3
4	.1733	.1781	.1823	.1858	.1888	.1912	.1931	.1944	.1951	.1954	4
5	.1075	.1140	.1203	.1264	.1322	.1377	.1429	.1477	.1522	.1563	5
6	.0555	.0608	.0662	.0716	.0771	.0826	.0881	.0936	.0989	.1042	6
7	.0246	.0278	.0312	.0348	.0385	.0425	.0466	.0508	.0551	.0595	7
8	.0095	.0111	.0129	.0148	.0169	.0191	.0215	.0241	.0269	.0298	8
9	.0033	.0040	.0047	.0056	.0066	.0076	.0089	.0102	.0116	.0132	9

Table F (continued)

Poisson Distribution — Individual Terms

$$f_P(x|\lambda,t) = \frac{(\lambda t)^x e^{-\lambda t}}{x!}$$

x	3.10	3.20	3.30	3.40	3.50	3.60	3.70	3.80	3.90	4.00	x
10	.0010	.0013	.0016	.0019	.0023	.0028	.0033	.0039	.0045	.0053	10
11	.0003	.0004	.0005	.0006	.0007	.0009	.0011	.0013	.0016	.0019	11
12	.0001	.0001	.0001	.0002	.0002	.0003	.0003	.0004	.0005	.0006	12
13	.0000	.0000	.0000	.0000	.0001	.0001	.0001	.0001	.0002	.0002	13
14	.0000	.0000	.0000	.0000	.0000	.0000	.0000	.0000	.0000	.0001	14
15	.0000	.0000	.0000	.0000	.0000	.0000	.0000	.0000	.0000	.0000	15

x	4.10	4.20	4.30	4.40	4.50	4.60	4.70	4.80	4.90	5.00	x
0	.0166	.0150	.0136	.0123	.0111	.0101	.0091	.0082	.0074	.0067	0
1	.0679	.0630	.0583	.0540	.0500	.0462	.0427	.0395	.0365	.0337	1
2	.1393	.1323	.1254	.1188	.1125	.1063	.1005	.0948	.0894	.0842	2
3	.1904	.1852	.1798	.1743	.1687	.1631	.1574	.1517	.1460	.1404	3
4	.1951	.1944	.1933	.1917	.1898	.1875	.1849	.1820	.1789	.1755	4
5	.1600	.1633	.1662	.1687	.1708	.1725	.1738	.1747	.1753	.1755	5
6	.1093	.1143	.1191	.1237	.1281	.1323	.1362	.1398	.1432	.1462	6
7	.0640	.0686	.0732	.0778	.0824	.0869	.0914	.0959	.1002	.1044	7
8	.0328	.0360	.0393	.0428	.0463	.0500	.0537	.0575	.0614	.0653	8
9	.0150	.0168	.0188	.0209	.0232	.0255	.0281	.0307	.0334	.0363	9
10	.0061	.0071	.0081	.0092	.0104	.0118	.0132	.0147	.0164	.0181	10
11	.0023	.0027	.0032	.0037	.0043	.0049	.0056	.0064	.0073	.0082	11
12	.0008	.0009	.0011	.0013	.0016	.0019	.0022	.0026	.0030	.0034	12
13	.0002	.0003	.0004	.0005	.0006	.0007	.0008	.0009	.0011	.0013	13
14	.0001	.0001	.0001	.0001	.0002	.0002	.0003	.0003	.0004	.0005	14
15	.0000	.0000	.0000	.0000	.0001	.0001	.0001	.0001	.0001	.0002	15
16	.0000	.0000	.0000	.0000	.0000	.0000	.0000	.0000	.0000	.0000	16

x	5.10	5.20	5.30	5.40	5.50	5.60	5.70	5.80	5.90	6.00	x
0	.0061	.0055	.0050	.0045	.0041	.0037	.0033	.0030	.0027	.0025	0
1	.0311	.0287	.0265	.0244	.0225	.0207	.0191	.0176	.0162	.0149	1
2	.0793	.0746	.0701	.0659	.0618	.0580	.0544	.0509	.0477	.0446	2
3	.1348	.1293	.1239	.1185	.1133	.1082	.1033	.0985	.0938	.0892	3
4	.1719	.1681	.1641	.1600	.1558	.1515	.1472	.1428	.1383	.1339	4
5	.1753	.1748	.1740	.1728	.1714	.1697	.1678	.1656	.1632	.1606	5
6	.1490	.1515	.1537	.1555	.1571	.1584	.1594	.1601	.1605	.1606	6
7	.1086	.1125	.1163	.1200	.1234	.1267	.1298	.1326	.1353	.1377	7
8	.0692	.0731	.0771	.0810	.0849	.0887	.0925	.0962	.0998	.1033	8
9	.0392	.0423	.0454	.0486	.0519	.0552	.0586	.0620	.0654	.0688	9
10	.0200	.0220	.0241	.0262	.0285	.0309	.0334	.0359	.0386	.0413	10
11	.0093	.0104	.0116	.0129	.0143	.0157	.0173	.0190	.0207	.0225	11
12	.0039	.0045	.0051	.0058	.0065	.0073	.0082	.0092	.0102	.0113	12
13	.0015	.0018	.0021	.0024	.0028	.0032	.0036	.0041	.0046	.0052	13
14	.0006	.0007	.0008	.0009	.0011	.0013	.0015	.0017	.0019	.0022	14
15	.0002	.0002	.0003	.0003	.0004	.0005	.0006	.0007	.0008	.0009	15
16	.0001	.0001	.0001	.0001	.0001	.0002	.0002	.0002	.0003	.0003	16
17	.0000	.0000	.0000	.0000	.0000	.0001	.0001	.0001	.0001	.0001	17
18	.0000	.0000	.0000	.0000	.0000	.0000	.0000	.0000	.0000	.0000	18

x	6.10	6.20	6.30	6.40	6.50	6.60	6.70	6.80	6.90	7.00	x
0	.0022	.0020	.0018	.0017	.0015	.0014	.0012	.0011	.0010	.0009	0
1	.0137	.0126	.0116	.0106	.0098	.0090	.0082	.0076	.0070	.0064	1
2	.0417	.0390	.0364	.0340	.0318	.0296	.0276	.0258	.0240	.0223	2
3	.0848	.0806	.0765	.0726	.0688	.0652	.0617	.0584	.0552	.0521	3
4	.1294	.1249	.1205	.1162	.1118	.1076	.1034	.0992	.0952	.0912	4
5	.1579	.1549	.1519	.1487	.1454	.1420	.1385	.1349	.1314	.1277	5
6	.1605	.1601	.1595	.1586	.1575	.1562	.1546	.1529	.1511	.1490	6
7	.1399	.1418	.1435	.1450	.1462	.1472	.1480	.1486	.1489	.1490	7
8	.1066	.1099	.1130	.1160	.1188	.1215	.1240	.1263	.1284	.1304	8
9	.0723	.0757	.0791	.0825	.0858	.0891	.0923	.0954	.0985	.1014	9
10	.0441	.0469	.0498	.0528	.0558	.0588	.0618	.0649	.0679	.0710	10
11	.0244	.0265	.0285	.0307	.0330	.0353	.0377	.0401	.0426	.0452	11
12	.0124	.0137	.0150	.0164	.0179	.0194	.0210	.0227	.0245	.0263	12
13	.0058	.0065	.0073	.0081	.0089	.0099	.0108	.0119	.0130	.0142	13
14	.0025	.0029	.0033	.0037	.0041	.0046	.0052	.0058	.0064	.0071	14
15	.0010	.0012	.0014	.0016	.0018	.0020	.0023	.0026	.0029	.0033	15
16	.0004	.0005	.0005	.0006	.0007	.0008	.0010	.0011	.0013	.0014	16
17	.0001	.0002	.0002	.0002	.0003	.0003	.0004	.0004	.0005	.0006	17

Table F (continued)

Poisson Distribution — Individual Terms

$$f_P(x|\lambda,t) = \frac{(\lambda t)^x e^{-\lambda t}}{x!}$$

x	6.10	6.20	6.30	6.40	6.50	6.60	6.70	6.80	6.90	7.00	x
					λt						
18	.0000	.0001	.0001	.0001	.0001	.0001	.0001	.0002	.0002	.0002	18
19	.0000	.0000	.0000	.0000	.0000	.0000	.0001	.0001	.0001	.0001	19
20	.0000	.0000	.0000	.0000	.0000	.0000	.0000	.0000	.0000	.0000	20

x	7.10	7.20	7.30	7.40	7.50	7.60	7.70	7.80	7.90	8.00	x
					λt						
0	.0008	.0007	.0007	.0006	.0006	.0005	.0005	.0004	.0004	.0003	0
1	.0059	.0054	.0049	.0045	.0041	.0038	.0035	.0032	.0029	.0027	1
2	.0208	.0194	.0180	.0167	.0156	.0145	.0134	.0125	.0116	.0107	2
3	.0492	.0464	.0438	.0413	.0389	.0366	.0345	.0324	.0305	.0286	3
4	.0874	.0836	.0799	.0764	.0729	.0696	.0663	.0632	.0602	.0573	4
5	.1241	.1204	.1167	.1130	.1094	.1057	.1021	.0986	.0951	.0916	5
6	.1468	.1445	.1420	.1394	.1367	.1339	.1311	.1282	.1252	.1221	6
7	.1489	.1486	.1481	.1474	.1465	.1454	.1442	.1428	.1413	.1396	7
8	.1321	.1337	.1351	.1363	.1373	.1381	.1388	.1392	.1395	.1396	8
9	.1042	.1070	.1096	.1121	.1144	.1167	.1187	.1207	.1224	.1241	9
10	.0740	.0770	.0800	.0829	.0858	.0887	.0914	.0941	.0967	.0993	10
11	.0478	.0504	.0531	.0558	.0585	.0613	.0640	.0667	.0695	.0722	11
12	.0283	.0303	.0323	.0344	.0366	.0388	.0411	.0434	.0457	.0481	12
13	.0154	.0168	.0181	.0196	.0211	.0227	.0243	.0260	.0278	.0296	13
14	.0078	.0086	.0095	.0104	.0113	.0123	.0134	.0145	.0157	.0169	14
15	.0037	.0041	.0046	.0051	.0057	.0062	.0069	.0075	.0083	.0090	15
16	.0016	.0019	.0021	.0024	.0026	.0030	.0033	.0037	.0041	.0045	16
17	.0007	.0008	.0009	.0010	.0012	.0013	.0015	.0017	.0019	.0021	17
18	.0003	.0003	.0004	.0004	.0005	.0006	.0006	.0007	.0008	.0009	18
19	.0001	.0001	.0001	.0002	.0002	.0002	.0003	.0003	.0003	.0004	19
20	.0000	.0000	.0001	.0001	.0001	.0001	.0001	.0001	.0001	.0002	20
21	.0000	.0000	.0000	.0000	.0000	.0000	.0000	.0000	.0001	.0001	21
22	.0000	.0000	.0000	.0000	.0000	.0000	.0000	.0000	.0000	.0000	22

x	8.10	8.20	8.30	8.40	8.50	8.60	8.70	8.80	8.90	9.00	x
					λt						
0	.0003	.0003	.0002	.0002	.0002	.0002	.0002	.0002	.0001	.0001	0
1	.0025	.0023	.0021	.0019	.0017	.0016	.0014	.0013	.0012	.0011	1
2	.0100	.0092	.0086	.0079	.0074	.0068	.0063	.0058	.0054	.0050	2
3	.0269	.0252	.0237	.0222	.0208	.0195	.0183	.0171	.0160	.0150	3
4	.0544	.0517	.0491	.0466	.0443	.0420	.0398	.0377	.0357	.0337	4
5	.0882	.0849	.0816	.0784	.0752	.0722	.0692	.0663	.0635	.0607	5
6	.1191	.1160	.1128	.1097	.1066	.1034	.1003	.0972	.0941	.0911	6
7	.1378	.1358	.1338	.1317	.1294	.1271	.1247	.1222	.1197	.1171	7
8	.1395	.1392	.1388	.1382	.1375	.1366	.1356	.1344	.1332	.1318	8
9	.1256	.1269	.1280	.1290	.1299	.1306	.1311	.1315	.1317	.1318	9
10	.1017	.1040	.1063	.1084	.1104	.1123	.1140	.1157	.1172	.1186	10
11	.0749	.0776	.0802	.0828	.0853	.0878	.0902	.0925	.0948	.0970	11
12	.0505	.0530	.0555	.0579	.0604	.0629	.0654	.0679	.0703	.0728	12
13	.0315	.0334	.0354	.0374	.0395	.0416	.0438	.0459	.0481	.0504	13
14	.0182	.0196	.0210	.0225	.0240	.0256	.0272	.0289	.0306	.0324	14
15	.0098	.0107	.0116	.0126	.0136	.0147	.0158	.0169	.0182	.0194	15
16	.0050	.0055	.0060	.0066	.0072	.0079	.0086	.0093	.0101	.0109	16
17	.0024	.0026	.0029	.0033	.0036	.0040	.0044	.0048	.0053	.0058	17
18	.0011	.0012	.0014	.0015	.0017	.0019	.0021	.0024	.0026	.0029	18
19	.0005	.0005	.0006	.0007	.0008	.0009	.0010	.0011	.0012	.0014	19
20	.0002	.0002	.0002	.0003	.0003	.0004	.0004	.0005	.0005	.0006	20
21	.0001	.0001	.0001	.0001	.0001	.0002	.0002	.0002	.0002	.0003	21
22	.0000	.0000	.0000	.0000	.0001	.0001	.0001	.0001	.0001	.0001	22
23	.0000	.0000	.0000	.0000	.0000	.0000	.0000	.0000	.0000	.0000	23

x	9.10	9.20	9.30	9.40	9.50	9.60	9.70	9.80	9.90	10.00	x
					λt						
0	.0001	.0001	.0001	.0001	.0001	.0001	.0001	.0001	.0001	.0000	0
1	.0010	.0009	.0009	.0008	.0007	.0007	.0006	.0005	.0005	.0005	1
2	.0046	.0043	.0040	.0037	.0034	.0031	.0029	.0027	.0025	.0023	2
3	.0140	.0131	.0123	.0115	.0107	.0100	.0093	.0087	.0081	.0076	3
4	.0319	.0302	.0285	.0269	.0254	.0240	.0226	.0213	.0201	.0189	4
5	.0581	.0555	.0530	.0506	.0483	.0460	.0439	.0418	.0398	.0378	5
6	.0881	.0851	.0822	.0793	.0764	.0736	.0709	.0682	.0656	.0631	6
7	.1145	.1118	.1091	.1064	.1037	.1010	.0982	.0955	.0928	.0901	7
8	.1302	.1286	.1269	.1251	.1232	.1212	.1191	.1170	.1148	.1126	8
9	.1317	.1315	.1311	.1306	.1300	.1293	.1284	.1274	.1263	.1251	9

Table F (continued)

Poisson Distribution — Individual Terms

$$f_P(x|\lambda,t) = \frac{(\lambda t)^x e^{-\lambda t}}{x!}$$

x	9.10	9.20	9.30	9.40	9.50	9.60	9.70	9.80	9.90	10.00	x
	λt										
10	.1198	.1210	.1219	.1228	.1235	.1241	.1245	.1249	.1250	.1251	10
11	.0991	.1012	.1031	.1049	.1067	.1083	.1098	.1112	.1125	.1137	11
12	.0752	.0776	.0799	.0822	.0844	.0866	.0888	.0908	.0928	.0948	12
13	.0526	.0549	.0572	.0594	.0617	.0640	.0662	.0685	.0707	.0729	13
14	.0342	.0361	.0380	.0399	.0419	.0439	.0459	.0479	.0500	.0521	14
15	.0208	.0221	.0235	.0250	.0265	.0281	.0297	.0313	.0330	.0347	15
16	.0118	.0127	.0137	.0147	.0157	.0168	.0180	.0192	.0204	.0217	16
17	.0063	.0069	.0075	.0081	.0088	.0095	.0103	.0111	.0119	.0128	17
18	.0032	.0035	.0039	.0042	.0046	.0051	.0055	.0060	.0065	.0071	18
19	.0015	.0017	.0019	.0021	.0023	.0026	.0028	.0031	.0034	.0037	19
20	.0007	.0008	.0009	.0010	.0011	.0012	.0014	.0015	.0017	.0019	20
21	.0003	.0003	.0004	.0004	.0005	.0006	.0006	.0007	.0008	.0009	21
22	.0001	.0001	.0002	.0002	.0002	.0002	.0003	.0003	.0004	.0004	22
23	.0000	.0001	.0001	.0001	.0001	.0001	.0001	.0001	.0002	.0002	23
24	.0000	.0000	.0000	.0000	.0000	.0000	.0000	.0001	.0001	.0001	24
25	.0000	.0000	.0000	.0000	.0000	.0000	.0000	.0000	.0000	.0000	25

x	11.00	12.00	13.00	14.00	15.00	16.00	17.00	18.00	19.00	20.00	x
	λt										
0	.0000	.0000	.0000	.0000	.0000	.0000	.0000	.0000	.0000	.0000	0
1	.0002	.0001	.0000	.0000	.0000	.0000	.0000	.0000	.0000	.0000	1
2	.0010	.0004	.0002	.0001	.0000	.0000	.0000	.0000	.0000	.0000	2
3	.0037	.0018	.0008	.0004	.0002	.0001	.0000	.0000	.0000	.0000	3
4	.0102	.0053	.0027	.0013	.0006	.0003	.0001	.0001	.0000	.0000	4
5	.0224	.0127	.0070	.0037	.0019	.0010	.0005	.0002	.0001	.0001	5
6	.0411	.0255	.0152	.0087	.0048	.0026	.0014	.0007	.0004	.0002	6
7	.0646	.0437	.0281	.0174	.0104	.0060	.0034	.0019	.0010	.0005	7
8	.0888	.0655	.0457	.0304	.0194	.0120	.0072	.0042	.0024	.0013	8
9	.1085	.0874	.0661	.0473	.0324	.0213	.0135	.0083	.0050	.0029	9
10	.1194	.1048	.0859	.0663	.0486	.0341	.0230	.0150	.0095	.0058	10
11	.1194	.1144	.1015	.0844	.0663	.0496	.0355	.0245	.0164	.0106	11
12	.1094	.1144	.1099	.0984	.0829	.0661	.0504	.0368	.0259	.0176	12
13	.0926	.1056	.1099	.1060	.0956	.0814	.0658	.0509	.0378	.0271	13
14	.0728	.0905	.1021	.1060	.1024	.0930	.0800	.0655	.0514	.0387	14
15	.0534	.0724	.0885	.0989	.1024	.0992	.0906	.0786	.0650	.0516	15
16	.0367	.0543	.0719	.0866	.0960	.0992	.0963	.0884	.0772	.0646	16
17	.0237	.0383	.0550	.0713	.0847	.0934	.0963	.0936	.0863	.0760	17
18	.0145	.0255	.0397	.0554	.0706	.0830	.0909	.0936	.0911	.0844	18
19	.0084	.0161	.0272	.0409	.0557	.0699	.0814	.0887	.0911	.0888	19
20	.0046	.0097	.0177	.0286	.0418	.0559	.0692	.0798	.0866	.0888	20
21	.0024	.0055	.0109	.0191	.0299	.0426	.0560	.0684	.0783	.0846	21
22	.0012	.0030	.0065	.0121	.0204	.0310	.0433	.0560	.0676	.0769	22
23	.0006	.0016	.0037	.0074	.0133	.0216	.0320	.0438	.0559	.0669	23
24	.0003	.0008	.0020	.0043	.0083	.0144	.0226	.0328	.0442	.0557	24
25	.0001	.0004	.0010	.0024	.0050	.0092	.0154	.0237	.0336	.0446	25
26	.0000	.0002	.0005	.0013	.0029	.0057	.0101	.0164	.0246	.0343	26
27	.0000	.0001	.0002	.0007	.0016	.0034	.0063	.0109	.0173	.0254	27
28	.0000	.0000	.0001	.0003	.0009	.0019	.0038	.0070	.0117	.0181	28
29	.0000	.0000	.0001	.0002	.0004	.0011	.0023	.0044	.0077	.0125	29
30	.0000	.0000	.0000	.0001	.0002	.0006	.0013	.0026	.0049	.0083	30
31	.0000	.0000	.0000	.0000	.0001	.0003	.0007	.0015	.0030	.0054	31
32	.0000	.0000	.0000	.0000	.0001	.0001	.0004	.0009	.0018	.0034	32
33	.0000	.0000	.0000	.0000	.0000	.0001	.0002	.0005	.0010	.0020	33
34	.0000	.0000	.0000	.0000	.0000	.0000	.0001	.0002	.0006	.0012	34
35	.0000	.0000	.0000	.0000	.0000	.0000	.0000	.0001	.0003	.0007	35
36	.0000	.0000	.0000	.0000	.0000	.0000	.0000	.0001	.0002	.0004	36
37	.0000	.0000	.0000	.0000	.0000	.0000	.0000	.0000	.0001	.0002	37
38	.0000	.0000	.0000	.0000	.0000	.0000	.0000	.0000	.0000	.0001	38
39	.0000	.0000	.0000	.0000	.0000	.0000	.0000	.0000	.0000	.0001	39

Table G

Poisson Distribution — Cumulative Terms

$$F_P(x|\lambda,t) = \sum_{y=0}^{x} \frac{(\lambda t)^y e^{-\lambda t}}{y!}$$

x	λt .001	.002	.003	.004	.005	.006	.007	.008	.009	0.010	x
0	.9990	.9980	.9970	.9960	.9950	.9940	.9930	.9920	.9910	.9900	0
1	1.0000	1.0000	1.0000	1.0000	1.0000	1.0000	1.0000	1.0000	1.0000	1.0000	1

x	λt .010	.020	.030	.040	.050	.060	.070	.080	.090	0.100	x
0	.9900	.9802	.9704	.9608	.9512	.9418	.9324	.9231	.9139	.9048	0
1	1.0000	.9998	.9996	.9992	.9988	.9983	.9977	.9970	.9962	.9953	1
2	1.0000	1.0000	1.0000	1.0000	1.0000	1.0000	.9999	.9999	.9999	.9998	2
3	1.0000	1.0000	1.0000	1.0000	1.0000	1.0000	1.0000	1.0000	1.0000	1.0000	3

x	λt 0.10	0.20	0.30	0.40	0.50	0.60	0.70	0.80	0.90	1.00	x
0	.9048	.8187	.7408	.6703	.6065	.5488	.4966	.4493	.4066	.3679	0
1	.9953	.9825	.9631	.9384	.9098	.8781	.8442	.8088	.7725	.7358	1
2	.9998	.9989	.9964	.9921	.9856	.9769	.9659	.9526	.9371	.9197	2
3	1.0000	.9999	.9997	.9992	.9982	.9966	.9942	.9909	.9865	.9810	3
4	1.0000	1.0000	1.0000	.9999	.9998	.9996	.9992	.9986	.9977	.9963	4
5	1.0000	1.0000	1.0000	1.0000	1.0000	1.0000	.9999	.9998	.9997	.9994	5
6	1.0000	1.0000	1.0000	1.0000	1.0000	1.0000	1.0000	1.0000	1.0000	.9999	6
7	1.0000	1.0000	1.0000	1.0000	1.0000	1.0000	1.0000	1.0000	1.0000	1.0000	7

x	λt 1.10	1.20	1.30	1.40	1.50	1.60	1.70	1.80	1.90	2.00	x
0	.3329	.3012	.2725	.2466	.2231	.2019	.1827	.1653	.1496	.1353	0
1	.6990	.6626	.6268	.5918	.5578	.5249	.4932	.4628	.4337	.4060	1
2	.9004	.8795	.8571	.8335	.8088	.7834	.7572	.7306	.7037	.6767	2
3	.9743	.9662	.9569	.9463	.9344	.9212	.9068	.8913	.8747	.8571	3
4	.9946	.9923	.9893	.9857	.9814	.9763	.9704	.9636	.9559	.9473	4
5	.9990	.9985	.9978	.9968	.9955	.9940	.9920	.9896	.9868	.9834	5
6	.9999	.9997	.9996	.9994	.9991	.9987	.9981	.9974	.9966	.9955	6
7	1.0000	1.0000	.9999	.9999	.9998	.9997	.9996	.9994	.9992	.9989	7
8	1.0000	1.0000	1.0000	1.0000	1.0000	1.0000	.9999	.9999	.9998	.9998	8
9	1.0000	1.0000	1.0000	1.0000	1.0000	1.0000	1.0000	1.0000	1.0000	1.0000	9

x	λt 2.10	2.20	2.30	2.40	2.50	2.60	2.70	2.80	2.90	3.00	x
0	.1225	.1108	.1003	.0907	.0821	.0743	.0672	.0608	.0550	.0498	0
1	.3796	.3546	.3309	.3084	.2873	.2674	.2487	.2311	.2146	.1991	1
2	.6496	.6227	.5960	.5697	.5438	.5184	.4936	.4695	.4460	.4232	2
3	.8386	.8194	.7993	.7787	.7576	.7360	.7141	.6919	.6696	.6472	3
4	.9379	.9275	.9162	.9041	.8912	.8774	.8629	.8477	.8318	.8153	4
5	.9796	.9751	.9700	.9643	.9580	.9510	.9433	.9349	.9258	.9161	5
6	.9941	.9925	.9906	.9884	.9858	.9828	.9794	.9756	.9713	.9665	6
7	.9985	.9980	.9974	.9967	.9958	.9947	.9934	.9919	.9901	.9881	7
8	.9997	.9995	.9994	.9991	.9989	.9985	.9981	.9976	.9969	.9962	8
9	.9999	.9999	.9999	.9998	.9997	.9996	.9995	.9993	.9991	.9989	9
10	1.0000	1.0000	1.0000	1.0000	.9999	.9999	.9999	.9998	.9998	.9997	10
11	1.0000	1.0000	1.0000	1.0000	1.0000	1.0000	1.0000	1.0000	.9999	.9999	11
12	1.0000	1.0000	1.0000	1.0000	1.0000	1.0000	1.0000	1.0000	1.0000	1.0000	12

x	λt 3.10	3.20	3.30	3.40	3.50	3.60	3.70	3.80	3.90	4.00	x
0	.0450	.0408	.0369	.0334	.0302	.0273	.0247	.0224	.0202	.0183	0
1	.1847	.1712	.1586	.1468	.1359	.1257	.1162	.1074	.0992	.0916	1
2	.4012	.3799	.3594	.3397	.3208	.3027	.2854	.2689	.2531	.2381	2
3	.6248	.6025	.5803	.5584	.5366	.5152	.4942	.4735	.4532	.4335	3
4	.7982	.7806	.7626	.7442	.7254	.7064	.6872	.6678	.6484	.6288	4
5	.9057	.8946	.8829	.8705	.8576	.8441	.8301	.8156	.8006	.7851	5
6	.9612	.9554	.9490	.9421	.9347	.9267	.9182	.9091	.8995	.8893	6
7	.9858	.9832	.9802	.9769	.9733	.9692	.9648	.9599	.9546	.9489	7
8	.9953	.9943	.9931	.9917	.9901	.9883	.9863	.9840	.9815	.9786	8
9	.9986	.9982	.9978	.9973	.9967	.9960	.9952	.9942	.9931	.9919	9
10	.9996	.9995	.9994	.9992	.9990	.9987	.9984	.9981	.9977	.9972	10
11	.9999	.9999	.9998	.9998	.9997	.9996	.9995	.9994	.9993	.9991	11
12	1.0000	1.0000	1.0000	.9999	.9999	.9999	.9999	.9998	.9998	.9997	12
13	1.0000	1.0000	1.0000	1.0000	1.0000	1.0000	1.0000	1.0000	.9999	.9999	13
14	1.0000	1.0000	1.0000	1.0000	1.0000	1.0000	1.0000	1.0000	1.0000	1.0000	14

Table G (continued)

Poisson Distribution — Cumulative Terms

$$F_P(x|\lambda,t) = \sum_{y=0}^{x} \frac{(\lambda t)^y e^{-\lambda t}}{y!}$$

x	λt 4.10	4.20	4.30	4.40	4.50	4.60	4.70	4.80	4.90	5.00	x
0	.0166	.0150	.0136	.0123	.0111	.0101	.0091	.0082	.0074	.0067	0
1	.0845	.0780	.0719	.0663	.0611	.0563	.0518	.0477	.0439	.0404	1
2	.2238	.2102	.1974	.1851	.1736	.1626	.1523	.1425	.1333	.1247	2
3	.4142	.3954	.3772	.3594	.3423	.3257	.3097	.2942	.2793	.2650	3
4	.6093	.5898	.5704	.5512	.5321	.5132	.4946	.4763	.4582	.4405	4
5	.7693	.7531	.7367	.7199	.7029	.6858	.6684	.6510	.6335	.6160	5
6	.8786	.8675	.8558	.8436	.8311	.8180	.8046	.7908	.7767	.7622	6
7	.9427	.9361	.9290	.9214	.9134	.9049	.8960	.8867	.8769	.8666	7
8	.9755	.9721	.9683	.9642	.9597	.9549	.9497	.9442	.9382	.9319	8
9	.9905	.9889	.9871	.9851	.9829	.9805	.9778	.9749	.9717	.9682	9
10	.9966	.9959	.9952	.9943	.9933	.9922	.9910	.9896	.9880	.9863	10
11	.9989	.9986	.9983	.9980	.9976	.9971	.9966	.9960	.9953	.9945	11
12	.9997	.9996	.9995	.9993	.9992	.9990	.9988	.9986	.9983	.9980	12
13	.9999	.9999	.9998	.9998	.9997	.9997	.9996	.9995	.9994	.9993	13
14	1.0000	1.0000	1.0000	.9999	.9999	.9999	.9999	.9999	.9998	.9998	14
15	1.0000	1.0000	1.0000	1.0000	1.0000	1.0000	1.0000	1.0000	.9999	.9999	15
16	1.0000	1.0000	1.0000	1.0000	1.0000	1.0000	1.0000	1.0000	1.0000	1.0000	16

x	λt 5.10	5.20	5.30	5.40	5.50	5.60	5.70	5.80	5.90	6.00	x
0	.0061	.0055	.0050	.0045	.0041	.0037	.0033	.0030	.0027	.0025	0
1	.0372	.0342	.0314	.0289	.0266	.0244	.0224	.0206	.0189	.0174	1
2	.1165	.1088	.1016	.0948	.0884	.0824	.0768	.0715	.0666	.0620	2
3	.2513	.2381	.2254	.2133	.2017	.1906	.1800	.1700	.1604	.1512	3
4	.4231	.4061	.3895	.3733	.3575	.3422	.3272	.3127	.2987	.2851	4
5	.5984	.5809	.5635	.5461	.5289	.5119	.4950	.4783	.4619	.4457	5
6	.7474	.7324	.7171	.7017	.6860	.6703	.6544	.6384	.6224	.6063	6
7	.8560	.8449	.8335	.8217	.8095	.7970	.7841	.7710	.7576	.7440	7
8	.9252	.9181	.9106	.9027	.8944	.8857	.8766	.8672	.8574	.8472	8
9	.9644	.9603	.9559	.9512	.9462	.9409	.9352	.9292	.9228	.9161	9
10	.9844	.9823	.9800	.9775	.9747	.9718	.9686	.9651	.9614	.9574	10
11	.9937	.9927	.9916	.9904	.9890	.9875	.9859	.9841	.9821	.9799	11
12	.9976	.9972	.9967	.9962	.9955	.9949	.9941	.9932	.9922	.9912	12
13	.9992	.9990	.9988	.9986	.9983	.9980	.9977	.9973	.9969	.9964	13
14	.9997	.9997	.9996	.9995	.9994	.9993	.9991	.9990	.9988	.9986	14
15	.9999	.9999	.9999	.9998	.9998	.9998	.9997	.9996	.9996	.9995	15
16	1.0000	1.0000	1.0000	.9999	.9999	.9999	.9999	.9999	.9999	.9998	16
17	1.0000	1.0000	1.0000	1.0000	1.0000	1.0000	1.0000	1.0000	1.0000	.9999	17
18	1.0000	1.0000	1.0000	1.0000	1.0000	1.0000	1.0000	1.0000	1.0000	1.0000	18

x	λt 6.10	6.20	6.30	6.40	6.50	6.60	6.70	6.80	6.90	7.00	x
0	.0022	.0020	.0018	.0017	.0015	.0014	.0012	.0011	.0010	.0009	0
1	.0159	.0146	.0134	.0123	.0113	.0103	.0095	.0087	.0080	.0073	1
2	.0577	.0536	.0498	.0463	.0430	.0400	.0371	.0344	.0320	.0296	2
3	.1425	.1342	.1264	.1189	.1118	.1052	.0988	.0928	.0871	.0818	3
4	.2719	.2592	.2469	.2351	.2237	.2127	.2022	.1920	.1823	.1730	4
5	.4298	.4141	.3988	.3837	.3690	.3547	.3406	.3270	.3137	.3007	5
6	.5902	.5742	.5582	.5423	.5265	.5108	.4953	.4799	.4647	.4497	6
7	.7301	.7160	.7017	.6873	.6728	.6581	.6433	.6285	.6136	.5987	7
8	.8367	.8259	.8148	.8033	.7916	.7796	.7673	.7548	.7420	.7291	8
9	.9090	.9016	.8939	.8858	.8774	.8686	.8596	.8502	.8405	.8305	9
10	.9531	.9486	.9437	.9386	.9332	.9274	.9214	.9151	.9084	.9015	10
11	.9776	.9750	.9723	.9693	.9661	.9627	.9591	.9552	.9510	.9467	11
12	.9900	.9887	.9873	.9857	.9840	.9821	.9801	.9779	.9755	.9730	12
13	.9958	.9952	.9945	.9937	.9929	.9920	.9909	.9898	.9885	.9872	13
14	.9984	.9981	.9978	.9974	.9970	.9966	.9961	.9956	.9950	.9943	14
15	.9994	.9993	.9992	.9990	.9988	.9986	.9984	.9982	.9979	.9976	15
16	.9998	.9997	.9997	.9996	.9996	.9995	.9994	.9993	.9992	.9990	16
17	.9999	.9999	.9999	.9999	.9998	.9998	.9998	.9997	.9997	.9996	17
18	1.0000	1.0000	1.0000	1.0000	.9999	.9999	.9999	.9999	.9999	.9999	18
19	1.0000	1.0000	1.0000	1.0000	1.0000	1.0000	1.0000	1.0000	1.0000	1.0000	19

x	λt 7.10	7.20	7.30	7.40	7.50	7.60	7.70	7.80	7.90	8.00	x
0	.0008	.0007	.0007	.0006	.0006	.0005	.0005	.0004	.0004	.0003	0
1	.0067	.0061	.0056	.0051	.0047	.0043	.0039	.0036	.0033	.0030	1
2	.0275	.0255	.0236	.0219	.0203	.0188	.0174	.0161	.0149	.0138	2

Table G (continued)

Poisson Distribution — Cumulative Terms

$$F_P(x|\lambda,t) = \sum_{y=0}^{x} \frac{(\lambda t)^y e^{-\lambda t}}{y!}$$

x	λt 7.10	7.20	7.30	7.40	7.50	7.60	7.70	7.80	7.90	8.00	x
3	.0767	.0719	.0674	.0632	.0591	.0554	.0518	.0485	.0453	.0424	3
4	.1641	.1555	.1473	.1395	.1321	.1249	.1181	.1117	.1055	.0996	4
5	.2881	.2759	.2640	.2526	.2414	.2307	.2203	.2103	.2006	.1912	5
6	.4349	.4204	.4060	.3920	.3782	.3646	.3514	.3384	.3257	.3134	6
7	.5838	.5689	.5541	.5393	.5246	.5100	.4956	.4812	.4670	.4530	7
8	.7160	.7027	.6892	.6757	.6620	.6482	.6343	.6204	.6065	.5925	8
9	.8202	.8097	.7988	.7877	.7764	.7649	.7531	.7411	.7290	.7166	9
10	.8942	.8867	.8788	.8707	.8622	.8535	.8445	.8352	.8257	.8159	10
11	.9420	.9371	.9319	.9265	.9208	.9148	.9085	.9020	.8952	.8881	11
12	.9703	.9673	.9642	.9609	.9573	.9536	.9496	.9454	.9409	.9362	12
13	.9857	.9841	.9824	.9805	.9784	.9762	.9739	.9714	.9687	.9658	13
14	.9935	.9927	.9918	.9908	.9897	.9886	.9873	.9859	.9844	.9827	14
15	.9972	.9969	.9964	.9959	.9954	.9948	.9941	.9934	.9926	.9918	15
16	.9989	.9987	.9985	.9983	.9980	.9978	.9974	.9971	.9967	.9963	16
17	.9996	.9995	.9994	.9993	.9992	.9991	.9989	.9988	.9986	.9984	17
18	.9998	.9998	.9998	.9997	.9997	.9996	.9996	.9995	.9994	.9993	18
19	.9999	.9999	.9999	.9999	.9999	.9999	.9998	.9998	.9998	.9997	19
20	1.0000	1.0000	1.0000	1.0000	1.0000	1.0000	.9999	.9999	.9999	.9999	20
21	1.0000	1.0000	1.0000	1.0000	1.0000	1.0000	1.0000	1.0000	1.0000	1.0000	21

x	λt 8.10	8.20	8.30	8.40	8.50	8.60	8.70	8.80	8.90	9.00	x
0	.0003	.0003	.0002	.0002	.0002	.0002	.0002	.0002	.0001	.0001	0
1	.0028	.0025	.0023	.0021	.0019	.0018	.0016	.0015	.0014	.0012	1
2	.0127	.0118	.0109	.0100	.0093	.0086	.0079	.0073	.0068	.0062	2
3	.0396	.0370	.0346	.0323	.0301	.0281	.0262	.0244	.0228	.0212	3
4	.0940	.0887	.0837	.0789	.0744	.0701	.0660	.0621	.0584	.0550	4
5	.1822	.1736	.1653	.1573	.1496	.1422	.1352	.1284	.1219	.1157	5
6	.3013	.2896	.2781	.2670	.2562	.2457	.2355	.2256	.2160	.2068	6
7	.4391	.4254	.4119	.3987	.3856	.3728	.3602	.3478	.3357	.3239	7
8	.5786	.5647	.5507	.5369	.5231	.5094	.4958	.4823	.4689	.4557	8
9	.7041	.6915	.6788	.6659	.6530	.6400	.6269	.6137	.6006	.5874	9
10	.8058	.7956	.7850	.7743	.7634	.7522	.7409	.7294	.7178	.7060	10
11	.8807	.8731	.8652	.8571	.8487	.8400	.8311	.8220	.8126	.8030	11
12	.9313	.9261	.9207	.9150	.9091	.9029	.8965	.8898	.8829	.8758	12
13	.9628	.9595	.9561	.9524	.9486	.9445	.9403	.9358	.9311	.9261	13
14	.9810	.9791	.9771	.9749	.9726	.9701	.9675	.9647	.9617	.9585	14
15	.9908	.9898	.9887	.9875	.9862	.9848	.9832	.9816	.9798	.9780	15
16	.9958	.9953	.9947	.9941	.9934	.9926	.9918	.9909	.9899	.9889	16
17	.9982	.9979	.9977	.9973	.9970	.9966	.9962	.9957	.9952	.9947	17
18	.9992	.9991	.9990	.9989	.9987	.9985	.9983	.9981	.9978	.9976	18
19	.9997	.9997	.9996	.9995	.9995	.9994	.9993	.9992	.9991	.9989	19
20	.9999	.9999	.9998	.9998	.9998	.9998	.9997	.9997	.9996	.9996	20
21	1.0000	1.0000	.9999	.9999	.9999	.9999	.9999	.9999	.9999	.9998	21
22	1.0000	1.0000	1.0000	1.0000	1.0000	1.0000	1.0000	1.0000	.9999	.9999	22
23	1.0000	1.0000	1.0000	1.0000	1.0000	1.0000	1.0000	1.0000	1.0000	1.0000	23

x	λt 9.10	9.20	9.30	9.40	9.50	9.60	9.70	9.80	9.90	10.00	x
0	.0001	.0001	.0001	.0001	.0001	.0001	.0001	.0001	.0001	.0000	0
1	.0011	.0010	.0009	.0009	.0008	.0007	.0007	.0006	.0005	.0005	1
2	.0058	.0053	.0049	.0045	.0042	.0038	.0035	.0033	.0030	.0028	2
3	.0198	.0184	.0172	.0160	.0149	.0138	.0129	.0120	.0111	.0103	3
4	.0517	.0486	.0456	.0429	.0403	.0378	.0355	.0333	.0312	.0293	4
5	.1098	.1041	.0986	.0935	.0885	.0838	.0793	.0750	.0710	.0671	5
6	.1978	.1892	.1808	.1727	.1649	.1574	.1502	.1433	.1366	.1301	6
7	.3123	.3010	.2900	.2792	.2687	.2584	.2485	.2388	.2294	.2202	7
8	.4426	.4296	.4168	.4042	.3918	.3796	.3676	.3558	.3442	.3328	8
9	.5742	.5611	.5479	.5349	.5218	.5089	.4960	.4832	.4705	.4579	9
10	.6941	.6820	.6699	.6576	.6453	.6329	.6205	.6080	.5955	.5830	10
11	.7932	.7832	.7730	.7626	.7520	.7412	.7303	.7193	.7081	.6968	11
12	.8684	.8607	.8529	.8448	.8364	.8279	.8191	.8101	.8009	.7916	12
13	.9210	.9156	.9100	.9042	.8981	.8919	.8853	.8786	.8716	.8645	13
14	.9552	.9517	.9480	.9441	.9400	.9357	.9312	.9265	.9216	.9165	14
15	.9760	.9738	.9715	.9691	.9665	.9638	.9609	.9579	.9546	.9513	15
16	.9878	.9865	.9852	.9838	.9823	.9806	.9789	.9770	.9751	.9730	16
17	.9941	.9934	.9927	.9919	.9911	.9902	.9892	.9881	.9870	.9857	17
18	.9973	.9969	.9966	.9962	.9957	.9952	.9947	.9941	.9935	.9928	18
19	.9988	.9986	.9985	.9983	.9980	.9978	.9975	.9972	.9969	.9965	19

Table G (continued)

Poisson Distribution — Cumulative Terms

$$F_P(x|\lambda,t) = \sum_{y=0}^{x} \frac{(\lambda t)^y e^{-\lambda t}}{y!}$$

x	λt = 9.10	9.20	9.30	9.40	9.50	9.60	9.70	9.80	9.90	10.00	x
20	.9995	.9994	.9993	.9992	.9991	.9990	.9989	.9987	.9986	.9984	20
21	.9998	.9998	.9997	.9997	.9996	.9996	.9995	.9995	.9994	.9993	21
22	.9999	.9999	.9999	.9999	.9999	.9998	.9998	.9998	.9997	.9997	22
23	1.0000	1.0000	1.0000	1.0000	.9999	.9999	.9999	.9999	.9999	.9999	23
24	1.0000	1.0000	1.0000	1.0000	1.0000	1.0000	1.0000	1.0000	1.0000	1.0000	24

x	λt = 11.00	12.00	13.00	14.00	15.00	16.00	17.00	18.00	19.00	20.00	x
0	.0000	.0000	.0000	.0000	.0000	.0000	.0000	.0000	.0000	.0000	0
1	.0002	.0001	.0000	.0000	.0000	.0000	.0000	.0000	.0000	.0000	1
2	.0012	.0005	.0002	.0001	.0000	.0000	.0000	.0000	.0000	.0000	2
3	.0049	.0023	.0011	.0005	.0002	.0001	.0000	.0000	.0000	.0000	3
4	.0151	.0076	.0037	.0018	.0009	.0004	.0002	.0001	.0000	.0000	4
5	.0375	.0203	.0107	.0055	.0028	.0014	.0007	.0003	.0002	.0001	5
6	.0786	.0458	.0259	.0142	.0076	.0040	.0021	.0010	.0005	.0003	6
7	.1432	.0895	.0540	.0316	.0180	.0100	.0054	.0029	.0015	.0008	7
8	.2320	.1550	.0998	.0621	.0374	.0220	.0126	.0071	.0039	.0021	8
9	.3405	.2424	.1658	.1094	.0699	.0433	.0261	.0154	.0089	.0050	9
10	.4599	.3472	.2517	.1757	.1185	.0774	.0491	.0304	.0183	.0108	10
11	.5793	.4616	.3532	.2600	.1848	.1270	.0847	.0549	.0347	.0214	11
12	.6887	.5760	.4631	.3585	.2676	.1931	.1350	.0917	.0606	.0390	12
13	.7813	.6815	.5730	.4644	.3632	.2745	.2009	.1426	.0984	.0661	13
14	.8540	.7720	.6751	.5704	.4657	.3675	.2808	.2081	.1497	.1049	14
15	.9074	.8444	.7636	.6694	.5681	.4667	.3715	.2867	.2148	.1565	15
16	.9441	.8987	.8355	.7559	.6641	.5660	.4677	.3751	.2920	.2211	16
17	.9678	.9370	.8905	.8272	.7489	.6593	.5640	.4686	.3784	.2970	17
18	.9823	.9626	.9302	.8826	.8195	.7423	.6550	.5622	.4695	.3814	18
19	.9907	.9787	.9573	.9235	.8752	.8122	.7363	.6509	.5606	.4703	19
20	.9953	.9884	.9750	.9521	.9170	.8682	.8055	.7307	.6472	.5591	20
21	.9977	.9939	.9859	.9712	.9469	.9108	.8615	.7991	.7255	.6437	21
22	.9990	.9970	.9924	.9833	.9673	.9418	.9047	.8551	.7931	.7206	22
23	.9995	.9985	.9960	.9907	.9805	.9633	.9367	.8989	.8490	.7875	23
24	.9998	.9993	.9980	.9950	.9888	.9777	.9594	.9317	.8933	.8432	24
25	.9999	.9997	.9990	.9974	.9938	.9869	.9748	.9554	.9269	.8878	25
26	1.0000	.9999	.9995	.9987	.9967	.9925	.9848	.9718	.9514	.9221	26
27	1.0000	.9999	.9998	.9994	.9983	.9959	.9912	.9827	.9687	.9475	27
28	1.0000	1.0000	.9999	.9997	.9991	.9978	.9950	.9897	.9805	.9657	28
29	1.0000	1.0000	1.0000	.9999	.9996	.9989	.9973	.9941	.9882	.9782	29
30	1.0000	1.0000	1.0000	.9999	.9998	.9994	.9986	.9967	.9930	.9865	30
31	1.0000	1.0000	1.0000	1.0000	.9999	.9997	.9993	.9982	.9960	.9919	31
32	1.0000	1.0000	1.0000	1.0000	1.0000	.9999	.9996	.9990	.9978	.9953	32
33	1.0000	1.0000	1.0000	1.0000	1.0000	.9999	.9998	.9995	.9988	.9973	33
34	1.0000	1.0000	1.0000	1.0000	1.0000	1.0000	.9999	.9998	.9994	.9985	34
35	1.0000	1.0000	1.0000	1.0000	1.0000	1.0000	1.0000	.9999	.9997	.9992	35
36	1.0000	1.0000	1.0000	1.0000	1.0000	1.0000	1.0000	.9999	.9998	.9996	36
37	1.0000	1.0000	1.0000	1.0000	1.0000	1.0000	1.0000	1.0000	.9999	.9998	37
38	1.0000	1.0000	1.0000	1.0000	1.0000	1.0000	1.0000	1.0000	1.0000	.9999	38
39	1.0000	1.0000	1.0000	1.0000	1.0000	1.0000	1.0000	1.0000	1.0000	.9999	39
40	1.0000	1.0000	1.0000	1.0000	1.0000	1.0000	1.0000	1.0000	1.0000	1.0000	40

Table H

Standard Normal Integral

$$F_N(z) = \int_{-\infty}^{z} \frac{1}{\sqrt{2\pi}} e^{-\frac{1}{2}t^2}\, dt$$

Z	-.09	-.08	-.07	-.06	-.05	-.04	-.03	-.02	-.01	.00	Z
-3.8	.0001	.0001	.0001	.0001	.0001	.0001	.0001	.0001	.0001	.0001	-3.8
-3.7	.0001	.0001	.0001	.0001	.0001	.0001	.0001	.0001	.0001	.0001	-3.7
-3.6	.0001	.0001	.0001	.0001	.0001	.0001	.0001	.0001	.0002	.0002	-3.6
-3.5	.0002	.0002	.0002	.0002	.0002	.0002	.0002	.0002	.0002	.0002	-3.5
-3.4	.0002	.0003	.0003	.0003	.0003	.0003	.0003	.0003	.0003	.0003	-3.4
-3.3	.0003	.0004	.0004	.0004	.0004	.0004	.0004	.0005	.0005	.0005	-3.3
-3.2	.0005	.0005	.0005	.0006	.0006	.0006	.0006	.0006	.0007	.0007	-3.2
-3.1	.0007	.0007	.0008	.0008	.0008	.0008	.0009	.0009	.0009	.0010	-3.1
-3.0	.0010	.0010	.0011	.0011	.0011	.0012	.0012	.0013	.0013	.0014	-3.0
-2.9	.0014	.0014	.0015	.0015	.0016	.0016	.0017	.0018	.0018	.0019	-2.9
-2.8	.0019	.0020	.0021	.0021	.0022	.0023	.0023	.0024	.0025	.0026	-2.8
-2.7	.0026	.0027	.0028	.0029	.0030	.0031	.0032	.0033	.0034	.0035	-2.7
-2.6	.0036	.0037	.0038	.0039	.0040	.0041	.0043	.0044	.0045	.0047	-2.6
-2.5	.0048	.0049	.0051	.0052	.0054	.0055	.0057	.0059	.0060	.0062	-2.5
-2.4	.0064	.0066	.0068	.0069	.0071	.0073	.0076	.0078	.0080	.0082	-2.4
-2.3	.0084	.0087	.0089	.0091	.0094	.0096	.0099	.0102	.0104	.0107	-2.3
-2.2	.0110	.0113	.0116	.0119	.0122	.0125	.0129	.0132	.0136	.0139	-2.2
-2.1	.0143	.0146	.0150	.0154	.0158	.0162	.0166	.0170	.0174	.0179	-2.1
-2.0	.0183	.0188	.0192	.0197	.0202	.0207	.0212	.0217	.0222	.0228	-2.0
-1.9	.0233	.0239	.0244	.0250	.0256	.0262	.0268	.0274	.0281	.0287	-1.9
-1.8	.0294	.0301	.0307	.0314	.0322	.0329	.0336	.0344	.0352	.0359	-1.8
-1.7	.0367	.0375	.0384	.0392	.0401	.0409	.0418	.0427	.0436	.0446	-1.7
-1.6	.0455	.0465	.0475	.0485	.0495	.0505	.0516	.0526	.0537	.0548	-1.6
-1.5	.0559	.0571	.0582	.0594	.0606	.0618	.0630	.0643	.0655	.0668	-1.5
-1.4	.0681	.0694	.0708	.0721	.0735	.0749	.0764	.0778	.0793	.0808	-1.4
-1.3	.0823	.0838	.0853	.0869	.0885	.0901	.0918	.0934	.0951	.0968	-1.3
-1.2	.0985	.1003	.1020	.1038	.1057	.1075	.1094	.1112	.1131	.1151	-1.2
-1.1	.1170	.1190	.1210	.1230	.1251	.1271	.1292	.1314	.1335	.1357	-1.1
-1.0	.1379	.1401	.1423	.1446	.1469	.1492	.1515	.1539	.1562	.1587	-1.0
-0.9	.1611	.1635	.1660	.1685	.1711	.1736	.1762	.1788	.1814	.1841	-0.9
-0.8	.1867	.1894	.1922	.1949	.1977	.2005	.2033	.2061	.2090	.2119	-0.8
-0.7	.2148	.2177	.2206	.2236	.2266	.2296	.2327	.2358	.2389	.2420	-0.7
-0.6	.2451	.2483	.2514	.2546	.2578	.2611	.2643	.2676	.2709	.2743	-0.6
-0.5	.2776	.2810	.2843	.2877	.2912	.2946	.2981	.3015	.3050	.3085	-0.5
-0.4	.3121	.3156	.3192	.3228	.3264	.3300	.3336	.3372	.3409	.3446	-0.4
-0.3	.3483	.3520	.3557	.3594	.3632	.3669	.3707	.3745	.3783	.3821	-0.3
-0.2	.3859	.3897	.3936	.3974	.4013	.4052	.4090	.4129	.4168	.4207	-0.2
-0.1	.4247	.4286	.4325	.4364	.4404	.4443	.4483	.4522	.4562	.4602	-0.1
0.0	.4641	.4681	.4721	.4761	.4801	.4840	.4880	.4920	.4960	.5000	0.0

Table H (continued)

Standard Normal Integral

$$F_N(z) = \int_{-\infty}^{z} \frac{1}{\sqrt{2\pi}} e^{-\frac{1}{2}t^2}\, dt$$

z	.00	.01	.02	.03	.04	.05	.06	.07	.08	.09	z
0.0	.5000	.5040	.5080	.5120	.5160	.5199	.5239	.5279	.5319	.5359	0.0
0.1	.5398	.5438	.5478	.5517	.5557	.5596	.5636	.5675	.5714	.5753	0.1
0.2	.5793	.5832	.5871	.5910	.5948	.5987	.6026	.6064	.6103	.6141	0.2
0.3	.6179	.6217	.6255	.6293	.6331	.6368	.6406	.6443	.6480	.6517	0.3
0.4	.6554	.6591	.6628	.6664	.6700	.6736	.6772	.6808	.6844	.6879	0.4
0.5	.6915	.6950	.6985	.7019	.7054	.7088	.7123	.7157	.7190	.7224	0.5
0.6	.7257	.7291	.7324	.7357	.7389	.7422	.7454	.7486	.7517	.7549	0.6
0.7	.7580	.7611	.7642	.7673	.7704	.7734	.7764	.7794	.7823	.7852	0.7
0.8	.7881	.7910	.7939	.7967	.7995	.8023	.8051	.8078	.8106	.8133	0.8
0.9	.8159	.8186	.8212	.8238	.8264	.8289	.8315	.8340	.8365	.8389	0.9
1.0	.8413	.8438	.8461	.8485	.8508	.8531	.8554	.8577	.8599	.8621	1.0
1.1	.8643	.8665	.8686	.8708	.8729	.8749	.8770	.8790	.8810	.8830	1.1
1.2	.8849	.8869	.8888	.8906	.8925	.8943	.8962	.8980	.8997	.9015	1.2
1.3	.9032	.9049	.9066	.9082	.9099	.9115	.9131	.9147	.9162	.9177	1.3
1.4	.9192	.9207	.9222	.9236	.9251	.9265	.9279	.9292	.9306	.9319	1.4
1.5	.9332	.9345	.9357	.9370	.9382	.9394	.9406	.9418	.9429	.9441	1.5
1.6	.9452	.9463	.9474	.9484	.9495	.9505	.9515	.9525	.9535	.9545	1.6
1.7	.9554	.9564	.9573	.9582	.9591	.9599	.9608	.9616	.9625	.9633	1.7
1.8	.9641	.9648	.9656	.9664	.9671	.9678	.9686	.9693	.9699	.9706	1.8
1.9	.9713	.9719	.9726	.9732	.9738	.9744	.9750	.9756	.9761	.9767	1.9
2.0	.9772	.9778	.9783	.9788	.9793	.9798	.9803	.9808	.9812	.9817	2.0
2.1	.9821	.9826	.9830	.9834	.9838	.9842	.9846	.9850	.9854	.9857	2.1
2.2	.9861	.9864	.9868	.9871	.9875	.9878	.9881	.9884	.9887	.9890	2.2
2.3	.9893	.9896	.9898	.9901	.9904	.9906	.9909	.9911	.9913	.9916	2.3
2.4	.9918	.9920	.9922	.9924	.9927	.9929	.9931	.9932	.9934	.9936	2.4
2.5	.9938	.9940	.9941	.9943	.9945	.9946	.9948	.9949	.9951	.9952	2.5
2.6	.9953	.9955	.9956	.9957	.9959	.9960	.9961	.9962	.9963	.9964	2.6
2.7	.9965	.9966	.9967	.9968	.9969	.9970	.9971	.9972	.9973	.9974	2.7
2.8	.9974	.9975	.9976	.9977	.9977	.9978	.9979	.9979	.9980	.9981	2.8
2.9	.9981	.9982	.9982	.9983	.9984	.9984	.9985	.9985	.9986	.9986	2.9
3.0	.9986	.9987	.9987	.9988	.9988	.9989	.9989	.9989	.9990	.9990	3.0
3.1	.9990	.9991	.9991	.9991	.9992	.9992	.9992	.9992	.9993	.9993	3.1
3.2	.9993	.9993	.9994	.9994	.9994	.9994	.9994	.9995	.9995	.9995	3.2
3.3	.9995	.9995	.9995	.9996	.9996	.9996	.9996	.9996	.9996	.9997	3.3
3.4	.9997	.9997	.9997	.9997	.9997	.9997	.9997	.9997	.9997	.9998	3.4
3.5	.9998	.9998	.9998	.9998	.9998	.9998	.9998	.9998	.9998	.9998	3.5
3.6	.9998	.9998	.9999	.9999	.9999	.9999	.9999	.9999	.9999	.9999	3.6
3.7	.9999	.9999	.9999	.9999	.9999	.9999	.9999	.9999	.9999	.9999	3.7
3.8	.9999	.9999	.9999	.9999	.9999	.9999	.9999	.9999	.9999	.9999	3.8

Table I

Squares and Square Roots

n	n^2	$\sqrt{n}$	$\sqrt{10n}$	n	n^2	$\sqrt{n}$	$\sqrt{10n}$
1	1	1.0000	3.1623	51	2601	7.1414	22.5832
2	4	1.4142	4.4721	52	2704	7.2111	22.8035
3	9	1.7320	5.4772	53	2809	7.2801	23.0217
4	16	2.0000	6.3246	54	2916	7.3485	23.2379
5	25	2.2361	7.0711	55	3025	7.4162	23.4521
6	36	2.4495	7.7460	56	3136	7.4833	23.6643
7	49	2.6458	8.3666	57	3249	7.5498	23.8747
8	64	2.8284	8.9443	58	3364	7.6158	24.0832
9	81	3.0000	9.4868	59	3481	7.6811	24.2899
10	100	3.1623	10.0000	60	3600	7.7460	24.4949
11	121	3.3166	10.4881	61	3721	7.8102	24.6982
12	144	3.4641	10.9545	62	3844	7.8740	24.8998
13	169	3.6056	11.4018	63	3969	7.9373	25.0998
14	196	3.7417	11.8322	64	4096	8.0000	25.2982
15	225	3.8730	12.2474	65	4225	8.0623	25.4951
16	256	4.0000	12.6491	66	4356	8.1240	25.6905
17	289	4.1231	13.0384	67	4489	8.1854	25.8844
18	324	4.2426	13.4164	68	4624	8.2462	26.0768
19	361	4.3589	13.7840	69	4761	8.3066	26.2678
20	400	4.4721	14.1421	70	4900	8.3666	26.4575
21	441	4.5826	14.4914	71	5041	8.4261	26.6458
22	484	4.6904	14.8324	72	5184	8.4853	26.8328
23	529	4.7958	15.1658	73	5329	8.5440	27.0185
24	576	4.8990	15.4919	74	5476	8.6023	27.2029
25	625	5.0000	15.8114	75	5625	8.6603	27.3861
26	676	5.0990	16.1245	76	5776	8.7178	27.5681
27	729	5.1962	16.4317	77	5929	8.7750	27.7489
28	784	5.2915	16.7332	78	6084	8.8318	27.9285
29	841	5.3852	17.0294	79	6241	8.8882	28.1069
30	900	5.4772	17.3205	80	6400	8.9443	28.2843
31	961	5.5678	17.6068	81	6561	9.0000	28.4605
32	1024	5.6569	17.8885	82	6724	9.0554	28.6356
33	1089	5.7446	18.1659	83	6889	9.1104	28.8097
34	1156	5.8310	18.4391	84	7056	9.1652	28.9827
35	1225	5.9161	18.7083	85	7225	9.2195	29.1548
36	1296	6.0000	18.9737	86	7396	9.2736	29.3257
37	1369	6.0828	19.2354	87	7569	9.3274	29.4958
38	1444	6.1644	19.4936	88	7744	9.3808	29.6648
39	1521	6.2450	19.7484	89	7921	9.4340	29.8329
40	1600	6.3246	20.0000	90	8100	9.4868	30.0000
41	1681	6.4031	20.2484	91	8281	9.5394	30.1662
42	1764	6.4807	20.4939	92	8464	9.5917	30.3315
43	1849	6.5574	20.7364	93	8649	9.6437	30.4959
44	1936	6.6332	20.9762	94	8836	9.6954	30.6594
45	2025	6.7082	21.2132	95	9025	9.7468	30.8221
46	2116	6.7823	21.4476	96	9216	9.7980	30.9839
47	2209	6.8557	21.6795	97	9409	9.8489	31.1448
48	2304	6.9282	21.9089	98	9604	9.8995	31.3049
49	2401	7.0000	22.1359	99	9801	9.9499	31.4643
50	2500	7.0711	22.3607	100	10000	10.0000	31.6228

Table J

Four-Place Mantissas for Common Logarithms

	.00	.01	.02	.03	.04	.05	.06	.07	.08	.09
1.0	.0000	.0043	.0086	.0128	.0170	.0212	.0253	.0294	.0334	.0374
1.1	.0414	.0453	.0492	.0531	.0569	.0607	.0645	.0682	.0719	.0755
1.2	.0792	.0828	.0864	.0899	.0934	.0969	.1004	.1038	.1072	.1106
1.3	.1139	.1173	.1206	.1239	.1271	.1303	.1335	.1367	.1399	.1430
1.4	.1461	.1492	.1523	.1553	.1584	.1614	.1644	.1673	.1703	.1732
1.5	.1761	.1790	.1818	.1847	.1875	.1903	.1931	.1959	.1987	.2014
1.6	.2041	.2068	.2095	.2122	.2148	.2175	.2201	.2227	.2253	.2279
1.7	.2304	.2330	.2355	.2380	.2405	.2430	.2455	.2480	.2504	.2529
1.8	.2553	.2577	.2601	.2625	.2648	.2672	.2695	.2718	.2742	.2765
1.9	.2788	.2810	.2833	.2856	.2878	.2900	.2923	.2945	.2967	.2989
2.0	.3010	.3032	.3054	.3075	.3096	.3118	.3139	.3160	.3181	.3201
2.1	.3222	.3243	.3263	.3284	.3304	.3324	.3345	.3365	.3385	.3404
2.2	.3424	.3444	.3464	.3483	.3502	.3522	.3541	.3560	.3579	.3598
2.3	.3617	.3636	.3655	.3674	.3692	.3711	.3729	.3747	.3766	.3784
2.4	.3802	.3820	.3838	.3856	.3874	.3892	.3909	.3927	.3945	.3962
2.5	.3979	.3997	.4014	.4031	.4048	.4065	.4082	.4099	.4116	.4133
2.6	.4150	.4166	.4183	.4200	.4216	.4232	.4249	.4265	.4281	.4298
2.7	.4314	.4330	.4346	.4362	.4378	.4393	.4409	.4425	.4440	.4456
2.8	.4472	.4487	.4502	.4518	.4533	.4548	.4564	.4579	.4594	.4609
2.9	.4624	.4639	.4654	.4669	.4683	.4698	.4713	.4728	.4742	.4757
3.0	.4771	.4786	.4800	.4814	.4829	.4843	.4857	.4871	.4886	.4900
3.1	.4914	.4928	.4942	.4955	.4969	.4983	.4997	.5011	.5024	.5038
3.2	.5051	.5065	.5079	.5092	.5105	.5119	.5132	.5145	.5159	.5172
3.3	.5185	.5198	.5211	.5224	.5237	.5250	.5263	.5276	.5289	.5302
3.4	.5315	.5328	.5340	.5353	.5366	.5378	.5391	.5403	.5416	.5428
3.5	.5441	.5453	.5465	.5478	.5490	.5502	.5514	.5527	.5539	.5551
3.6	.5563	.5575	.5587	.5599	.5611	.5623	.5635	.5647	.5658	.5670
3.7	.5682	.5694	.5705	.5717	.5729	.5740	.5752	.5763	.5775	.5786
3.8	.5798	.5809	.5821	.5832	.5843	.5855	.5866	.5877	.5888	.5899
3.9	.5911	.5922	.5933	.5944	.5955	.5966	.5977	.5988	.5999	.6010
4.0	.6021	.6031	.6042	.6053	.6064	.6075	.6085	.6096	.6107	.6117
4.1	.6128	.6138	.6149	.6159	.6170	.6180	.6191	.6201	.6212	.6222
4.2	.6232	.6243	.6253	.6263	.6274	.6284	.6294	.6304	.6314	.6325
4.3	.6335	.6345	.6355	.6365	.6375	.6385	.6395	.6405	.6415	.6425
4.4	.6435	.6444	.6454	.6464	.6474	.6484	.6493	.6503	.6513	.6522
4.5	.6532	.6542	.6551	.6561	.6571	.6580	.6590	.6599	.6609	.6618
4.6	.6628	.6637	.6646	.6656	.6665	.6675	.6684	.6693	.6702	.6712
4.7	.6721	.6730	.6739	.6749	.6758	.6767	.6776	.6785	.6794	.6803
4.8	.6812	.6821	.6830	.6839	.6848	.6857	.6866	.6875	.6884	.6893
4.9	.6902	.6911	.6920	.6928	.6937	.6946	.6955	.6964	.6972	.6981
5.0	.6990	.6998	.7007	.7016	.7024	.7033	.7042	.7050	.7059	.7067
5.1	.7076	.7084	.7093	.7101	.7110	.7118	.7126	.7135	.7143	.7152
5.2	.7160	.7168	.7177	.7185	.7193	.7202	.7210	.7218	.7226	.7235
5.3	.7243	.7251	.7259	.7267	.7275	.7284	.7292	.7300	.7308	.7316
5.4	.7324	.7332	.7340	.7348	.7356	.7364	.7372	.7380	.7388	.7396

Table J (continued)

Four-Place Mantissas for Common Logarithms

	.00	.01	.02	.03	.04	.05	.06	.07	.08	.09
5.5	.7404	.7412	.7419	.7427	.7435	.7443	.7451	.7459	.7466	.7474
5.6	.7482	.7490	.7497	.7505	.7513	.7520	.7528	.7536	.7543	.7551
5.7	.7559	.7566	.7574	.7582	.7589	.7597	.7604	.7612	.7619	.7627
5.8	.7634	.7642	.7649	.7657	.7664	.7672	.7679	.7686	.7694	.7701
5.9	.7709	.7716	.7723	.7731	.7738	.7745	.7752	.7760	.7767	.7774
6.0	.7782	.7789	.7796	.7803	.7810	.7818	.7825	.7832	.7839	.7846
6.1	.7853	.7860	.7868	.7875	.7882	.7889	.7896	.7903	.7910	.7917
6.2	.7924	.7931	.7938	.7945	.7952	.7959	.7966	.7973	.7980	.7987
6.3	.7993	.8000	.8007	.8014	.8021	.8028	.8035	.8041	.8048	.8055
6.4	.8062	.8069	.8075	.8082	.8089	.8096	.8102	.8109	.8116	.8122
6.5	.8129	.8136	.8142	.8149	.8156	.8162	.8169	.8176	.8182	.8189
6.6	.8195	.8202	.8209	.8215	.8222	.8228	.8235	.8241	.8248	.8254
6.7	.8261	.8267	.8274	.8280	.8287	.8293	.8299	.8306	.8312	.8319
6.8	.8325	.8331	.8338	.8344	.8351	.8357	.8363	.8370	.8376	.8382
6.9	.8388	.8395	.8401	.8407	.8414	.8420	.8426	.8432	.8439	.8445
7.0	.8451	.8457	.8463	.8470	.8476	.8482	.8488	.8494	.8500	.8506
7.1	.8513	.8519	.8525	.8531	.8537	.8543	.8549	.8555	.8561	.8567
7.2	.8573	.8579	.8585	.8591	.8597	.8603	.8609	.8615	.8621	.8627
7.3	.8633	.8639	.8645	.8651	.8657	.8663	.8669	.8675	.8681	.8686
7.4	.8692	.8698	.8704	.8710	.8716	.8722	.8727	.8733	.8739	.8745
7.5	.8751	.8756	.8762	.8768	.8774	.8779	.8785	.8791	.8797	.8802
7.6	.8808	.8814	.8820	.8825	.8831	.8837	.8842	.8848	.8854	.8859
7.7	.8865	.8871	.8876	.8882	.8887	.8893	.8899	.8904	.8910	.8915
7.8	.8921	.8927	.8932	.8938	.8943	.8949	.8954	.8960	.8965	.8971
7.9	.8976	.8982	.8987	.8993	.8998	.9004	.9009	.9015	.9020	.9025
8.0	.9031	.9036	.9042	.9047	.9053	.9058	.9063	.9069	.9074	.9079
8.1	.9085	.9090	.9096	.9101	.9106	.9112	.9117	.9122	.9128	.9133
8.2	.9138	.9143	.9149	.9154	.9159	.9165	.9170	.9175	.9180	.9186
8.3	.9191	.9196	.9201	.9206	.9212	.9217	.9222	.9227	.9232	.9238
8.4	.9243	.9248	.9253	.9258	.9263	.9269	.9274	.9279	.9284	.9289
8.5	.9294	.9299	.9304	.9309	.9315	.9320	.9325	.9330	.9335	.9340
8.6	.9345	.9350	.9355	.9360	.9365	.9370	.9375	.9380	.9385	.9390
8.7	.9395	.9400	.9405	.9410	.9415	.9420	.9425	.9430	.9435	.9440
8.8	.9445	.9450	.9455	.9460	.9465	.9469	.9474	.9479	.9484	.9489
8.9	.9494	.9499	.9504	.9509	.9513	.9518	.9523	.9528	.9533	.9538
9.0	.9542	.9547	.9552	.9557	.9562	.9566	.9571	.9576	.9581	.9586
9.1	.9590	.9595	.9600	.9605	.9609	.9614	.9619	.9624	.9628	.9633
9.2	.9638	.9643	.9647	.9652	.9657	.9661	.9666	.9671	.9675	.9680
9.3	.9685	.9689	.9694	.9699	.9703	.9708	.9713	.9717	.9722	.9727
9.4	.9731	.9736	.9741	.9745	.9750	.9754	.9759	.9764	.9768	.9773
9.5	.9777	.9782	.9786	.9791	.9795	.9800	.9805	.9809	.9814	.9818
9.6	.9823	.9827	.9832	.9836	.9841	.9845	.9850	.9854	.9859	.9863
9.7	.9868	.9872	.9877	.9881	.9886	.9890	.9894	.9899	.9903	.9908
9.8	.9912	.9917	.9921	.9926	.9930	.9934	.9939	.9943	.9948	.9952
9.9	.9956	.9961	.9965	.9969	.9974	.9978	.9983	.9987	.9991	.9996

Table K

Natural (Napierian) Logarithms
Note: All logarithms in this table are negative

	.000	.001	.002	.003	.004	.005	.006	.007	.008	.009
0.00	∞	6.9078	6.2146	5.8091	5.5215	5.2983	5.1160	4.9618	4.8283	4.7105
0.01	4.6052	4.5099	4.4228	4.3428	4.2687	4.1997	4.1352	4.0745	4.0174	3.9633
0.02	3.9120	3.8632	3.8167	3.7723	3.7297	3.6889	3.6497	3.6119	3.5756	3.5405
0.03	3.5066	3.4738	3.4420	3.4112	3.3814	3.3524	3.3242	3.2968	3.2702	3.2442
0.04	3.2189	3.1942	3.1701	3.1466	3.1236	3.1011	3.0791	3.0576	3.0366	3.0159
0.05	2.9957	2.9759	2.9565	2.9375	2.9188	2.9004	2.8824	2.8647	2.8473	2.8302
0.06	2.8134	2.7969	2.7806	2.7646	2.7489	2.7334	2.7181	2.7031	2.6882	2.6736
0.07	2.6593	2.6451	2.6311	2.6173	2.6037	2.5903	2.5770	2.5640	2.5510	2.5383
0.08	2.5257	2.5133	2.5010	2.4889	2.4769	2.4651	2.4534	2.4418	2.4304	2.4191
0.09	2.4079	2.3969	2.3860	2.3752	2.3645	2.3539	2.3434	2.3330	2.3228	2.3126
0.10	2.3026	2.2926	2.2828	2.2730	2.2634	2.2538	2.2443	2.2349	2.2256	2.2164
0.11	2.2073	2.1982	2.1893	2.1804	2.1716	2.1628	2.1542	2.1456	2.1371	2.1286
0.12	2.1203	2.1120	2.1037	2.0956	2.0875	2.0794	2.0715	2.0636	2.0557	2.0479
0.13	2.0402	2.0326	2.0250	2.0174	2.0099	2.0025	1.9951	1.9878	1.9805	1.9733
0.14	1.9661	1.9590	1.9519	1.9449	1.9379	1.9310	1.9241	1.9173	1.9105	1.9038
0.15	1.8971	1.8905	1.8839	1.8773	1.8708	1.8643	1.8579	1.8515	1.8452	1.8389
0.16	1.8326	1.8264	1.8202	1.8140	1.8079	1.8018	1.7958	1.7898	1.7838	1.7779
0.17	1.7720	1.7661	1.7603	1.7545	1.7487	1.7430	1.7373	1.7316	1.7260	1.7204
0.18	1.7148	1.7093	1.7037	1.6983	1.6928	1.6874	1.6820	1.6766	1.6713	1.6660
0.19	1.6607	1.6555	1.6503	1.6451	1.6399	1.6348	1.6296	1.6246	1.6195	1.6145
0.20	1.6094	1.6045	1.5995	1.5945	1.5896	1.5847	1.5799	1.5750	1.5702	1.5654
0.21	1.5606	1.5559	1.5512	1.5465	1.5418	1.5371	1.5325	1.5279	1.5233	1.5187
0.22	1.5141	1.5096	1.5051	1.5006	1.4961	1.4917	1.4872	1.4828	1.4784	1.4740
0.23	1.4697	1.4653	1.4610	1.4567	1.4524	1.4482	1.4439	1.4397	1.4355	1.4313
0.24	1.4271	1.4230	1.4188	1.4147	1.4106	1.4065	1.4024	1.3984	1.3943	1.3903
0.25	1.3863	1.3823	1.3783	1.3744	1.3704	1.3665	1.3626	1.3587	1.3548	1.3509
0.26	1.3471	1.3432	1.3394	1.3356	1.3318	1.3280	1.3243	1.3205	1.3168	1.3130
0.27	1.3093	1.3056	1.3020	1.2983	1.2946	1.2910	1.2874	1.2837	1.2801	1.2765
0.28	1.2730	1.2694	1.2658	1.2623	1.2588	1.2553	1.2518	1.2483	1.2448	1.2413
0.29	1.2379	1.2344	1.2310	1.2276	1.2242	1.2208	1.2174	1.2140	1.2107	1.2073
0.30	1.2040	1.2006	1.1973	1.1940	1.1907	1.1874	1.1842	1.1809	1.1777	1.1744
0.31	1.1712	1.1680	1.1648	1.1616	1.1584	1.1552	1.1520	1.1489	1.1457	1.1426
0.32	1.1394	1.1363	1.1332	1.1301	1.1270	1.1239	1.1209	1.1178	1.1147	1.1117
0.33	1.1087	1.1056	1.1026	1.0996	1.0966	1.0936	1.0906	1.0877	1.0847	1.0818
0.34	1.0788	1.0759	1.0729	1.0700	1.0671	1.0642	1.0613	1.0584	1.0556	1.0527
0.35	1.0498	1.0470	1.0441	1.0413	1.0385	1.0356	1.0328	1.0300	1.0272	1.0244
0.36	1.0217	1.0189	1.0161	1.0134	1.0106	1.0079	1.0051	1.0024	0.9997	0.9970
0.37	0.9943	0.9916	0.9889	0.9862	0.9835	0.9808	0.9782	0.9755	0.9729	0.9702
0.38	0.9676	0.9650	0.9623	0.9597	0.9571	0.9545	0.9519	0.9493	0.9468	0.9442
0.39	0.9416	0.9390	0.9365	0.9339	0.9314	0.9289	0.9263	0.9238	0.9213	0.9188
0.40	0.9163	0.9138	0.9113	0.9088	0.9063	0.9039	0.9014	0.8989	0.8965	0.8940
0.41	0.8916	0.8892	0.8867	0.8843	0.8819	0.8795	0.8771	0.8747	0.8723	0.8699
0.42	0.8675	0.8651	0.8628	0.8604	0.8580	0.8557	0.8533	0.8510	0.8486	0.8463
0.43	0.8440	0.8416	0.8393	0.8370	0.8347	0.8324	0.8301	0.8278	0.8255	0.8233
0.44	0.8210	0.8187	0.8164	0.8142	0.8119	0.8097	0.8074	0.8052	0.8030	0.8007
0.45	0.7985	0.7963	0.7941	0.7919	0.7897	0.7875	0.7853	0.7831	0.7809	0.7787
0.46	0.7765	0.7744	0.7722	0.7700	0.7679	0.7657	0.7636	0.7614	0.7593	0.7572
0.47	0.7550	0.7529	0.7508	0.7487	0.7465	0.7444	0.7423	0.7402	0.7381	0.7361
0.48	0.7340	0.7319	0.7298	0.7277	0.7257	0.7236	0.7215	0.7195	0.7174	0.7154
0.49	0.7134	0.7113	0.7093	0.7072	0.7052	0.7032	0.7012	0.6992	0.6972	0.6951

Table K (continued)

Natural (Napierian) Logarithms
Note: All logarithms in this table are negative

	.000	.001	.002	.003	.004	.005	.006	.007	.008	.009
0.50	0.6931	0.6911	0.6892	0.6872	0.6852	0.6832	0.6812	0.6792	0.6773	0.6753
0.51	0.6733	0.6714	0.6694	0.6675	0.6655	0.6636	0.6616	0.6597	0.6578	0.6559
0.52	0.6539	0.6520	0.6501	0.6482	0.6463	0.6444	0.6425	0.6406	0.6387	0.6368
0.53	0.6349	0.6330	0.6311	0.6292	0.6274	0.6255	0.6236	0.6218	0.6199	0.6180
0.54	0.6162	0.6143	0.6125	0.6106	0.6088	0.6070	0.6051	0.6033	0.6015	0.5997
0.55	0.5978	0.5960	0.5942	0.5924	0.5906	0.5888	0.5870	0.5852	0.5834	0.5816
0.56	0.5798	0.5780	0.5763	0.5745	0.5727	0.5709	0.5692	0.5674	0.5656	0.5639
0.57	0.5621	0.5604	0.5586	0.5569	0.5551	0.5534	0.5516	0.5499	0.5482	0.5465
0.58	0.5447	0.5430	0.5413	0.5396	0.5379	0.5361	0.5344	0.5327	0.5310	0.5293
0.59	0.5276	0.5259	0.5242	0.5226	0.5209	0.5192	0.5175	0.5158	0.5142	0.5125
0.60	0.5108	0.5092	0.5075	0.5058	0.5042	0.5025	0.5009	0.4992	0.4976	0.4959
0.61	0.4943	0.4927	0.4910	0.4894	0.4878	0.4861	0.4845	0.4829	0.4813	0.4797
0.62	0.4780	0.4764	0.4748	0.4732	0.4716	0.4700	0.4684	0.4668	0.4652	0.4636
0.63	0.4620	0.4604	0.4589	0.4573	0.4557	0.4541	0.4526	0.4510	0.4494	0.4479
0.64	0.4463	0.4447	0.4432	0.4416	0.4401	0.4385	0.4370	0.4354	0.4339	0.4323
0.65	0.4308	0.4292	0.4277	0.4262	0.4246	0.4231	0.4216	0.4201	0.4186	0.4170
0.66	0.4155	0.4140	0.4125	0.4110	0.4095	0.4080	0.4065	0.4050	0.4035	0.4020
0.67	0.4005	0.3990	0.3975	0.3960	0.3945	0.3930	0.3916	0.3901	0.3886	0.3871
0.68	0.3857	0.3842	0.3827	0.3813	0.3798	0.3783	0.3769	0.3754	0.3740	0.3725
0.69	0.3711	0.3696	0.3682	0.3667	0.3653	0.3638	0.3624	0.3610	0.3595	0.3581
0.70	0.3567	0.3552	0.3538	0.3524	0.3510	0.3496	0.3481	0.3467	0.3453	0.3439
0.71	0.3425	0.3411	0.3397	0.3383	0.3369	0.3355	0.3341	0.3327	0.3313	0.3299
0.72	0.3285	0.3271	0.3257	0.3243	0.3230	0.3216	0.3202	0.3188	0.3175	0.3161
0.73	0.3147	0.3133	0.3120	0.3106	0.3092	0.3079	0.3065	0.3052	0.3038	0.3025
0.74	0.3011	0.2998	0.2984	0.2971	0.2957	0.2944	0.2930	0.2917	0.2904	0.2890
0.75	0.2877	0.2863	0.2850	0.2837	0.2824	0.2810	0.2797	0.2784	0.2771	0.2758
0.76	0.2744	0.2731	0.2718	0.2705	0.2692	0.2679	0.2666	0.2653	0.2640	0.2627
0.77	0.2614	0.2601	0.2588	0.2575	0.2562	0.2549	0.2536	0.2523	0.2510	0.2497
0.78	0.2485	0.2472	0.2459	0.2446	0.2433	0.2421	0.2408	0.2395	0.2383	0.2370
0.79	0.2357	0.2345	0.2332	0.2319	0.2307	0.2294	0.2282	0.2269	0.2256	0.2244
0.80	0.2231	0.2219	0.2206	0.2194	0.2182	0.2169	0.2157	0.2144	0.2132	0.2120
0.81	0.2107	0.2095	0.2083	0.2070	0.2058	0.2046	0.2033	0.2021	0.2009	0.1997
0.82	0.1985	0.1972	0.1960	0.1948	0.1936	0.1924	0.1912	0.1900	0.1887	0.1875
0.83	0.1863	0.1851	0.1839	0.1827	0.1815	0.1803	0.1791	0.1779	0.1767	0.1755
0.84	0.1744	0.1732	0.1720	0.1708	0.1696	0.1684	0.1672	0.1661	0.1649	0.1637
0.85	0.1625	0.1613	0.1602	0.1590	0.1578	0.1567	0.1555	0.1543	0.1532	0.1520
0.86	0.1508	0.1497	0.1485	0.1473	0.1462	0.1450	0.1439	0.1427	0.1416	0.1404
0.87	0.1393	0.1381	0.1370	0.1358	0.1347	0.1335	0.1324	0.1312	0.1301	0.1290
0.88	0.1278	0.1267	0.1256	0.1244	0.1233	0.1222	0.1210	0.1199	0.1188	0.1177
0.89	0.1165	0.1154	0.1143	0.1132	0.1120	0.1109	0.1098	0.1087	0.1076	0.1065
0.90	0.1054	0.1043	0.1031	0.1020	0.1009	0.0998	0.0987	0.0976	0.0965	0.0954
0.91	0.0943	0.0932	0.0921	0.0910	0.0899	0.0888	0.0877	0.0866	0.0856	0.0845
0.92	0.0834	0.0823	0.0812	0.0801	0.0790	0.0780	0.0769	0.0758	0.0747	0.0736
0.93	0.0726	0.0715	0.0704	0.0694	0.0683	0.0672	0.0661	0.0651	0.0640	0.0629
0.94	0.0619	0.0608	0.0598	0.0587	0.0576	0.0566	0.0555	0.0545	0.0534	0.0523
0.95	0.0513	0.0502	0.0492	0.0481	0.0471	0.0460	0.0450	0.0440	0.0429	0.0419
0.96	0.0408	0.0398	0.0387	0.0377	0.0367	0.0356	0.0346	0.0336	0.0325	0.0315
0.97	0.0305	0.0294	0.0284	0.0274	0.0263	0.0253	0.0243	0.0233	0.0222	0.0212
0.98	0.0202	0.0192	0.0182	0.0171	0.0161	0.0151	0.0141	0.0131	0.0121	0.0111
0.99	0.0101	0.0090	0.0080	0.0070	0.0060	0.0050	0.0040	0.0030	0.0020	0.0010

Answers To Selected Problems

CHAPTER 2

2.2 Subjective

2.4 (a) 1/13; (b) 3/13; (c) 1/52; (d) 1/4; (e) 1/2; (f) 1; (g) 9/13; (h) 3/52; (i) 1/2

2.6 (a) .294; (b) .092; (c) .018; (d) .283

2.8 .00013

CHAPTER 3

3.1 72

3.3 24

3.5 5,040

3.7 720

3.8 (a) 720; (b) 360; (c) 144; (d) 36; (e) 72

3.9 1,663,200

3.11 1,287

3.13 $.49592 \times 10^{15}$

3.15 1,028,160

3.17 16.9 millions

3.19 (a) .04; (b) .48

3.20 (a) 32; (b) 12; (c) 4; (d) 6

3.21 15,504

3.23 3,718 millions

3.24 (a) 2,380; (c) 408; (e) 425

3.25 635 billions

3.27 98,794,080

3.29 (a) 420; (b) 4

3.30 (a) (i) $\binom{5}{2}\binom{10}{3}\binom{5}{2}$, (ii) $\binom{40}{2}\binom{100}{5}\binom{60}{3}$,

(iii) (1) $\binom{120}{6}\binom{59}{3}\binom{20}{2}$, (iv) $\dfrac{\binom{200}{5}\binom{50}{4}}{\binom{250}{9}}$

(b) (i) $\dfrac{50!\ 200!\ 242!\ 8!}{47!\ 3!\ 195!\ 5!\ 250!}$, (ii) $\dfrac{50!\ 125!\ 75!\ 239!\ 11!}{48!\ 2!\ 119!\ 6!\ 72!\ 3!\ 250!}$,

(iii) $\sum_{x=3}^{10} \dfrac{25!\ 99!\ 10!\ 114!}{x!\ (25-x)!\ (10-x)!\ (89+x)!\ 124!}$

CHAPTER 4

4.2 $S_1 = \{x|x \text{ is one of the last three months of the year}\}$
$S_2 = \{x|x \text{ is an English vowel}\}$
$S_3 = \{x|x \text{ is a positive-even integer}\}$

4.4 (a) 252; (b) 1,024

4.6 (b) (i) .57, (ii) .43, (iii) .18, (iv) .33, (v) .24, (vi) .49, (vii) .43, (viii) .14, (ix) .57, (x) .10, (xi) .20, (xii) .31

4.8 $H = A \cap B = \{5,6,7\}$ $N = \{7\}$
$I = A \cap C = \{3,4,5,6,7\}$ $O = \emptyset$
$J = A \cap D = \{4,5,6\}$ $P = \{p|7 < p < 8\}$
$K = C \cap D = \{k|3 < k < 7\}$ $Q = \{3,4,5,6,7,8,9\}$
$L = \emptyset$ $R = \{r|3 < r \leq 8\}$
$M = \{m|7 < m < 8\}$ $S = \{s|3 \leq s \leq 9\}$

4.10 (a), (d), (f), and (h) are partitions

4.12 (a) $X \cap Y = \{2\}$;
(b) $X' \cap Y' = \{5,7,9\}$;
(c) $X' \cap Z' = \{4,8\}$;
(d) $X \cap Y \cap Z' = \emptyset$
(e) $(X \cap Y')' = \{2,4,5,6,7,8,9\}$;
(f) $(Y \cap Z)' = \{1,3,4,5,7,8,9\}$;
(g) $(X \cap Y \cap Z)' = \{1,3,4,5,6,7,8,9\}$;
(h) $(X' \cap Y' \cap Z)' = \{1,2,3,4,6,8\}$

4.14 (a) $J = \{a,b,c,d\}$; $K = \{e\}$; $L = \{a,c,h\}$;
$M = \{e\}$; $N = \{c,f,h\}$; $O = \{a,b,f,h\}$
(b) $P = L \cap N$; $Q = J \cap O$; $R = J' \cap O'$; $S = L \cup M$;
$T = J' \cup (M \cup N)'$
(c) $P = \{c,h\}$; $Q = \{a,b\}$; $R = \{e,g\}$; $S = \{a,c,e,h\}$; $T = \{a,b,d,e,f,g,h\}$

CHAPTER 5

5.1 .75

5.3 .50

5.5 .10

5.7 (b) (i) .75, (ii) .20, (iii) .40, (iv) .33, (v) .20
(c) (i) $P(M' \cap T)$, (ii) $P(M' \cap T')$, (iii) $P(M|T')$, (iv) $P(T'|M)$, (v) $P(M'|T')$

5.9 (a) .32; (b) .48; (c) .08; (d) .12; (e) .80; (f) .80; (g) .20; (h) .20

5.11 18

5.13 (a) 2/3; (b) no

5.15 (a) .80; (b) no

5.17 .60

5.19 $P(X) = .40$, $P(Y) = .40$, $P(Z) = .20$

5.21 (a) .90; (b) .50

5.22 .80

5.23 .30

5.25 (a) .75; (b) .64

5.27 (a) .56; (b) .06; (c) .94; (d) .44; (e) .70; (f) .20; (g) .24

5.29 (a) .10; (b) .90; (c) .30; (d) .50; (e) .83; (f) .83; (g) .75

5.31 (a) .200; (b) .229; (c) .789; (d) .800; (e) .7875; (f) .290; (g) .967

5.33 (a) .02; (b) .35; (c) .75; (d) .83; (e) .60; (f) .90; (g) .26

5.34 (a) .04; (b) .81; (c) .06; (d) .12; (e) .64; (f) .09; (g) .10;
(h) .05; (i) .15; (j) .85; (k) .95

5.35 .9656

5.37 (a) 3; (b) 9

5.39 (a) 5

5.41 2/3

5.43 .043

5.45 (a) .72, .20, .08; (b) .20, .76, .04; (c) .10, .34, .56

CHAPTER 6

6.2 (a) $s = f(c) = 1.25c$; (b) $s = f(c) = \dfrac{c}{.6}$

6.4 (a) 4; (b) Domain: $\{0,1,2,3,4,5\}$, Range: $\{4,6,10\}$

6.6 (a) $T = f(H) = \$20 + \$5H$
(b) (i) $P = f(T) = \$50 - T$, (ii) $P = g(H) = 30 - 5H$

6.8 (b)

$$Q_p = \begin{cases} 2X & \text{if } X \leq \dfrac{s_p}{2} \\ s_p & \text{if } X > \dfrac{s_p}{2} \end{cases}$$

(d)

$$R_b = \begin{cases} X & \text{if } X \leq \dfrac{s_b}{4} \\ .25s_b & \text{if } X > \dfrac{s_b}{4} \end{cases}$$

(f)

$$R_p = \begin{cases} .30X & \text{if } X \leq \dfrac{s_p}{2} \\ .15s_p & \text{if } X > \dfrac{s_p}{2} \end{cases}$$

(i)

$$R_t = \begin{cases} X + .30X & \text{if } X \leq \dfrac{s_b}{4} \text{ and } X \leq \dfrac{s_p}{2} \\ X + .15s_p & \text{if } X \leq \dfrac{s_b}{4} \text{ and } X > \dfrac{s_p}{2} \\ .25s_b + .30X & \text{if } X > \dfrac{s_b}{4} \text{ and } X \leq \dfrac{s_p}{2} \\ .25s_b + .15s_p & \text{if } X > \dfrac{s_b}{4} \text{ and } X > \dfrac{s_p}{2} \end{cases}$$

(m) $G_t = R_b + R_p - .10s_b - .05s_p$

6.10 (a) 12; (c) $S = \{5,7,8,11,12,14\}$;
(e) $E_1 = \{(2,3), (2,5), (3,2), (5,2)\}$
$E_2 = \{(2,5), (2,9), (3,5), (5,2), (5,3), (9,2)\}$
$E_3 = \{(2,9), (3,5), (5,3), (9,2)\}$

CHAPTER 7

7.1 (a)

x	$f(x)$
1	.12
2	.16
3	.20
4	.24
5	.28

(b)

$$F(x) = \begin{cases} 0 & \text{if } x < 1 \\ \sum_{t=1}^{x} (t+2)/25 & \text{if } x = 1,2,3,4,5 \\ 1 & \text{if } x > 5 \end{cases}$$

(c)

x	$F(x)$
1	.12
2	.28
3	.48
4	.72
5	1.00

7.3 (a) .16; (b) .18; (c) .34; (d) 1; (e) .60

7.5 (a)

$$F(z) = \begin{cases} 0 & \text{if } z < 0 \\ \sum_{t=0}^{z} \dfrac{1}{t!\, e} & \text{if } z = 0, 1, 2, \ldots \end{cases}$$

(b) .2605

7.7 (a) .4096; (b) .9728

7.9 (a) $k = .1$; (b) .33

7.11 $\frac{1}{1{,}440} = .000694$

7.13 (c) is the corresponding p.n.f.

7.15 (a) By inspection, $f(x) \geq 0$ for all real values.

Also, $\int_{-\infty}^{\infty} f(x)\,dx = \int_0^1 2g\,dg = g^2\Big]_0^1 = 1$

(c) $F(g) = \begin{cases} 0 & \text{if } g < 0 \\ g^2 & \text{if } 0 \leq g \leq 1 \\ 1 & \text{if } g > 1 \end{cases}$

(d) (i) .25, (ii) .64, (iii) .33

7.17 (a)

$$F(w) = \begin{cases} 0 & \text{if } w < 0 \\ \frac{w^2}{16} & \text{if } 0 \leq w \leq 4 \\ 1 & \text{if } w > 4 \end{cases}$$

(b) (i) .1875, (ii) .5625, (iii) .6094

7.19 (c)

$$F(h) = \begin{cases} 0 & \text{if } h < 0 \\ 3h^2 - 2h^3 & \text{if } 0 \leq h \leq 1 \\ 1 & \text{if } h > 1 \end{cases}$$

(d) (i) .104, (ii) .0607, (iii) .6874

7.21 (a)

$$F(h) = \begin{cases} 0 & \text{if } h < 0 \\ .3\left(\frac{h^2}{2} - \frac{h^3}{3} + 2h\right) & \text{if } 0 \leq h \leq 2 \\ 1 & \text{if } h > 2 \end{cases}$$

(b) .65; (c) .10

7.23 (a)

$$F(m) = \begin{cases} 0 & \text{if } m < 1 \\ m^2 - 2m + 1 & \text{if } 1 \leq m \leq 2 \\ 1 & \text{if } m > 2 \end{cases}$$

(b) (i) 0, (ii) 1, (iii) .39

7.25 (b)

$$F(v) = \begin{cases} 0 & \text{if } v < 2 \\ \frac{v^2}{2} - 2v + 2 & \text{if } 2 \leq v < 3 \\ 4v - \frac{v^2}{2} - 7 & \text{if } 3 \leq v \leq 4 \\ 1 & \text{if } v > 4 \end{cases}$$

(d) (i) .125, (ii) .875, (iii) .32, (iv) .84

7.27 (b)

$$F(x) = \begin{cases} 0 & \text{if } x < 0 \\ \dfrac{x^3}{2} & \text{if } 0 \le x < 1 \\ 1.5 - \dfrac{1}{x} & \text{if } 1 \le x \le 2 \\ 1 & \text{if } x > 2 \end{cases}$$

(d) (i) .50, (iii) .9375, (v) .119

7.29 (a)

$$F(x) = \begin{cases} 0 & \text{if } x < 0 \\ \dfrac{x^2}{16} & \text{if } 0 \le x \le 4; \\ 1 & \text{if } x > 4 \end{cases}$$

(b) (i) 1/4, (ii) 3/16, (iii) 5/16, (iv) 15/16, (v) 5/12;
(c) No, since $P(A_2 \cap A_3) \neq P(A_2)P(A_3)$

7.31 (a) $F(x) = \begin{cases} 0 & \text{if } x < 0 \\ 1 - e^{-2x} & \text{if } x \ge 0 \end{cases}$;

(b) (i) .1170, (ii) .6295

7.33 .9029

7.35 (a) $f(x) = \begin{cases} e^x & \text{if } x \le 0 \\ 0 & \text{if } x > 0 \end{cases}$;

(c) .2326

7.37 (b)

$$F(x) = \begin{cases} \dfrac{1}{2} e^x & \text{if } x < 0 \\ 1 - \dfrac{1}{2} e^{-x} & \text{if } x \ge 0 \end{cases}$$

(d) (i) .3352, (ii) .6967, (iii) .2202

CHAPTER 8

8.2 $\mu = 3.4$, Median = 4, Mode = 5

8.4 \$5.00

8.6 $\mu = 1.33$, Median = 1.41

8.8 $\mu = .5$, Mode = .5

8.10 5/3

8.11 (a) $a = 1.2$, $b = -.4$; (b) .95; (c) 1.27

8.12 $E(W) = 1/2$, Median = .35

8.14 (a) ∞; (b) .216

8.16 (a) .70; (b) 11.0; (c) \$11,100

8.18 (a) 5; (b) \$27,000

8.20 \$3,667

8.22 (a) 5 inches; (b) \$10,000; (c) 1.10 inches of rain

8.23 (a) 6.67 hours; (b) \$173.00

8.24 A.D. = 1.064, $\sigma^2 = 1.64$, $\sigma = 1.28$

8.26 A.D. = 1.80, $\sigma^2 = 4.66$, $\sigma = 2.16$

8.28 A.D. = .395, $\sigma^2 = 2/9$, $\sigma = .471$

8.30 $\sigma^2 = .05$, $\sigma = .224$

8.31 A.D. = .3679, $\sigma^2 = 1/4$, $\sigma = 1/2$

8.34 (b) $\mu = 2.5$, $\sigma^2 = 1.33$

8.36 150,000

8.38 (a) 14, .67; (b) \$32; (c) \$32; (d) 96.48

CHAPTER 9

9.1 .1475

9.3 (a) .1591; (b) .9955;

(c) (i) $f_h(x|N = 12, n = 3, D = 4) = \dfrac{\binom{4}{x}\binom{8}{3-x}}{\binom{12}{3}}$,

(ii)

x	$f(x)$
0	.2545
1	.5091
2	.2182
3	.0182

9.5 (a)

0	.1937
1	.4402
2	.2935
3	.0677
4	.0048
5	.0001
	1.0000

(b) 1.25

9.7 (a) .2051; (b) .2051; (c) .9893; (d) .0547; (e) .0107

9.9 (a) .5852; (b) 8; (c) .0609; (d) .7858

9.11 .0435

9.13 (a) .1501; (b) .0970; (c) \$23.4 million

9.15 (a) .0586; (b) 17.5; (c) .9081; (d) .1440

9.17 (a) .8737; (b) .9601; (c) 22.2

9.19 .1904

9.21 .0016

9.23 (a) .2225; (b) .5704; (c) .4013; (d) .8442; (e) 14; (f) 14, 3.74

CHAPTER 10

10.2 (a) .9772; (b) .0228; (c) .3085; (d) .0401; (e) .5319

10.5 .5631

10.6 (a) 190,880; (b) It is only a rough approximation since economic expenditures tend to be skewed to the right.

10.8 (a) Process B; (b) Process B

10.9 (a) 16.20; (b) 23.80

10.10 1.16

10.12 (a) 19.05 years; (b) 4.65 years; (c) 67.14%

10.13 .0866

10.14 (a) .1972; (b) .4024

10.16 (a) .6626; (b) 40

10.18 .1205

10.20 (a) .7619; (b) 6 hours; (c) \$47.62

10.22 .0025

10.23 .5488

10.24 (b) $F(h) = \begin{cases} 1 - e^{-.25h} & \text{if } h \geq 0 \\ 0 & \text{if } h < 0 \end{cases}$

(d) $\sigma = 4$ hours; (e) .368; (f) .2568

10.26 (a) (i) $\frac{1}{3}$, (ii) $\frac{1}{3}$, (iii) $\frac{1}{3}$; (b) 3.5; (c) .75

10.28 .70

CHAPTER 11

11.1 (a) .0030; (c) .0013; (e) .9881

11.3

	(a)	(b)	(c)
(i)	.0012	.0012	.0000
(ii)	.0758	.0756	+.0002
(iii)	.1839	.1858	−.0019
(iv)	.2565	.2611	−.0046
(v)	.0842	.0779	+.0063
(vi)	.0023	.0011	+.0012

11.5 (a) .9982; (b) .9984; (c) −.0002

11.7 (a) .5418; (b) .1753; (c) .2793; (d) .8384

11.9 (a) .1024; (b) .0028; (c) .1248; (d) .9095

11.11 (a) .0558; (b) .0237; (c) .0430; (d) .7123; (e) .2327

11.13 (a) .1038; (b) .8962; (c) .0240; (d) .0000+; (e) .2709; (f) .9988

11.15 (a) .0360; (b) .9857; (c) .6046; (d) .6554

11.17 (a) .0068
(b) Normal approximation was used because p is close to .50 and n is large.

11.19 .1781

11.21 (a) .8171; (b) .3642

CHAPTER 12

12.2 (a) No; (b) Yes; (c) No; (d) Yes

12.4 (a) Payoff Table

	a_1	a_2	a_3	a_4
θ_1	2	−3	−8	−13
θ_2	2	4	−1	−6
θ_3	2	4	6	1
θ_4	2	4	6	8

(b) Stock 2 melons; (c) Stock 3 melons.

12.6 (a) Payoff Table

	a_1	a_2
θ_1	100	−10
θ_2	−20	90

(c) COL Table

	a_1	a_2
θ_1	0	110
θ_2	110	0

(d) Snow shoes (a_1)
(e) Yes, because the sum of EOL and EMV is constant for all acts.

12.8 (a) 25; (b) 30; (c) 30; (d) 28 or 29; (e) 27; (f) 27; (g) 28; (h) 28

12.10 .60

12.12 19 trees

12.13 892 calendars

12.14 456 kilograms

12.16 (a) 1,051 blouses; (b) 1,105 blouses

12.18 (a) 600 pounds;
(b) (ii) $\begin{cases} 1{,}000x - 1{,}800 & \text{if } x \leq 6 \\ 6{,}000 - 1{,}800 & \text{if } x \geq 6 \end{cases}$
(c) 2 bags
(d) (i) \$2,089.58; (ii) \$3,533.33; (iii) \$3,289.58

CHAPTER 13

13.1 (a) \$3.30; (b) \$5.20; (c) \$1.90; (d) Stock 2 melons;
(e) Yes. This will always be the case, since the EVPI is the EOL of the optimal act.
(f) The sum is constant for all acts (\$5.20)
(g) The sum is equal to the EMV of the decision with perfect information.

13.3 \$44.00

13.5 (a) \$54.90; (b) \$1.60

13.7 (a) Issue; (b) Issue;
(c) Yes. The two criteria always lead to the same decision because the sum of EMV and EOL is constant for all acts.
(d) (i) $P(\theta_1|F) = 21/23$; $P(\theta_2|F) = 2/23$
(ii) $P(\theta_1|F') = 1/3$; $P(\theta_2|F') = 2/3$
(e) (i) $\text{EMV}(a_1|F) = \$440$; $\text{EMV}(a_2|F) = \$0$
(ii) $\text{EMV}(a_1|F') = \$40$; $\text{EMV}(a_2|F') = \$0$
(f) \$224;
(g) Not reasonable to pay \$40 because the EVII = \$0.
(h) \$76

CHAPTER 14

14.1 (a) The major points on the utility curve are:

$$u(-\$20{,}000) = 0,\ u(\$40{,}000) = .25,\ u(\$70{,}000) = .50,$$
$$u(\$90{,}000) = .75,\ u(\$100{,}000) = 1.00$$

14.4 .60

14.5 (a) $\text{EMV}(a_1) = \$1{,}175$, $\text{EMV}(a_2) = \$1{,}825$; should not adopt the EMV criterion.
(b) $\text{EU}(a_1) = 5$, $\text{EU}(a_2) = -12.5$; choose a_1.
(c) $\text{CE}(a_1) = \$25$, $\text{CE}(a_2) = -\$12.50$; choose a_1.
(d) Yes. This is a general phenomenon since the act which has a higher utility must have a higher certainty equivalent.
(e) Yes.

Index

D

E

M

N

O

P

R

W